AF663435

STATICS

STATICS

[Written according to the syllabi of BA and BSc (General and Honours) of all Indian universities and all Competitive Examinations of equivalent standard]

Dr M M Rahman
Double Gold Medallist
Associate Professor, P. G. Dept. of Mathematics
A. N. College, Patna, Bihar

Edited by

Dr Binoy Bhattacharyya, Ph.D [Moscow], MIE

New Central Book Agency (P) Ltd
LONDON
HYDERABAD ERNAKULAM BHUBANESWAR
NEW DELHI KOLKATA PUNE GUWAHATI

NCBA

REGD OFFICE

8/1 Chintamoni Das Lane, Kolkata 700 009, India

email: ncbapvtltd@eth.net

OVERSEAS

NCBA (UK) Ltd, 149 Park Avenue North, Northampton, NN3 2HY, UK

email: ncbauk@yahoo.com

EXPORT

NCBA Exports Pvt. Ltd, 212 Shahpur Jat, New Delhi 110 049

email: ncbaexports@ncbapvtltd.com

BRANCHES

2/27 Ansari Road, 1st Floor
Daryaganj, **New Delhi** 110 002
email: ncbadel@ncbapvtltd.com

House No. 3-1-315, 1st Floor
Nimboliadda, Kachiguda, **Hyderabad** 500 027
email: ncbahydb@ncbapvtltd.com

Shop Nos. 3 & 4, Vinayak Towers
681/B Budhwar Peth, Appa Balwant Chowk
Pune 411 002
email: ncbapune@ncbapvtltd.com

GSS Shopping Complex, 1st Floor
Opposite College Post Office, Convent Road
Ernakulam 682 035
0484-4021054 / 53

Shop No. 15, Bharati Towers, Ground Floor
Forest Park
Bhubaneswar 751 001
email: ncbabub@ncbapvtltd.com

Radhamayee Complex [Second Floor]
Opp. Axis Bank, Panbazar, Jaswanta Road
Guwahati 781 001
email: ncbaguwahati@ncbapvtltd.com

STATICS ● Dr M M Rahman

First Published: 1994
Reprinted: 1998, 2002
Second Edition: 2004
Thoroughly Revised, Enlarged and Updated Third Edition: 2014
Reprinted: 2015, 2016, 2019

PUBLISHER AND TYPESETTER : New Central Book Agency (P) Ltd
8/1 Chintamoni Das Lane, Kolkata 700 009

PRINTER : New Central Book Agency (P) Ltd
Web-Offset Division, Dhulagarh, Sankrail, Howrah

PROJECT SUPERVISOR : Abhijit Sen

TECHNICAL EDITOR : Dipan Roy

PROJECT TEAM : Pradip Biswas and Prodip Baidya

ISBN: 978 81 7381 027 5

Price ₹ 295.00

Contents

Preface to the Third Edition

This book has been throughly revised. Several new examples from various examination papers of different universities have been added and almost all of them have been solved. This book has been revised with the intention to make it more user-friendly to the students. Two new chapters, namely *Bending Moment* and *Objective Type Questions* have been added in this edition, as they have been included in the syllabi of some universities. Latest University Question Papers have been included as well.

I am indebted to the teachers and students without whose generous patronage the book would not have appeared in its third edition. I am also thankful to my wife Mrs Nelofar Jahan and sons Tauseef, Tauqeer, and Ayan for their constant encouragement and inspiration.

Every effort has been made to avoid the errors and misprints of previous editions and to improve the text. Even then, there always remains some scope of improvement, and every suggestion for further improvement of the book as well as pointing out of errors by the observant readers will be thankfully acknowledged.

Last, but not the least, I am thankful to Mr Amitabha Sen, the Director of New Central Book Agency Pvt. Ltd., for his cordial cooperation during the preparation of the third edition of this book.

Patna
18 October 2013

Md. Motiur Rahman

Publisher's Note to the Revised Second Edition

We are very happy to publish the **Revised Second Edition** of this book for the benefit of the students. This book covers the syllabi of BSc (Honours) of all Indian Universities as well as of all competitive examinations. A special feature of this book is the vector treatment of theorems.

Illustration of each principle has been given by means of solved problems, so that students may solve unsolved problems independently. The selected and important problems have been worked out nicely.

This book will be of immense benefit to the students of all Indian Universities and also candidates aspiring to sit for the competitive examinations.

We shall consider our endeavour successful if this revised edition is well received by the esteemed teachers and the student community all over India.

We welcome suggestions and criticisms from all quarters which will enable us in the subsequent editions to make the title more helpful to the interested students.

11 November 2004 **Publisher**

Preface to the First Edition

This book has been written for the BA and BSc (Honours) students of all Indian Universities as well as competitive examinations. After teaching the Honours students more than ten years, I found that students lack the basic concepts of Statics. So, I have given stress on the fundamentals so that the Students may grasp it easily. A special feature of the book is the vector treatment of the theorems. Each principle is illustrated by means of solved the examples, so that students may solve the unsolved problems easily.

I cannot claim that the book is free from mistakes, though every care has been taken to avoid them. I shall appreciate the pointing out of mistakes and suggestions to improve the book.

Lastly but not the least I am thankful to Mr Amitabha Sen, Director of New Central Book Agency (P) Ltd, due to whose cooperation the book is in this form.

28 August 1994 **Md. Motiur Rahman**

List of Important Formulae and Theorems

Chapter 1: Coplanar Forces

1. A system of forces acting in one plane at different points $(x_1, y_1), (x_2, y_2), \ldots$ of a rigid body can be reduced to a single force R together with a couple.
   If X and Y stands for the sum of the resolved parts of the forces along any two perpendicular directions, then magnitude of the force $R = \sqrt{X^2 + Y^2}$ and its inclination θ to the direction of X is $\tan^{-1}\frac{Y}{X}$. Also the moment of the couple $G = \Sigma(x_1 Y_1 - y_1 X_1)$.
2. Equation of the line of action of the single force R is $xY - yX = G$.
   Remember. G is the algebraic sum of the moments of forces about the origin O and *not about any point.*
3. The moment of a system about any point $(x_1, y_1) = Xy_1 - Yx_1 + G$.
4. If three forces acting on a rigid body keep it in equilibrium, they must be coplanar.
5. If three forces acting on a rigid body keep it in equilibrium, they must either be parallel or meet in a point.
6. General conditions of equilibrium: (a) In usual notations $X = 0$, $Y = 0$ and $G = 0$; (b) The algebraic sum of moments of all the forces about each of three non-collinear points vanishes.

Chapter 2: Virtual Work

1. Virtual work done by a force = force × virtual displacement in the direction of the force = force × projection of virtual displacement on the direction of the force = virtual displacement × resolved part of the force in the direction of virtual displacement.
2. The algebraic sum of the virtual work done by a system of coplanar forces acting on a particle is equal to the virtual work done by their resultant.
3. The virtual work done by a force in moving a particle from one position to another imaginary position is equal to the algebraic sum of the virtual works done by the resolved part of the forces.
4. *Principle of virtual work for a system of coplanar forces acting on a particle.* The necessary and sufficient condition that a particle acted upon by a number of coplanar forces to be in equilibrium is that the algebraic sum of the virtual works done by the forces in any small virtual displacement, consistent with the geometrical conditions of the system, is zero.
5. *Principle of virtual work for a system of coplanar forces acting at different points of a rigid body.* The necessary and sufficient condition that a system of coplanar forces acting at

different points of a rigid body to be in equilibrium is that the algebraic sum of the virtual work done by the forces in any small virtual displacement consistent with the geometrical conditions of the system is zero.

6. The string is replaced by two forces T and T acting inwards and the work done by the tension is $-T\delta l$.
7. The rod is replaced by two forces T acting outwards and the work done by the thrust is $T\delta l$.

Convention of signs. If the distances of various points from a fixed point are measured in the same direction in which the force is acting, then the work done is positive and if they are measured in a direction opposite to the direction of force, the work done is negative.

Chapter 3: Equilibrium of Strings and Chains

1. Intrinsic equation of catenary is $s = c\tan\psi$.
2. Cartesian equation of centenary is $y = c\cosh\frac{x}{c}$.
3. *Various other relations*
   (i) $y = c\sec\psi$
   (ii) $s = c\sinh\frac{x}{c}$
   (iii) $y^2 = c^2 + s^2$
   (iv) $x = c\log(\sec\psi + \tan\psi)$
   (v) $T_0 = wc$
   (vi) $T = wy$, i.e., tension at any point varies as its height above the directrix.
   (vii) $T\cos\psi = T_0 = wc$, i.e., the horizontal component of tension at any point = tension at the lowest point.
   (viii) $T\sin\psi = ws$, i.e., the vertical component of tension at any point = weight of the length of the chain measured from the vertex to the point under consideration.
   (ix) Span $= 2x$
   (x) Sag $= y - c$.
4. *Equilibrium of a string under any given forces in one plane.* In usual notations, the equations of equilibrium are

$$mX + \frac{d}{ds}\left(T\frac{dx}{ds}\right) = 0,\ mY + \frac{d}{ds}\left(T\frac{dy}{ds}\right) = 0.$$

5. *Light string on a smooth plane curve:* $T = \text{constant} = c, R = c/\rho$.
6. Heavy string on a smooth plane curve: $T = wy + c, R = \frac{T}{\rho} - w\cos\psi, T_2 - T_1 = w(y_1 - y_2)$.
7. *Light string on a rough plane curve:* $T = T_0 e^{\mu\psi}$, where T_0 is the lesser tension and T the greater tension and ψ is the angle through which the tangent turns from T_0 to T unwrapping the whole string from the curve.
8. *Heavy string on a rough plane curve:* $Te^{\mu\psi} = \int w\cdot\rho\cdot e^{-\mu\psi}(\sin\psi - \mu\cos\psi)\,d\psi + c$.

   In case of circle, $\rho = a$.

Chapter 4: Stability of Equilibrium

1. *Stable equilibrium.* A body is said to be in stable equilibrium when if it be slightly displaced from its position of equilibrium the forces acting on it tend to bring it back to its original position. For stable equilibrium, CG *of the body should be at its lowest height or its depth below a certain fixed horizontal should be greatest.*
2. *Unstable equilibrium.* A body is said to be in unstable equilibrium when if it be slightly displaced from its position of equilibrium the forces acting on it tend to move still from its original position of equilibrium. For unstable equilibrium, the CG of the body should be at its greatest height or its depth below a certain fixed horizontal line should be least.
3. *Test for the determination of the nature of stability.* Express the height z of the CG of the body above a fixed point or a fixed horizontal plane in terms of available θ.
   The body will be in equilibrium if z is either a maximum or minimum, i.e., if $\frac{dz}{d\theta}=0$. The value(s) of θ obtained from this equation will give the positions of equilibrium.
   (a) For stable equilibrium, $\frac{d^2z}{d\theta^2}>0$.
   (b) For unstable equilibrium, $\frac{d^2z}{d\theta^2}<0$.
   If z be the depth of the CG below a fixed point, or a fixed horizontal plane, then
   (a) For stable equilibrium, $\frac{d^2z}{d\theta^2}<0$.
   (b) For unstable equilibrium, $\frac{d^2z}{d\theta^2}>0$.
4. *Stable and unstable equilibrium when a body of radius r is resting on a body of radius R and h is the height of* CG *of upper body above the point of contact.*
   (a) For stable equilibrium, $\frac{1}{h}>\frac{1}{r}+\frac{1}{R}$.
   (b) For unstable equilibrium, $\frac{1}{h}\le\frac{1}{r}+\frac{1}{R}$.
5. *If the lower body of radius R near the point of contact be a plane, then put $R=\infty$, i.e., $\frac{1}{R}=0$ in the above,*
   then for stable equilibrium, $\frac{1}{h}>\frac{1}{r}$ or $h<r$
   and for unstable equilibrium, $\frac{1}{h}<\frac{1}{r}$, i.e., $h>r$.
6. If the upper surface of radius r has a plane face in contact with lower body, then put $r=\infty$, i.e., $\frac{1}{r}=0$ in the above. Hence, for stable equilibrium $\frac{1}{h}>\frac{1}{R}$ or $h<R$ and for unstable equilibrium $\frac{1}{h}<\frac{1}{R}$ or $h>R$.
7. *A body of radius r resting inside another concave body of radius R and h is the height of* CG *of the body resting above the point of contact,* then
   (a) For stable equilibrium, $\frac{1}{h}>\frac{1}{r}-\frac{1}{R}$.
   (b) For unstable equilibrium, $\frac{1}{h}<\frac{1}{r}-\frac{1}{R}$.
   (c) If $\frac{1}{h}=\frac{1}{r}-\frac{1}{R}$, then equilibrium is stable for $R>2r$ and unstable for $R<2r$.
8. *General case.* Equilibrium will be stable or unstable according as $\frac{1}{h}>$ or $<\left(\frac{1}{\rho_1}+\frac{1}{\rho_2}\right)$ $\sec\alpha$, where ρ_1,ρ_2 are the radii of curvatures and α the angle which the common normal makes with the vertical.
9. *Energy test.* For stable equilibrium potential energy is minimum and it is maximum for unstable equilibrium.

Chapter 5: Forces in Three Dimensions

1. $R^2 = X^2 + Y^2 + Z^2$ and it acts along a line whose d.c.'s are $\frac{X}{R}, \frac{Y}{R}, \frac{Z}{R}$.
   $G^2 = L^2 + M^2 + N^2$, d.c.'s of its axis are $\frac{L}{G}, \frac{M}{G}, \frac{N}{G}$.
2. A given system of forces can be reduced to a single force acting through an arbitrarily chosen point and a couple whose axis passes through that point.
3. *General conditions of equilibrium.* $R=0, G=0$, i.e., $X=0, Y=0, Z=0$; $L=0, M=0, N=0$, where X, Y, Z are sums of the resolved parts of the forces parallel to any three axis of coordinates and L, M, N are the sums of the moments of the forces about these axis.
4. *Elements of system.* $X, Y, Z; L, M, N$ are called elements of system. Let (X_1, Y_1, Z_1) be the components of a force at (x_1, y_1, z_1), then $X = \Sigma Y_1, Y = \Sigma Y_1, Z = \Sigma Z_1$.
   The values of L_r, M_r, N_r are given by the determinant

$$L_r\vec{i} + M_r\vec{j} + N_r\vec{k} = \begin{vmatrix} \vec{i} & \vec{j} & \vec{k} \\ x_r & y_r & z_r \\ X_r & Y_r & Z_r \end{vmatrix}.$$

   By equating the coefficients of $\vec{i}, \vec{j}, \vec{k}$ on both sides, we get L_r, M_r, N_r; then $L = \Sigma L_r$, $M = \Sigma M_r$, $N = \Sigma N_r$.

   Another form. If a force P acts along a line,

$$\frac{x - x_1}{l} = \frac{y - y_1}{m} = \frac{z - z_1}{n},$$

   where l, m, n are d.c.'s of the line, then $X_1 = Pl$, $Y_1 = Pm$, $Z_1 = Pn$.
   The values of L_r, M_r, N_r are obtained from the determinant

$$L_r\vec{i} + M_r\vec{j} + N_r\vec{k} = \begin{vmatrix} \vec{i} & \vec{j} & \vec{k} \\ x_r & y_r & z_r \\ Pl & Pm & Pn \end{vmatrix}.$$

   By equating the coefficients of $\vec{i}, \vec{j}, \vec{k}$ on both sides, we get the values of L_r, M_r, N_r; then $L = \Sigma L_r, M = \Sigma M_r, N = \Sigma N_r$.
5. *Central axis.* Any system of forces acting on a rigid body can be reduced to a single force together with a couple whose axis is along the direction of the force. This axis of the couple which is the line of action of single force is called *Poinsot's central axis.*
6. *Wrench.* A single force R together with a couple k whose axis coincides with the direction of the force when taken together are called a wrench of the system. Force R is called intensity of wrench and the ratio K/R is called the *pitch of the wrench.*
7. *Screw.* The straight line along which the single force acts considered together with the pitch is called a screw.
8. Condition for a given system of forces to be compounded to a single force is that $LX + MY + NZ = 0$.
9. *Equation of central axis is*

$$\frac{L - yZ + zY}{X} = \frac{M - zX + xZ}{Y} = \frac{N - xY + yX}{Z} = \frac{K}{R}.$$

10. Moment of a force about a line is defined as the product of force, shortest distance and sine of included angle.

11. *Null lines, planes and point.*

Null lines. Corresponding to a given system of forces referred to a base or origin O, all those lines which radiate from O', such that the moment of the given system of forces about these lines is zero are called null lines at O'.

Null plane. The plane in which all these null lines lie is called the *null plane of point* O' and the point O' is called *null point.*

12. *Null plane of a given point* (f, g, h) *is*

$$x(L-gZ+hY)+y(M-hX+fZ)+z(N-fY+gX)=(fL+gM+hN)$$

$$\text{or,}\quad \begin{vmatrix} x & y & z \\ f & g & h \\ X & Y & Z \end{vmatrix} = L(x-f)+M(y-g)+N(z-h).$$

13. *Null point of a given plane* $lx+my+nz=1$. If (f, g, h) be the point of the above plane, then

$$\frac{f}{X-\begin{vmatrix} M & N \\ m & n \end{vmatrix}} = \frac{g}{Y-\begin{vmatrix} N & L \\ n & l \end{vmatrix}} = \frac{h}{Z-\begin{vmatrix} L & M \\ l & m \end{vmatrix}} = \frac{1}{lX+mY+nZ}.$$

14. *Condition for a given line* $\frac{x-f}{l}=\frac{y-g}{m}=\frac{z-h}{n}$ *to be a null line for a given system X, Y, Z; L, M, N.*

$$\begin{vmatrix} X & Y & Z \\ l & m & n \\ f & g & h \end{vmatrix} = lL+mM+nN.$$

Chapter 6: Centre of Gravity (CG)

1. CG of a number of particles arranged in a straight line

$$\bar{x} = \frac{\Sigma w_r x_r}{\Sigma w_r}.$$

2. CG of a number of weights placed at points in a plane

$$\bar{x} = \frac{\Sigma w_r x_r}{\Sigma w_r},\ \bar{y} = \frac{\Sigma w_r y_r}{\Sigma w_r},\ \bar{x} = \frac{\int x\,dm}{\int dm} \text{ and } \bar{y} = \frac{\int y\,dm}{\int dm}.$$

CG will be on the line of symmetry.

3. CG *of a compound body.*

(a) CG of a body consisting of two parts of weights $w_1\, w_2$ is given by $\bar{x} = \frac{w_1x_1+w_2x_2}{w_1+w_2}$, where $\bar{x}, x_1, x_2$ are the distances of G, G_1, G_2 from a fixed point O on G_1G_2.

(b) CG of a body consisting of several parts of weight $w_1, w_2, \ldots, w_n$ is given by

$$\bar{x} = \frac{w_1x_1 + w_2x_2 + \cdots + w_nx_n}{w_1 + w_2 + \cdots + w_n} = \frac{\Sigma w_r x_r}{\Sigma w_r}$$
$$\bar{y} = \frac{w_1y_1 + w_2y_2 + \cdots + w_ny_n}{w_1 + w_2 + \cdots + w_n} = \frac{\Sigma w_r y_r}{\Sigma w_r},$$

where $(\bar{x}, \bar{y})$ and (x_r, y_r) $(r = 1, 2, \ldots, n)$ are the coordinates of $G, G_1, G_2, G_3, \ldots, G_n$ respectively referred to two rectangular axes in their planes.

4. CG *of the remainder.* If w be the weight of the whole body and w_1 the weight of the portion removed, then the CG of the remainder of weight $(w - w_1)$ is given by $\bar{x} = \frac{wx - w_1x_1}{w - w_1}$, where x, x_1 and x_2 are the distances of the CGs of the whole body, the portion removed and the portion left from a fixed point on OG_1.

5. CG *of an arc.* $\bar{x} = \dfrac{\int x\,ds}{\int ds}$ and $\bar{y} = \dfrac{\int y\,ds}{\int ds}$. Replace ds by $\dfrac{ds}{dx}dx, \dfrac{ds}{dt}dt, \dfrac{ds}{d\theta}d\theta$, etc.

6. CG *of an area bounded by the curve $y = f(x)$, axis of x and ordinates at $x = a$ and $x = b$.*

$$\bar{x} = \frac{\int_a^b x(y\,dx)}{\int_a^b y\,dx}, \quad \bar{y} = \frac{\int_a^b \frac{y}{2}(y\,dx)}{\int_a^b y\,dx}$$

and in case the area be symmetrical about x-axis, $\bar{y} = 0$.

7. CG of an area bounded by the curve $x = f(y)$, axis of y and abscissas at $y = a$ and $y = b$.

$$\bar{x} = \frac{\int_a^b \frac{x}{2}(x\,dy)}{\int_a^b x\,dy}, \quad \bar{y} = \frac{\int_a^b y(x\,dy)}{\int_a^b x\,dy}$$

and in case the area be symmetrical about y-axis, $\bar{x} = 0$.

8. CG of an area bounded between the curve $r = f(\theta)$ and radii vectors at $\theta = \alpha$ and $\theta = \beta$.

$$\bar{x} = \frac{\frac{2}{3}\int_\alpha^\beta r^3 \cos\theta\,d\theta}{\int_\alpha^\beta r^3\,d\theta}, \quad \bar{y} = \frac{\frac{2}{3}\int_\alpha^\beta r^3 \sin\theta\,d\theta}{\int_\alpha^\beta r^3\,d\theta}.$$

In case the area be symmetrical about the initial line, $\bar{y} = 0$.

9. CG *of area between two curves.*

$$\bar{x}=\frac{\int x\,(y_1-y_2)\,dx}{\int (y_1-y_2)\,dx},\ \bar{y}=\frac{\int \frac{y_1-y_2}{2}\,(y_1-y_2)\,dx}{\int (y_1-y_2)\,dx}.$$

10. CG *of area between two curves.*

$$\bar{x}=\frac{\iint x\rho\,dx\,dy}{\iint \rho\,dx\,dy},\ \bar{y}=\frac{\iint y\rho\,dx\,dy}{\iint \rho\,dx\,dy}.$$

$$\bar{x}=\frac{\iint r\cos\theta\,\rho\,r\,d\theta\,dr}{\iint \rho\,r\,d\theta\,dr},\ \bar{y}=\frac{\iint r\sin\theta\,\rho\,r\,d\theta\,dr}{\iint \rho\,r\,d\theta\,dr}.$$

11. CG *of a solid of revolution.*

$$\bar{x}=\frac{\int_a^b x\left(y^2dx\right)}{\int_a^b y^2dx}\ \text{and}\ \bar{y}=0\ \text{or,}\ \bar{x}=\frac{\int_a^b x\left(y_1^2-y_2^2\right)dx}{\int \left(y_1^2-y_2^2\right)}\ \text{and}\ \bar{y}=0.$$

When revolved about y-axis,

$$\bar{x}=0\ \text{and}\ \bar{y}=\frac{\int_a^b y\left(x^2dy\right)}{\int_a^b x^2dy}.$$

12. CG *of surface of revolution.*

When revolved about x-axis, $\bar{x}=\dfrac{\int xy\,ds}{\int y\,ds}$ and $\bar{y}=0.$

When revolved about y-axis, $\bar{x}=0$ and $\bar{y}=\dfrac{\int yx\,ds}{\int x\,ds}.$

Note. In all the above formulae, ρ the density is taken to be uniform. If density varies, then we should write ρ in Nr and Dr of both $\bar{x}$ and $\bar{y}$.

13. *Polar coordinates.*

$$\bar{x} = \frac{\iiint x\rho r d\theta dr (r \sin\theta) d\phi}{\iiint \rho r d\theta dr (r\sin\theta) d\phi} \text{ etc.,}$$

where $r = r\sin\theta\cos\phi, y = r\sin\theta\sin\phi, z = r\cos\theta$.

Chapter 7: Friction

1. *Friction.* If two bodies be in contact with one another, the property of the two bodies, by virtue of which a force is exerted between them at their point of contact to prevent one body sliding over the other is called *friction* and the force exerted is called the *force of friction.*
2. Friction is a passive and self adjusting force.
3. *Limiting friction.* When one body is on the verge of sliding on another body, the force of friction called into play attains its maximum value and is called *limiting friction.* The equilibrium then is called *limiting equilibrium.*
4. *Statical friction.* When a body in contact with another body is in any position of equilibrium but not limiting equilibrium, then the friction exerted is called statical friction. Thus, statical friction is less than the limiting friction.
5. *Laws of statical friction.*
   (a) The direction of statical friction is opposite to the direction in which the body moves or tends to move.
   (b) The magnitude of statical friction is just sufficient to prevent the body from moving.
6. *Laws of limiting friction.*
   (a) Limiting friction is equal in magnitude and opposite in direction to the force which tends to produce motion.
   (b) The magnitude of limiting friction at the point of contact between two bodies bears a constant ratio to the normal reaction at the point.
   (c) The constant ratio depends entirely on the nature of the material of which the surface in contact are composed of and is independent of their extent and shape.
7. *Angle of friction.* The angle between the resultant of limiting friction and normal reaction is called the angle of friction and is generally denoted by λ.
8. *Coefficient of friction.* Tangent of the angle of friction is called *coefficient of friction* and is generally denoted by μ. Thus $\mu = \tan\lambda$.
9. *Cone of friction.* A cone whose vertex is the point of contact of two rough bodies and whose axis lies along the common normal and whose semi-vertical angle is equal to λ, i.e., the angle of friction, is called the *cone of friction.*

$$\mu = \tan\lambda \text{ and } \sin 2\lambda = \frac{2\tan\lambda}{1+\tan^2\lambda} = \frac{2\mu}{1+\mu^2}.$$

$$\cos 2\lambda = \frac{1-\tan^2\lambda}{1+\tan^2\lambda} = \frac{1-\mu^2}{1+\mu^2}; \quad \tan 2\lambda = \frac{2\tan\lambda}{1-\tan^2\lambda} = \frac{2\mu}{1-\mu^2}.$$

10. *Resultant reaction.* The resultant of the force of friction and the normal reaction at the point of contact is called the *resultant reaction.*
11. If F be the friction and R, the normal reaction between two bodies when the equilibrium is limiting, then $F = \mu R$.
12. Since friction is maximum, when the equilibrium is limiting, μR is the maximum value of the friction.

Chapter 0

Some Useful Geometrical Results

0.1 Areas

Triangle: (I) If h be the height and a the length of the base, then its area $= \frac{1}{2}ah$. (II) If a and b be the two sides having angle θ between them, then its area $= \frac{1}{2}ab\sin\theta$.

Hexagon: If a be the length of any side of a regular hexagon, then its area $= \frac{3\sqrt{3}}{2}a^2$.

Octagon: If a be the length of any side of a regular octagon, then its area $= 2\left(1+\sqrt{2}\right)a^2$.

Circle: If r be the radius of a circle, then its area $= \pi r^2$ and circumference $= 2\pi r$.

Sector of a circle: The area of a sector of a circle of radius r subtending an angle 2θ at the centre $= \frac{1}{2}r^2\theta$.

Ellipse: If $2a$ and $2b$ be the lengths of the major and minor axes of an ellipse, then its area $= \pi ab$.

0.2 Volumes and Surfaces

Rectangular solid: If a, b and c be the length, breadth and height of a rectangular solid, then

its volume $= abc$,

its whole surface $= 2(ab+bc+ca)$

and its diagonal $= \sqrt{a^2+b^2+c^2}$.

Cube: If a be the length of each edge of a cube, then

its volume $= a^3$,

its whole surface $= 6a^2$,

its diagonal $= \sqrt{3}a$.

Prism: If the base of a prism be a regular polygon, then its volume = (area of the base) × height,

its lateral face = (perimeter of the base) × height.

Cylinder: If r be the radius of the base of a cylinder of height h, then

its curved surface $= 2\pi rh$,

its whole surface $= 2\pi r(h+r)$ and

its volume $= \pi r^2 h$.

Pyramid: Its volume = $\frac{1}{3}$ (area of base) × height,

its slant height = $\frac{1}{2}$× (perimeter of its base) × slant height.

Cone: If h be the height, r the radius, l the slant height, α the semi-vertical angle of the cone, then $r = h\tan\alpha$, $l = h\sec\alpha$,

its volume = $\frac{1}{3}\pi r^2 h = \frac{1}{3}\pi h^3 \tan^2\alpha$,

its curved surface = $\pi r l = \pi r^2 \tan\alpha \sec\alpha$.

Frustum of a cone: If r_1, r_2 be the radii of its circular ends and d be the perpendicular distance between them, then

its volume = $\frac{1}{3}\pi d\left(r_1^2 + r_1 r_2 + r_2^2\right)$.

Sphere: If r be the radius of the sphere, then

its volume = $\frac{4}{3}\pi r^3$,

its curved surface = $4\pi r^2$.

0.3 Zone or Portion of a Sphere

If r be the radius of the sphere of which the zone is given between two parallel planes of distances h_1 and h_2 from the centre of the sphere,

its volume = $\frac{\pi}{3}(h_1 - h_2)\{3r^2 + (h_1^2 + h_1 h_2 + h_2^2)\}$

its curved surface = $2\pi r d$, where d is the perpendicular distance between the two parallel planes.

0.4 Paraboloid of Revolution

Solid formed by the revolution of a parabola about its axis is known as *paraboloid of revolution*.

Its volume cut off by a plane perpendicular to its axis = $\frac{1}{2}$ (area of its plane base) × its height.

0.5 Centre of Gravity

CG *of a cylinder* is the mid-point of its axis.

CG *of a solid cone* of height h is the point on the axis at a distance $\frac{3h}{4}$ from the vertex.

CG *of a curved surface of a circular cone* of height h is the point on axis at a distance $\frac{3h}{4}$ from the vertex.

CG *of a solid half cone* where base is of radius r, is a point whose distance from the axis of the whole cone = $\frac{r}{\pi}$.

CG *of a solid hemisphere* of radius r, is a point on its axis at a distance $\frac{3r}{8}$ from the centre of the base.

CG *of the curved surface of a hemisphere of radius r*, is a point on its axis at a distance $\frac{r}{2}$ from the centre of the base.

CG *of a semicircle* of radius r, is a point on the radius perpendicular to the bounding diameter at a distance $\frac{4r}{3\pi}$ from the centre of the circle.

CG *of a sector of a circle* of radius r subtending an angle 2θ at the centre of the circle is a point on the bisector of the angle 2θ, whose distance from the centre = $\frac{2r}{3}\frac{\sin\theta}{\theta}$.

CG *of a circular arc of radius* r subtending an angle 2θ at the centre is a point on the bisector of the angle 2θ, whose distance from the centre $= \frac{r\sin\theta}{\theta}$.

CG *of a semicircular arc* of radius r is a point on the radius perpendicular to the bounding diameter at a distance $\frac{2r}{\pi}$ from the centre of the circle.

Chapter 1

Preliminary Concepts

1.1 Definitions

Mechanics is the study of the physical behaviour of material bodies with interest generally focussed on the motion induced by forces. In other words, it is a branch of physical science studying the motion or rest of a body under the action of a system of forces.

Primarily in mechanics, we study the external effects of a system of forces acting on a rigid body. The external effect of a force on a body either produces acceleration or creates resisting forces (reaction) in the body. When the internal effects of a system of forces acting on the body are considered or the alterations in the shape of the body are taken into account, the mechanics is known as the *mechanics of materials.*

The principle of mechanics are simple and less in number but the applications are countless in number, various and also complicated too.

1.2 The Principles of Laws of Mechanics

The principles of the laws of mechanics are the principles of the laws of conservation. In mechanics there are certain quantities or characteristics which remain invariant during motion, such as mass, momentum, energy, etc. This property of invariance of characteristics of motion is put under the conservation principle. The conservation laws are highly potent in solving problems of mechanics on the following grounds:

1. The conservation laws do not depend on the path of motion and details of a particular force. The equations of motions are consequences of this property.
2. The impossibility of some aspects of motion are also judged by the conservation laws and thus, it helps us discard such motions.
3. The conservation laws tell us a lot about the motion even if the forces affecting the motion are not known.
4. The failure of conservation laws may lead to the discovery of new and not yet explained phenomenon.
5. The conservation laws provide us a convenient tool to deal with the motion of a system of particles.

Mechanics is divided into two branches, namely Statics and Dynamics.

1.3 Statics

It is an old branch of Mechanics. If the system of forces acting on a body is balanced, the body experiences no external effect and is in equilibrium and the problem is one of Statics. Thus, Statics is that branch of mechanics which deals with equilibrium of bodies, that is, the forces acting on the body produces no motion in the body.

The small distortions produced in a body by the forces acting on the body do not generally alter appreciably the external effects of the forces acting on the body. Thus, such changes in the shape of the body is not taken into account because that would simply complicate the problems for no good. The bodies, we take, are therefore, rigid bodies. A rigid body is that in which all particles are at constant distances from each other. Actually, no real body can be absolutely rigid but since the changes in the shape of the body have negligible effect upon the acceleration produced by a system of forces or upon reactions required to maintain equilibrium, a body may be assumed to be a rigid one if the changes produced in the distances between the particles of the body are negligibly small.

1.4 Force

A force is a cause which changes, or tends to change the state of rest or uniform motion of a body.

A force is characterised by its point of application, its magnitude and its direction. A force is represented by a vector. Thus, the force can be represented by a directed line-segment.

1.5 The Parallelogram Law of Forces

This law states that two forces acting on a particle at a point be represented in magnitude and direction by the sides of a parallelogram drawn from the point. Their resultant is represented both in magnitude and direction by the diagonal of the parallelogram drawn through the point.

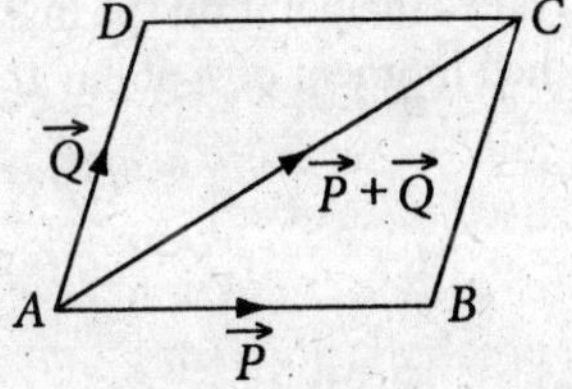

Fig. 1.1

Let two forces $\vec{P}$ and $\vec{Q}$ act at a point A of the particle. Let $\vec{P}$ and $\vec{Q}$ be represented in magnitude and direction by the adjacent sides AB and AD of the parallelogram $ABCD$.

Then the resultant $\vec{P}+\vec{Q}$ of the forces $\vec{P}$ and $\vec{Q}$ will be represented by $\overrightarrow{AC}$.

This law can be verified experimentally. We, however, assume the law as a fundamental principle of Statics.

1.6 The Principle of Transmissibility of Forces

This law states that the conditions of equilibrium or of motion of a rigid body remain unchanged if a force acting at a given point of the rigid body is replaced by another force of the same magnitude and same direction, but acting at a different point of the body, provided that the two forces have the same line of action.

1.7 Moment of a Force about a Point

It is the tendency of rotation of a force about a point and it is measured by the product of the magnitude of the force and the perpendicular distance from the point to the line of action of the force. It is conventionally taken as positive or negative according as the sense of rotation of force about the point is anticlockwise or clockwise.

If F be the force and p the perpendicular distance of its line of action AB from the fixed point O, then moment of F about $O = Fp$.

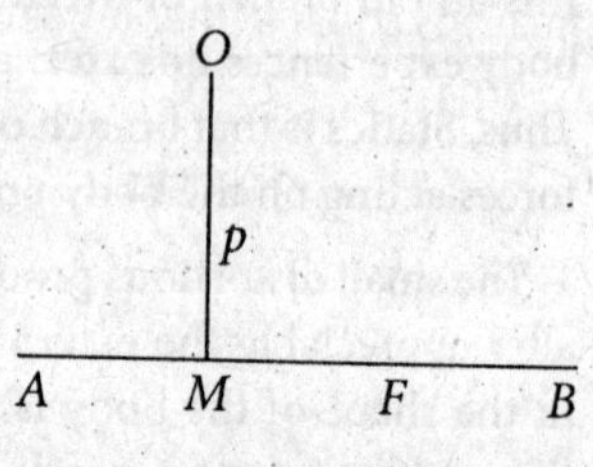

Fig. 1.2

1.8 Vector Moment of a Force about a Point

Let $\vec{F}$ be a force acting along the line AB and O be the point about which moment of $\vec{F}$ is to be found. From O we draw ON perpendicular to AB.

Then vector moment of $\vec{F}$ about O is defined as $\overrightarrow{OA} \times \vec{F}$.

In ΔOAN,

$$\sin\theta = \frac{ON}{OA}, \; ON = OA \cdot \sin\theta.$$

Now

$$\left|\overrightarrow{OA} \times \vec{F}\right| = OA \cdot F \sin\theta = F \cdot ON,$$

which is scalar moment of $\vec{F}$ about O.

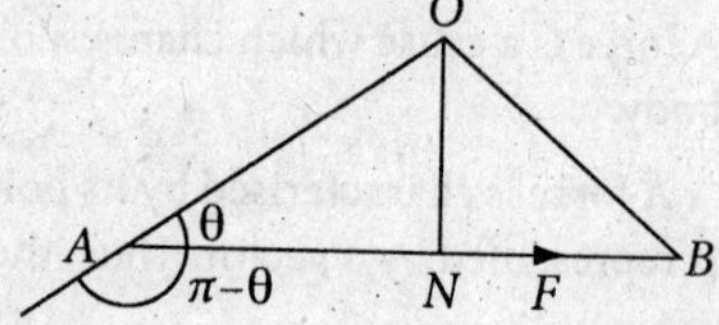

Fig. 1.3

It can be easily seen that the above vector moment is independent of choice of the point A (i.e., the point on the line AB), for if we take any other point B on the line of action of $\vec{F}$, then moment of $\vec{F}$ about B is defined as

$$\overrightarrow{OB} \times \vec{F} = \left(\overrightarrow{OA} + \overrightarrow{AB}\right) \times \vec{F} = \overrightarrow{OA} \times \vec{F} + \overrightarrow{AB} \times \vec{F}$$

$$= \overrightarrow{OA} \times \vec{F} \left(\because \overrightarrow{AB} \text{ and } \vec{F} \text{ are collinear}\right).$$

1.9 Couple

Two equal and unlike parallel forces are said to make a couple, if their lines of action are not the same.

Arm of a Couple

The perpendicular distance between the parallel forces forming a couple is called the *arm of a couple.* In Fig. 1.4, $AB = p$ is the arm of the couple.

Moment of a Couple

It is defined as the product of one of the forces forming the couple and the length of arm of the couple. Thus, the moment of the couple in Fig. 1.4 is Pp.

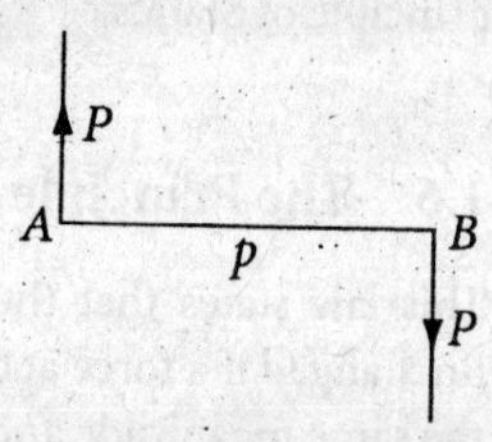

Fig. 1.4

1.10 Varignon's Theorem

The algebraic sum of moments of any number of coplanar forces about any point in their plane is equal to the moment of their resultant about the same point.

Chapter 2

Coplanar Forces

2.1 Introduction

A system of forces whose lines of action lie in a plane is known as a *coplanar system of forces.* In our daily life we usually come across coplanar forces, e.g., if a man goes up a hill he is under the action of three coplanar forces, namely his weight acting vertically downward, the reaction of the hill surface inclined at some angle to the surface, and the force of friction down the hill surface. Similarly, if a rod be placed in a smooth hemispherical bowl, it is in equilibrium under the action of three coplanar forces, namely the reactions at the ends of the rod along two radii of the bowl and the weight of the rod acting vertically downward, etc.

Theorem 1. *To show that any system of coplanar forces acting upon a rigid body can be reduced to either a single force or a single couple unless it is in equilibrium.* **[PU(H) 1996]**

Proof. Let P, Q, R, S, T, etc. be a system of coplanar forces acting on a rigid body.

We take any three forces P, Q, R, of the system. The two forces P and Q can be combined into a single resultant by parallelogram of forces, if they meet or by the method of combining parallel forces when they are parallel, like or unlike, excepting in the case when they form a couple. In case P and Q form a couple, we can combine P and R into a single force unless P forms a couple with R also. Now, if P forms a couple with Q as well as with R, Q and R must be like parallel forces which can be combined into a single force. In any case, the three forces can be reduced to two.

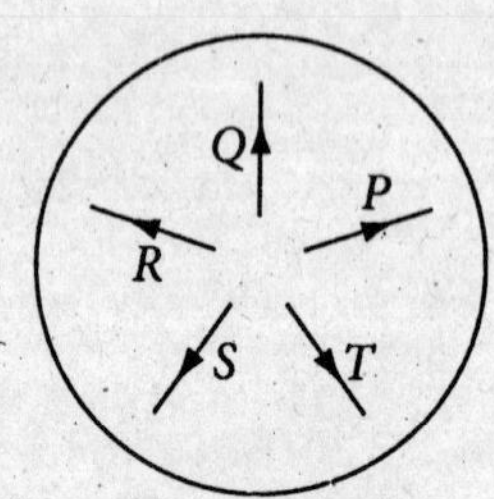

Fig. 2.1

With these two we take another force S of the system. These three forces just as before can be reduced to two. Proceeding in the same way, when all the forces of the system are exhausted, we get ultimately two forces. These two, if they are equal and opposite acting along the same line, will produce equilibrium. Otherwise, if they are equal and unlike parallel forces, they form a couple. In case they do not form a couple, we can finally combine them into a single resultant.

Theorem 2. *To show that a system of coplanar forces acting in one plane at different points of a rigid body can be reduced to a single force through any given point and a single couple.*

[PU(H) 2008, TMBU(H) 2010, LNMU(H) 1995, MU(H) 2009, BVABU(H) 2010]

Proof. Let a number of coplanar forces $P_1, P_2, P_3, \ldots$ act at the points $A_1, A_2, \ldots$ of the rigid body respectively. The coordinates of $A_1, A_2, A_3, \ldots$ are $(x_1, y_1), (x_2, y_2), (x_3, y_3), \ldots$.

Let the resolved parts of the forces in the directions parallel to axes be $(X_1, Y_1), (X_2, Y_2), \ldots$

Now, at O we introduce equal and opposite forces $X_1, X_2, \ldots$ along Ox and Ox' and $Y_1, Y_2, \ldots$ along Oy and Oy'. These forces along the axes acting in opposite direction balances each other.

Now, the forces X_1 at A_1 and X_1 at O along Ox' constitute couple of moment $-X_1 . A_1L = -X_1y_1$. Here negative sign is taken as the couple has the tendency to rotate the body in clockwise direction. Also the forces Y_1 at A_1 and Y_1 at O along Oy' form a couple of moment $Y_1 . A_1M = Y_1x_1$.

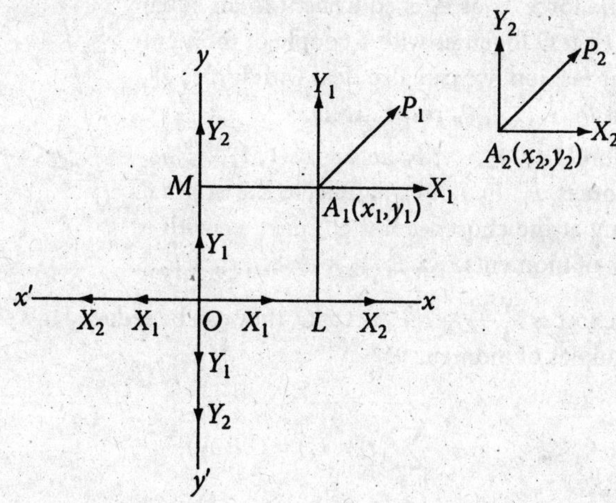

Fig. 2.2

Hence, the forces X_1 and Y_1 at A_1 are equivalent to forces X_1, Y_1 at O along Ox and Oy and a couple of moment $x_1Y_1 - y_1X_1$.

Let us proceed in the same way with the force P_2 at A_2 and so on.

Finally, we find that system of forces are equivalent to the forces

$$X = \Sigma X_1 = X_1 + X_2 + \ldots \text{ along } Ox$$
$$Y = \Sigma Y_1 = Y_1 + Y_2 + \ldots \text{ along } Oy$$

and a couple of moment

$$G = (x_1Y_1 - y_1X_1) + (x_2Y_2 - y_2X_2) + \ldots$$
$$= \Sigma(x_1Y_1 - y_1X_1) \text{ about } O.$$

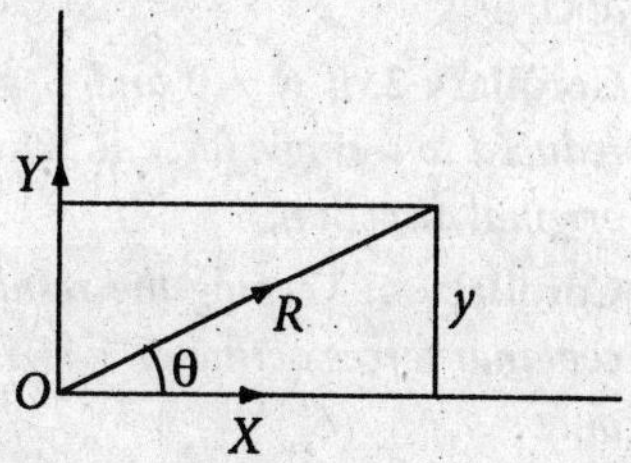

Fig. 2.3

If R be the resultant of the forces X and Y inclined at an angle θ to Ox then

$$X = X_1 + X_2 + \cdots = R\cos\theta$$
$$Y = Y_1 + Y_2 + \cdots = R\sin\theta.$$

$$R^2 = X^2 + Y^2.$$

$$\Rightarrow R = \sqrt{X^2 + Y^2} \text{ and } \tan\theta \Rightarrow Y/X \text{ or, } \theta = \tan^{-1} Y/X. \quad (1)$$

Thus, the system reduces to a force R through O whose magnitude and directions are given by (1) together with a couple, whose moment $G = \Sigma(x_1 Y_1 - y_1 X_1)$.

Alternative Method (Vector Method)

Let $\vec{F}_1, \vec{F}_2, \ldots, \vec{F}_n$ be any system of coplanar non-concurrent forces acting at points $P_1, P_2, \ldots, P_n$ respectively. Let O be an arbitrary point. Let O be origin of references and the position vectors of $P_1, P_2, \ldots, P_n$ be $\vec{r}_1, \vec{r}_2, \ldots, \vec{r}_n$ respectively.

We know that the force $\vec{F}_1$ at P_1 is equivalent to an equal like parallel force $\vec{F}_1$ at O together with a couple of moment $\vec{r}_1 \times \vec{F}_1$. In a similar fashion, we can also deal with forces $\vec{F}_2, \vec{F}_3, \ldots, \vec{F}_n$ acting at $P_2, P_3, \ldots, P_n$ respectively.

Thus the given forces $\vec{F}_1, \vec{F}_2, \ldots, \vec{F}_n$ acting at $P_1, P_2, \ldots, P_n$ are equivalent to forces. $\vec{F}_1, \vec{F}_2, \ldots, \vec{F}_n$ parallel to the original directions all acting at the chosen point O, together with a number of couples of moments $\vec{r}_1 \times \vec{F}_1, \vec{r}_2 \times \vec{F}_2, \ldots, \vec{r}_n \times \vec{F}_n$.

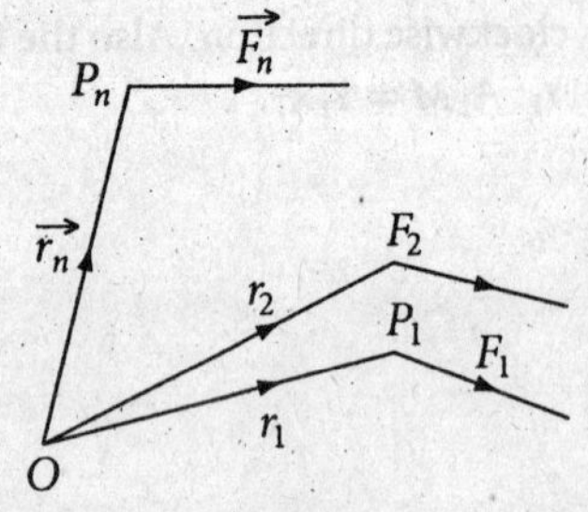

Fig. 2.4

Now since all the forces $\vec{F}_1, \vec{F}_2, \ldots, \vec{F}_n$ act at O, they can be reduced to a single force $\vec{R} = \Sigma F_i$ at O. Again, the couples of moment

$$\sum_{i=1}^{n} (r_i \times F_i) = G \text{ (say)},$$

which is equal to the algebraic sum of the moments of the given forces about O.

Thus the given system of forces is equivalent to the resultant force R at O together with the couple of moment G, where $\vec{R} = \Sigma \vec{F}_i$ and $G = \Sigma \vec{r}_i \times \vec{F}_i$.

Corollary 1. *If $\vec{R} = \vec{0}$ and $\vec{G} \neq \vec{0}$, then the given system of forces reduces to a couple.*

Corollary 2. *If $\vec{G} = \vec{0}$ and $\vec{R} \neq \vec{0}$, then the given system of forces reduces to a single force only at O.*

Corollary 3. *If $\vec{R} \neq 0$ and $\vec{G} \neq 0$, then the resultant force $\vec{R}$ and the couple $\vec{G}$ can again be reduced to a single force $\vec{R}$ acting parallel to the original direction at a distance G/R from its original direction.*

Corollary 4. *Keeping the results of the above three corollaries, we find that any number of coplanar forces acting on a rigid body can be reduced either to a single force or a single couple only.*

2.2 Moment of a System of Coplanar Forces about any point

Theorem 1. *If X and Y be the algebraic sum of resolved parts of the system of coplanar forces parallel to the axes and G their moments about the origin O, show that the moment G' of the system about any other point (h, k) in its plane is given by $G' = G - hY + kX$.*

Proof. Let X_1, Y_1 be the components of one of the coplanar forces acting at the point (x_1, y_1) of the rigid body parallel to the axes, then

$$X = \Sigma X_1,\ Y = \Sigma Y_1,\ G = \Sigma(x_1 Y_1 - y_1 X_1).$$

Now, G' = sum of moments of the system about any point (h, k)

$$\begin{aligned}
&= \Sigma[-X_1 \cdot A_1N + Y_1 \cdot O'N] \\
&= \Sigma[Y_1 \cdot ML - X_1 \cdot (A_1L - NL)] \\
&= \Sigma[Y_1 \cdot (x_1 - h) - X_1(y_1 - k)] \\
&= \Sigma[(x_1 Y_1 - y_1 X_1) + kX_1 - hY_1] \\
&= \Sigma(x_1 Y_1 - y_1 X_1) + k\Sigma X_1 - h\Sigma Y_1 \\
&= G + kX - hY.
\end{aligned}$$

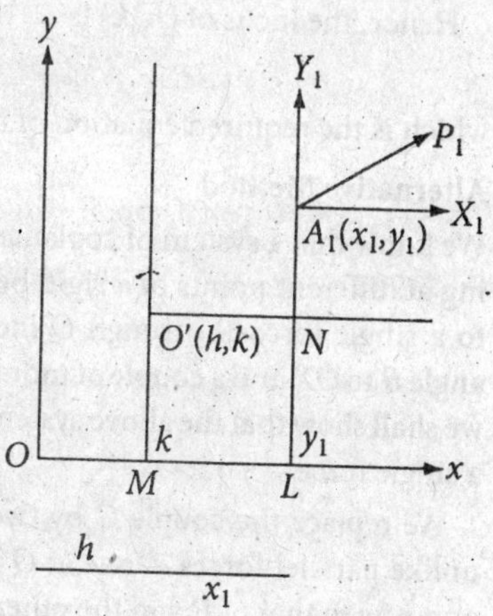

Fig. 2.5

Alternative Method

Since a system of coplanar forces not in equilibrium acting on a rigid body can be reduced to a single force acting at an arbitrary point in the plane of the forces together with a couple, let the system reduce to a force R through O together with a couple of moment G.

Let X, Y be the resolved parts of forces along the axes which are two mutually perpendicular lines through O.

Moment of system about (x_1, y_1)

= moment of R about $(x_1, y_1) + G$

($\because$ moment of a couple is constant)

= algebraic sum of moments of X and Y about $(x_1, y_1) + G$

$= Xy_1 - Yx_1 + G.$

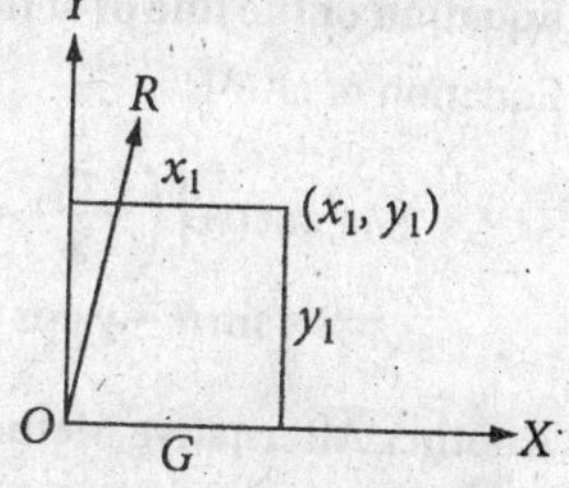

Fig. 2.6

Theorem 2. *To find the equation of line of action of the resultant of a system of coplanar forces acting at different points of a rigid body.*

[PU(H) 2011, MU(H) 2011, LNMU(H) 2011, BVABU(H) 2009, TMBU(H) 2011, RU(H) 1994, JPU(H) 2011]

Proof. We know that a system of coplanar forces acting on a rigid body can be reduced to a single force, R, acting at an arbitrary chosen point O, in the plane of the forces together with a couple G.

Let $P(h, k)$ be any point on the line of action of resultant of the given system. Then by Varignon's theorem by taking moment about P

$$G + X \cdot PQ - Y \cdot OQ = 0,$$

i.e., $G + Xk - Yh = 0.$

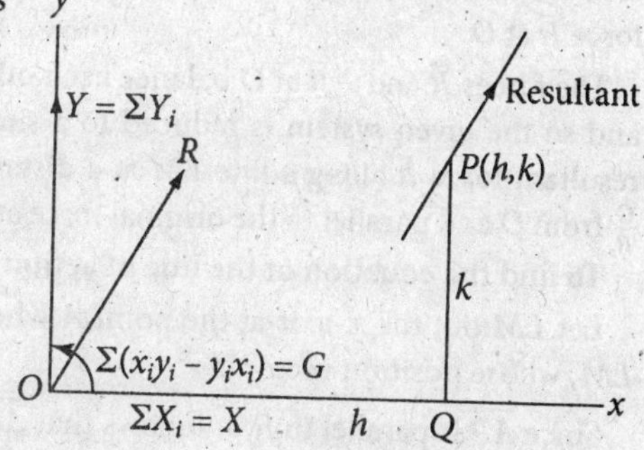

Fig. 2.7

Hence, the locus of (h, k) is

$$G + Xy - Yx = 0,$$

which is the required equation of the line of action of the resultant force.

Alternative Method

We know that a system of coplanar forces acting at different points of a rigid body reduces to a single force R through O inclined at an angle θ to OX and a couple of moment G. Now we shall show that the above system reduces to a single force.

We replace the couple G by two equal and unlike parallel forces — one at O in direction opposite to that of R and the other at O' along perpendicular to OO', where OO' is such that $R \cdot OO' = G$ or, $OO' = G/R$.

Now two equal and opposite forces at O balance each other and we are left with a force R through O'.

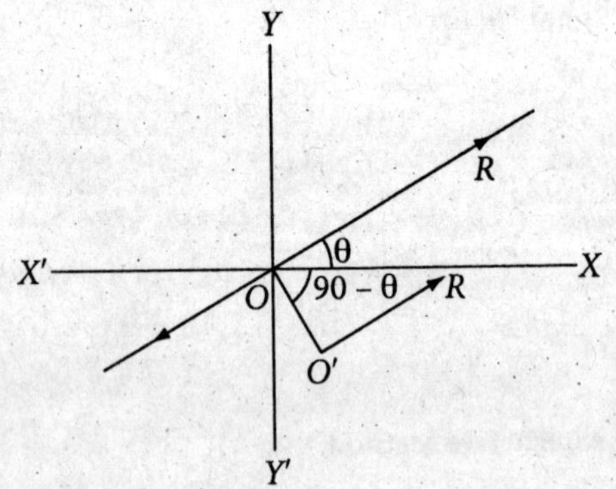

Fig. 2.8

Equation of the line of action of this single force *R*, i.e., *O′R*

Equation of $O'R$ is

$$x\cos\left\{\left(-\frac{\pi}{2}-\theta\right)\right\}+y\sin\left\{-\left(\frac{\pi}{2}-\theta\right)\right\}=OO'=\frac{G}{R}$$

$$\Rightarrow x\sin\theta - y\cos\theta = G/R \Rightarrow xR\sin\theta - yR\cos\theta = G \Rightarrow xY - yX = G.$$

Another Alternative Method (Vector Method)

Let the gives system of coplanar forces be reduced to a single force $\vec{R}$ at the origin O together with a couple of moment $\vec{G}$.

The couple of moment $\vec{G}$ can be replaced by two equal and unlike parallel forces $-\vec{R}$ at O and $\vec{R}$ at a distance G/R from O and parallel to the force $\vec{R}$ at O.

The forces $\vec{R}$ and $-\vec{R}$ at O balance each other and so the given system is reduced to a single resultant force $\vec{R}$ along a line LM at a distance $\frac{G}{R}$ from O and parallel to the original force at O.

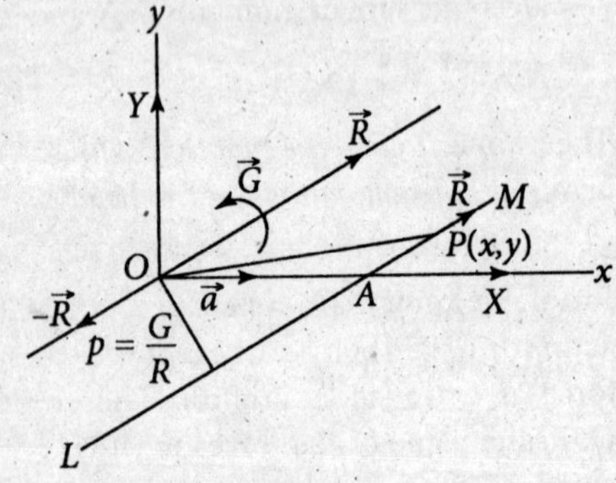

Fig. 2.9

To find the equation of the line of action *LM* of the resultant force $\vec{R}$.

Let LM out the x-axis at the point A whose position vector is $\vec{a}$. Let P be any point on LM, whose position vector is $\vec{r}$.

Since $\overrightarrow{AP}$ is parallel to $\vec{R}$, it follows that $\overrightarrow{AP} = \lambda\vec{R}$, where λ is scalar.

Now

$$\overrightarrow{OP} - \overrightarrow{OA} + \overrightarrow{AP} \Rightarrow \vec{r} = \vec{a} + \lambda\vec{R}. \quad (1)$$

Let $\vec{i}$ and $\vec{j}$ be the unit vector along the x-axis and y-axis respectively. If X and Y are the components of $\vec{R}$, parallel to the x-axis and y-axis respectively, then we can write

$$\vec{R} = X\vec{i} + Y\vec{j}.$$

Also on taking moments about O, we have

$$aY = G \Rightarrow a = \frac{G}{Y}.$$

So from (1), we have

$$\vec{r} = a\vec{i} + \lambda\vec{R} \quad \left(\because \vec{a} = a\vec{i}\right)$$

$$\Rightarrow \vec{r} = \frac{G}{Y}\vec{i} + \lambda\vec{R}$$

Thus the vector equation of the line of action LM of $\vec{R}$, the resultant is

$$\vec{r} = \left(\frac{G}{Y}\right)\vec{i} + \lambda\vec{R},$$

since this gives the position vector of any point P on LM.

Cartesian Equivalent or Cartesian Form

The above equation can be written as

$$x\vec{i} + y\vec{j} = \frac{G}{Y}\vec{i} + s(X\vec{i} + Y\vec{j}) \quad \left(\because \vec{r} = x\vec{i} + y\vec{j}\right).$$

Equating coefficients of $\vec{i}$ and $\vec{j}$ on both sides, we get

$$x - \frac{G}{Y} = sX \text{ and } y = sY.$$

Eliminating s, we have

$$x - \frac{G}{Y} = \frac{y}{Y}X \Rightarrow xY - G = yX \Rightarrow xY - yX = G.$$

This is the equation of the resultant in Cartesian form.

Theorem 3. *To find a necessary and sufficient condition for the equilibrium of a system of coplanar forces acting on a rigid body.*

[PU(H) 2001, MU(H) 2009, LNMU(H) 1998, BVABU(H) 1998, TMBU(H) 2009, RU(H) 1991]

Proof. The necessary and sufficient condition for a system of a coplanar forces acting on a rigid body to be in equilibrium are that the sums of the resolved parts of the system of forces along any two perpendicular line must vanish separately and the sum of the moments of the forces about any point in their plane must be zero.

We know that a system of coplanar forces acting at different points of a rigid body can be reduced to a single force R acting at an arbitrary point O and a couple whose moment is equal to the sum of the moments of the given forces about O, i.e., the system is equivalent to the single force

$$R=\sqrt{X^2+Y^2}, \tag{1}$$

and a couple of moment

$$G=\Sigma(x_1Y_1-y_1X_1)$$

where X, Y are the algebraic sum of the there solved parts of the system of forces parallel to the axes.

The given conditions are necessary

Let us assume that the given forces are in equilibrium.

Since a force and a couple cannot balance each other, it is necessary that R and G should be separately zero.

$\therefore$ $R=0$ and $G=0$.

So from (1),

$$X^2+Y^2=0 \Rightarrow X=0 \text{ and } Y=0.$$

Let G be the sum of moments of all the forces about any point (x_1, y_1), in their plane, then

$$G_1=G+y_1X-x_1Y \Rightarrow G_1=0+y_1\cdot 0-x_1\cdot 0=0.$$

Hence, the necessary condition for equilibrium of a system of coplanar forces acting on a rigid body are that $X=0$, $Y=0$ and $G_1=0$, i.e., the sum of the resolved parts of the forces in any two perpendicular directions should be zero separately and the sum of the moments of all the forces of the system about any arbitrary point is zero.

The given conditions are sufficient

Here we assume that $X=0$, $Y=0$ and $G_1=0$.

$\therefore$ we shall now show that the given system of forces are in equilibrium.

Now,

$$R=\sqrt{X^2+Y^2}=0.$$

This shows that the body has no motion of translation and

$$G_1=G+y_1X-x_1Y \Rightarrow 0=G+y_1\cdot 0-x_1\cdot 0 \Rightarrow 0=G \Rightarrow G=0.$$

This show that the body has no motion of rotation. Hence, forces are in equilibrium.

Alternative or Vector Method

The condition is necessary. Let a rigid body under the action of a system of coplanar forces $\vec{F}_1, \vec{F}_2, \ldots, \vec{F}_n$ acting at different points $P_1, P_2, \ldots, P_n$ of the rigid body whose position vectors referred to any origin O are $\vec{r}_1, \vec{r}_2, \ldots, \vec{r}_n$ be in equilibrium.

We know that the above system of forces is equivalent to a single force

$$\vec{R}=\sum_{i=1} F_i$$

acting at O and a single couple of moment

$$G=\sum_{i=1}^{n}\vec{r}_i\times\vec{F}_i.$$

The single force $\vec{R}$ and the couple G, have respectively the tendency of motion of translation and the motion of rotation of the body. In case the forces acting on the body keep it in equilibrium, the forces, have no tendency of motion of translation or of motion of rotation of the body. Hence, the necessary conditions for the equilibrium of the body are that $\vec{R}=\vec{0}$ and $\vec{G}=0$.

Let (x_i, y_i) be the coordinates of the point P_i, referred to some rectangular axes through the point O, then $\vec{r}_i=x_i\vec{i}+y_i\vec{j}$.

Also if X_i, Y_i are the resolved parts parallel to the axes of the forces $\vec{F}_i$ acting at the point P_i, then

$$\vec{F}_i=X_i\vec{i}+Y_i\vec{j}.$$

$$\therefore\ \vec{R}=\sum_{i=1}^{n}\vec{F}_i=\sum_{i=1}^{n}(X_i\vec{i}+Y_i\vec{j})=\left(\sum_{i=1}^{n}X_i\right)\vec{i}+\left(\sum_{i=1}^{n}Y_i\right)\vec{j}\Rightarrow\vec{R}=X\vec{i}+Y\vec{j},$$

where

$$X=\sum_{i=1}^{n}X_i,\ Y=\sum_{i=1}^{n}Y_i$$

and

$$\vec{G}=\sum_{i=1}^{n}\vec{r}_i\times\vec{F}_i=\sum_{i=1}^{n}\{(x_i\vec{i}+y_i\vec{j})\times(X_i\vec{i}+Y_i\vec{j})\}$$

$$=\sum_{i=1}^{n}(x_iY_i-y_iX_i)(\vec{i}\times\vec{j})\ (\because\ \vec{i}\times\vec{j}=-\vec{j}\times\vec{i};\ \vec{i}\times\vec{i}=\vec{0}=\vec{j}\times\vec{j}).$$

Now if $\vec{R}=\vec{0}$, then

$$X=\sum_{i=1}^{n}X_i=0 \text{ and } Y=\sum_{i=1}^{n}Y_i=0$$

and if $\vec{G}=\vec{0}$, then

$$G=\sum_{i=1}^{n}(x_iY_i-y_iX_i)=0.$$

Clearly, G is the algebraic sum of the scalar moments of the forces about the point O.

Hence, the necessary conditions of equilibrium of the body are that, the sum of the resolved parts of the forces in any two mutually perpendicular directions vanishes separately and the sum of the moments of the forces about any point in the plane of the forces vanishes.

The condition is sufficient

If the sum of the resolved parts of the forces in any two mutually perpendicular directions vanishes separately and the sum of the moments of the forces about any point in the plane of the forces vanishes, i.e., $X=0$, $Y=0$ and $G=0$, then $\vec{R}=X\vec{i}+Y\vec{j}=\vec{0}$ and $\vec{G}=\vec{0}$.

Thus, there is neither the motion of translation nor the motion of rotation of the body, therefore, the body is in equilibrium.

Hence, the necessary and sufficient conditions for the equilibrium of a body are that $X=0$, $Y=0$ and $G=0$.

Theorem 4. *If the algebraic sum of moments of a system of coplanar forces about any three non-collinear points on their plane vanishes separately, show that the forces of the system are in equilibrium.* **[BVABU(H) 2011]**

Proof. Let the three given non-collinear points be A, B, C.

Let A be origin and any two perpendicular lines through A as the coordinate axes. Let the coordinates of B and C be (x_1, y_1) and (x_2, y_2) respectively, referred to above axes.

Let X, Y be the algebraic sum of resolved parts of the system of forces parallel to the axes. Let the moments of the forces about A, B and C be G, G_1 and G_2 respectively, then we have

$$G_1 = G - x_1 Y + y_1 X \tag{1}$$

$$\text{and } G_2 = G - x_2 Y + y_2 X. \tag{2}$$

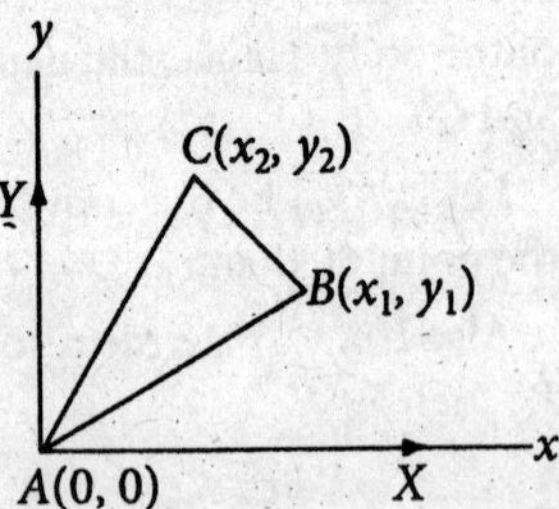

Fig. 2.10

It is given that

$$G = 0,\ G_1 = 0,\ G_2 = 0. \tag{3}$$

So from (1) and (2), with the help of (3), we get

$$y_1 X - x_1 Y = 0 \tag{4}$$

$$y_2 X - x_2 Y = 0. \tag{5}$$

Since A, B, C are non-collinear,

$$\begin{vmatrix} 0 & 0 & 1 \\ x_1 & y_1 & 1 \\ x_2 & y_2 & 1 \end{vmatrix} \neq 0 \text{ or, } (x_1 y_2 - x_2 y_1) \neq 0. \tag{6}$$

Multiplying (5) by x_1 and (4) by x_2,

$$x_2 y_1 X - x_1 x_2 Y = 0$$

$$x_1 y_2 X - x_1 x_2 Y = 0.$$

Subtracting, these two we get

$$X(x_2 y_1 - x_1 y_2) = 0$$

$$X = 0 \text{ [from (6)]}.$$

So from (4), we have

$$-x_1 Y = 0$$

$$\Rightarrow Y = 0.$$

Thus, we have shown, $X = 0$, $Y = 0$ and we are given that $G = 0$, which are the required conditions of equilibrium. Hence, the system is in equilibrium.

2.3 Astatic Equilibrium

When a given system of coplanar forces acts on a particle and keeps it at rest, the particle will remain in equilibrium if we turn each force about its point of application through the same angle. But when the coplanar forces act at different points of a rigid body and keep it in equilibrium, it is not in general true that the body will still be in equilibrium, when the forces are rotated by the same constant angle about their points of application. When the equilibrium is maintained by such a rotation, it is said to be Astatic Equilibrium.

The following theorem deals with such a situation.

Theorem 1. *If all the forces in a coplanar system are rotated about their points of application through the same angle in their own plane. show that their resultant passes through a fixed point in the body.* [LNMU(H) 1996; MU(H) 2003, BVABU(H) 2008, TMBU(H) 2000]

Proof. Let P_1 be one of the forces of the system acting at a point $A_1(x_1, y_1)$ and making an angle θ_1 with the x-axis.

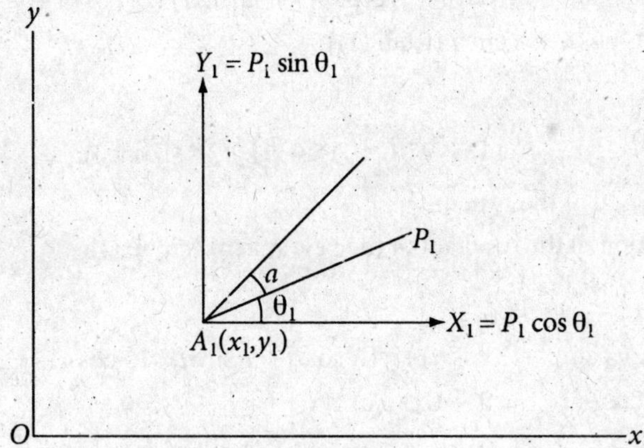

Fig. 2.11

Let the components of P_1 parallel to axes be X_1, Y_1, so for the other forces. Also if X, Y be the components of the system along the axes and G the sum of moments of the forces about the origin, then we known that

$$\left.\begin{aligned} X &= \Sigma X_1 = \Sigma P_1 \cos\theta_1 \\ Y &= \Sigma Y_1 = \Sigma P_1 \sin\theta_1 \\ \text{and } G &= \Sigma(x_1 Y_1 - y_1 X_1) \\ &= \Sigma(x_1 P_1 \sin\theta_1 - y_1 P_1 \cos\theta_1) \\ &= \Sigma P_1 (x_1 \sin\theta_1 - y_1 \cos\theta_1) \end{aligned}\right\} \tag{1}$$

Also the equation to the resultant of the systems is

$$xY - yX - G = 0. \tag{2}$$

Now let all the forces of the system be rotated through an angle α about their points of application.

If X', Y', G' be the new position of X, Y, G, then as in (1), we have

$$\begin{aligned} X' &= \Sigma P_1 \cos(\theta_1 + \alpha) = \Sigma P_1 (\cos\theta_1 \cos\alpha - \sin\theta_1 \sin\alpha) \\ &= \cos\alpha \Sigma P_1 \cos\theta_1 - \sin\alpha \Sigma P_1 \sin\theta_1 = X\cos\alpha - Y\sin\alpha \text{ [from (1)]} \quad (3) \\ Y' &= \Sigma P_1 \sin(\theta_1 + \alpha) = \Sigma P_1 (\sin\theta_1 \cos\alpha + \cos\theta_1 \sin\alpha) \\ &= \cos\alpha \Sigma P_1 \sin\theta_1 + \sin\alpha \Sigma P_1 \cos\theta_1 = Y\cos\alpha + X\sin\alpha \text{ [from (1)].} \quad (4) \end{aligned}$$

and

$$\begin{aligned} G' &= \Sigma P_1 [x_1 \sin(\theta_1 + \alpha) - y_1 \cos(\theta_1 + \alpha)] \\ &= \Sigma P_1 [x_1 (\sin\theta_1 \cos\alpha + \cos\theta_1 \sin\alpha) - y_1 (\cos\theta_1 \cos\alpha - \sin\theta_1 \sin\alpha)] \\ &= \Sigma P_1 [(x_1 \sin\theta_1 - y_1 \cos\theta_1)\cos\alpha + (x_1 \cos\theta_1 + y_1 \sin\theta_1)\sin\alpha] \\ &= \cos\alpha \Sigma P_1 (x_1 \sin\theta_1 - y_1 \cos\theta_1) + \sin\alpha \Sigma P_1 (x_1 \cos\theta_1 + y_1 \sin\theta_1) \\ \Rightarrow G' &= G\cos\alpha + V\sin\alpha \text{ [from (1)]} \quad (5) \end{aligned}$$

where

$$V = \Sigma P_1 (x_1 \cos\theta_1 + y_1 \sin\theta_1) = \Sigma (x_1 X_1 + y_1 Y_1). \quad (6)$$

is known as the viral of the system.

Also the equation of the resultant of the new system will be

$$\begin{aligned} & xY' - yX' = G', \\ & \Rightarrow x(X\sin\alpha + Y\cos\alpha) - y(X\cos\alpha - Y\sin\alpha) = G\cos\alpha + V\sin\alpha \\ & \Rightarrow \cos\alpha(xY - yX - G) + \sin\alpha(xX + yY - V) = 0. \quad (7) \end{aligned}$$

Whatever be the value of α, (7) always passes through the point whose coordinates are given by the point of intersection of straight lines

$$xY - yX - G = 0$$
$$\text{and } xX + yY - V = 0.$$

Solving these two equations by cross-multiplication, we get

$$\frac{x}{XV + GY} = \frac{y}{VY - GX} = \frac{1}{Y^2 + X^2}.$$

$$x = \frac{GY + VX}{X^2 + Y^2}, \; y = \frac{VY - GX}{X^2 + Y^2}.$$

Since the coordinates

$$\left(\frac{GY + VX}{X^2 + Y^2}, \frac{VY - GX}{X^2 + Y^2}\right)$$

are independent of α, therefore, this point is fixed in the body. This point is called *Astatic Centre.*

Corollary 1. *If the forces were in equilibrium before displacement, then* $X = 0$, $Y = 0$, $G = 0$.

Hence, after displacement $X' = 0$, $Y' = 0$, $G' = V\sin\alpha$.

This shows that if the forces were originally in equilibrium and if each force turned through the same angle α about their points of application, then they are equivalent to a couple whose moment is $V\sin\alpha$, where V is independent of α. Thus, for astatic equilibrium we must have $V = 0$.

Theorem 2. *Show that if the algebraic sum of moments of a system of coplanar forces is the same and of same sign, when taken about each of three non-collinear points, it reduces to a couple.*

Proof. Let three non-collinear points be $O(0,0)$, $A(x_1, y_1)$ and $B(x_2, y_2)$.

Let G, G_1 and G_2 be the algebraic sum of moments of the coplanar forces acting on the rigid body respectively.

Given $G = G_1 = G_2$.

Now

$$G_1 = G + y_1 X - x_1 Y$$
$$\text{and } G_2 = G + y_2 X - x_2 Y.$$

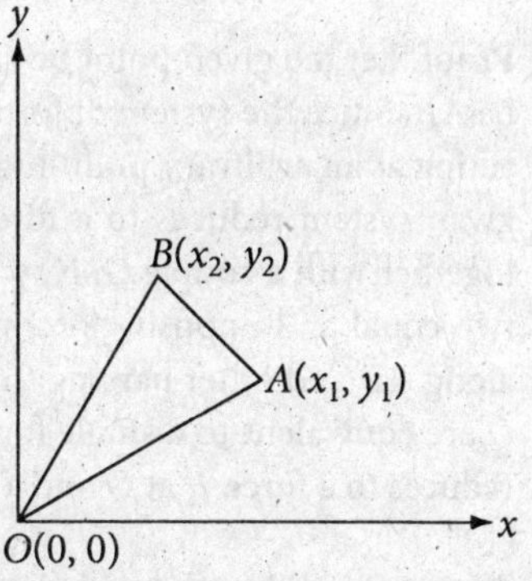

Now

$$G = G_1 \Rightarrow G = G + y_1 X - x_1 Y$$
$$\Rightarrow y_1 X - x_1 Y = 0. \quad (1)$$

Also

$$G = G_2 \Rightarrow G = G + y_2 X - x_2 Y$$
$$\Rightarrow y_2 X - x_2 Y = 0. \quad (2)$$

$(1) \times x_2$ gives

$$x_2 y_1 X - x_1 x_2 Y = 0. \quad (3)$$

Fig. 2.12

Also $(2) \times x_1$ gives

$$x_1 y_2 X - x_1 x_2 Y = 0. \quad (4)$$

(3) and (4) gives

$$(x_2 y_1 - x_1 y_2) X = 0. \quad (5)$$

Since O, A, B are non-collinear,

$$\therefore\ 0 + x_1(y_2 - 0) + x_2(0 - y_1) \neq 0 \Rightarrow x_1 y_2 - x_2 y_1 \neq 0.$$

Hence, from (5), we get $X = 0$.

So from (1), we get $Y = 0$.

Now

$$R = \sqrt{X^2 + y^2} = 0.$$

But $G \neq 0$.

Hence, the system reduces to a couple of moment G.

Theorem 3. *Prove that a single force and a couple acting in the same plane upon a rigid body cannot produce equilibrium, but are equivalent to a single force acting in a direction parallel to its original direction.* [MU(H) 2012; LNMU(H) 2000]

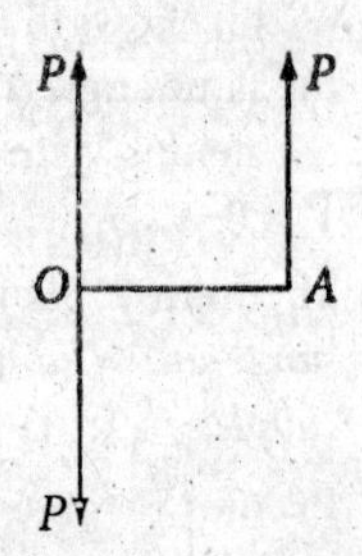

Fig. 2.13

Proof. Let P be the given force acting at O and G the moment of the couples, reckoned positive in the counter-clockwise direction. We replace the couple by two forces of magnitude P at a distance p apart such that $G = P \cdot p$ and place them so that one of them balances the given force P acting at O. Then the resultant of P and the couple G is a single parallel force P, displaced to a distance $OA = G/P$ from its original position. Hence, a force and a couple acting in the same plane cannot produce equilibrium.

Theorem 4. *Prove that in all cases a system of coplanar forces not in equilibrium can be replaced by two forces, one of which acts through a given point and the other along a given straight line.*

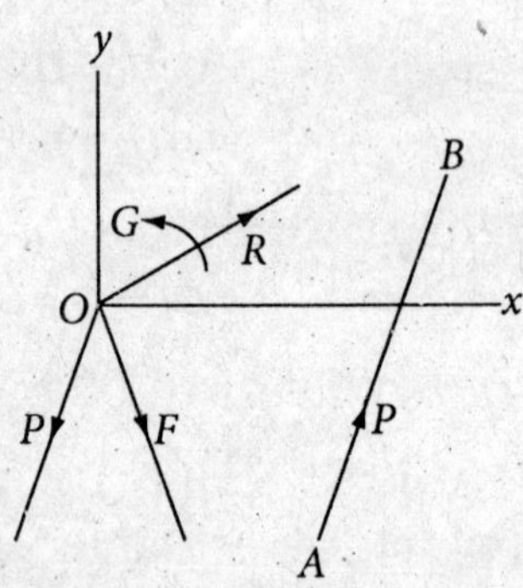

Fig. 2.14

Proof. Let the given point be O and the given straight line be AB. Since the system of forces always reduces to a force acting at an arbitrary point together with a couple, let the given system reduces to a force R acting at the origin O together with a couple G. Now we replace the couple G by two equal and opposite forces of magnitude P (say), one along AB and other passing through O. Again, R and P at O are equivalent to a single force F at O. Thus, the system reduces to a force F at O and a force P along AB.

Theorem 5. *Prove that a given system of forces, not in equilibrium, can be reduced to two forces acting one at each of the two given points A and B, the force at A making a given angle (not zero) with AB.*

Proof. Take the forces to be coplanar. The system can be reduced to a single force R acting at B together with a couple G.

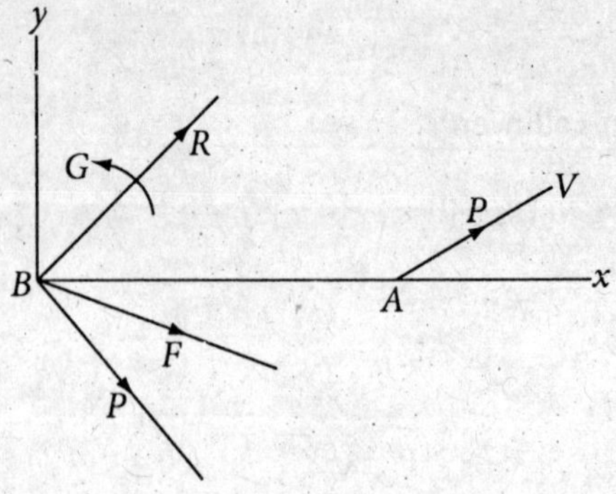

Fig. 2.15

Now we replace the couple G by two equal and opposite forces of magnitude P, one acting at A making a given angle with AB and the other acting at B.

Again, R and P at B are equivalent to a single force F at B. Thus, the system reduces to a force F at B and a force P at A making a given angle with AB.

2.4 Equilibriums of a Rigid Body under Three Forces

Theorem 1. *If three forces acting on a rigid body be in equilibrium, show that they must be* (i) *coplanar* (ii) *must either meet at a point or be all parallel.* [LNMU(H) 1993]

Proof. Let the three forces $\vec{P}$, $\vec{Q}$ and $\vec{R}$ acting at the points A, B and C respectively of a rigid body, keep it in equilibrium.

Since the forces are in equilibrium,

$$\therefore \vec{P}+\vec{Q}+\vec{R}=\vec{0}. \qquad (1)$$

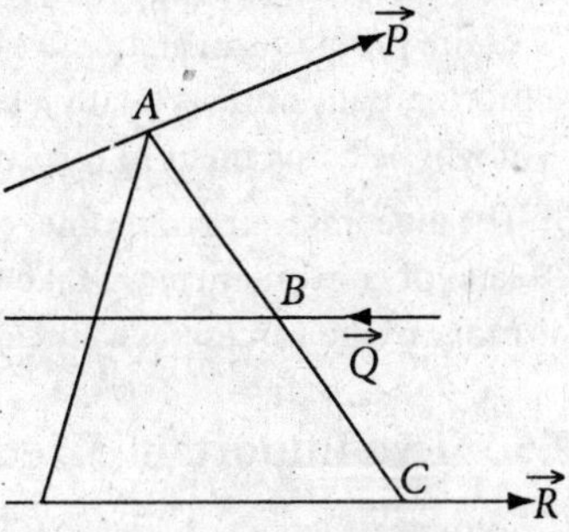

Fig. 2.16

Also the sum of the moments of the forces about any point must be zero.

$\therefore$ taking moments of the forces about the point A, we have

$$\overrightarrow{AB}\times\vec{Q}+\overrightarrow{AC}\times\vec{R}=\vec{0}\Rightarrow\overrightarrow{AB}\times\vec{Q}=-\overrightarrow{AC}\times\vec{R}\Rightarrow\overrightarrow{AB}\times\vec{Q}=\overrightarrow{CA}\times\vec{R}=\hat{n} \text{ (say).}$$

Thus, $\hat{n}$ is a vector perpendicular to $\overrightarrow{AB}$, $\vec{Q}$, $\overrightarrow{CA}$ and $\vec{R}$, i.e., the vector $\hat{n}$ is perpendicular to the plane ABC and the vectors Q and R passing through the point B and C respectively. Hence, the forces $\vec{Q}$ and $\vec{R}$ must lie in the plane ABC. Similarly, by taking moment about B, we get that $\vec{P}$ at $\vec{A}$ and $\vec{R}$ at C lie in the plane ABC.

Thus, $\vec{P}$, $\vec{Q}$, $\vec{R}$ acting at A, B, C respectively, are coplanar and lie in the plane ABC.

Since the forces $\vec{Q}$ and $\vec{R}$ are coplanar, hence, their lines of action must either intersect or be parallel.

Case I. The lines of action of $\vec{P}$ at A and $\vec{Q}$ at B are intersecting. Let D be the point of intersection. By the parallelogram law of forces the resultant of $\vec{P}$ at A, $\vec{Q}$ at B is $\vec{P}+\vec{Q}$ through D. Since the forces $\vec{P}$, $\vec{Q}$, $\vec{R}$ at A, B, C respectively, are in equilibrium, we must have $\vec{P}+\vec{Q}$ through D and $\vec{R}$ at C equal in magnitude, opposite in direction and passing through D. Thus, the lines of action of $\vec{P}$, $\vec{Q}$, $\vec{R}$ at A, B, C respectively, are concurrent at D.

Case II. The lines of action of $\vec{P}$ at A and $\vec{Q}$ at B are parallel. The resultant of these forces is $\vec{P}+\vec{Q}$ whose line of action is in the plane ABC and parallel to that of $\vec{P}$ and $\vec{Q}$. But $\vec{P}$, $\vec{Q}$, $\vec{R}$ are in equilibrium, so the third force $\vec{R}$ should be equal and opposite to $\vec{P}+\vec{Q}$ and as such its line of action is the same as that of $\vec{R}$ but opposite and hence, it is parallel to $\vec{P}$ and $\vec{Q}$. So the three forces are parallel.

2.5 The Conditions for Three Forces Acting upon a Rigid Body to be in Equilibrium

In view of the last two theorems, it follows that if three forces keep a body in equilibrium, then

1. They must be coplanar.
2. Their lines of action must meet at a point or they must be parallel.
3. (a) If the three forces are concurrent, then Lami's theorem holds and they must satisfy the converse of triangle law of forces.
   (b) If all the forces are parallel, then each force is proportional to the distance between the other two, i.e., the theorem on parallel forces hold.
4. The algebraic sum of resolved parts of the forces along any straight line is zero. In most of the problems in statics, we resolve the forces in any two perpendicular directions (i.e., horizontally and vertically), so that $X=0$, $Y=0$ separately. But in case of inclined plane, it will be convenient to resolve the forces along and perpendicular to the plane.
5. The algebraic sum of moments of the forces about any point in their plane is zero. For the sake of convenience, we take moments of the forces about a point through which one or more of the forces pass, since the moment of these forces about that point will be zero.

2.6 Two Important Trigonometrical Theorems

In any ΔABC, if D divides AB in the ratio $m:n$ and if $\angle CDB=\theta$, $\angle ACD=\alpha$ and $\angle BCD=\beta$, then

1. $(m+n)\cot\theta = m\cot\alpha - n\cot\beta$;
2. $(m+n)\cot\theta = n\cot A - m\cot B$.

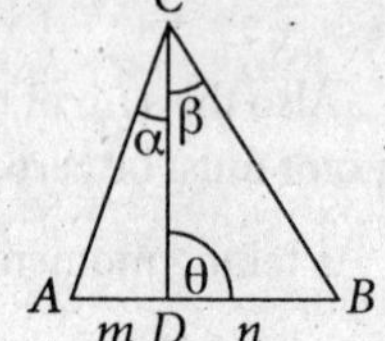

Fig. 2.17

Theorem 1. *To obtain condition for a particle to be at rest in contact with a smooth curve (in a plane).*

Proof. Let a particle of mass m be at rest at any point $P(x, y)$ on a given smooth plane curve. Let X, Y be the components of the forces acting on the particle parallel to the axes and R be the normal reaction at P. Let PT be the tangent at P to the curve and let PT make angle ψ with x-axis, then at P we have

$$\tan\psi = \frac{dy}{dx}, \ \sin\psi = \frac{dy}{ds} \text{ and } \cos\psi = \frac{dx}{ds}, \tag{1}$$

where arc $AP = s$.

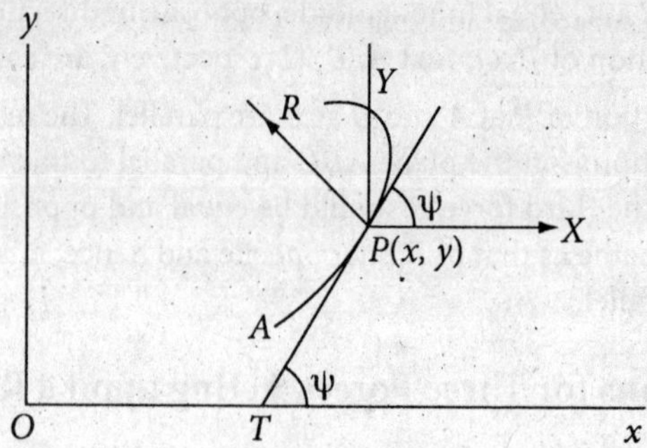

Fig. 2.18

Resolving along and perpendicular to the tangent, we get for equilibrium

$$X\cos\psi + Y\sin\psi = 0, \text{ i.e., } X\frac{dx}{ds} + Y\frac{dy}{ds} = 0 \text{ [from (1)]} \tag{2}$$

and

$$X\sin\psi - Y\cos\psi = R, \text{ i.e., } X\frac{dy}{ds} - Y\frac{dx}{ds} = R. \tag{3}$$

Equation (2) gives the condition that a particle be at rest in contact with the smooth curve and (3) gives the reaction at any point in contact will the curve.

Theorem 2. *Prove that a system of coplanar forces will be in equilibrium if the algebraic sum of the moments about each of any two points is zero and the algebraic sum of the resolved parts of the forces in any given direction, not perpendicular to the line joining the given points is zero.*

Proof. Let us take the given direction of resolution of the forces as x-axis.

Let G and G' be the moments of the forces of the system about any two points O and $A(x_1, y_1)$.

Also OA is not perpendicular to Ox.

Now, we have

$$G' = G + y_1X - x_1Y. \tag{1}$$

As given, X = algebraic sum of resolved parts of the forces along x-axis = 0.

Given $G = 0$, $G' = 0$.

So from (1), we get

$$-x_1Y = 0 \Rightarrow Y = 0 \; (\because x_1 \neq 0).$$

Fig. 2.19

Hence, for the given system, we have $X = 0$, $Y = 0$ and $G = 0$.

Hence, the given system is in equilibrium.

Theorem 3. *A system of coplanar forces is such that the sum of their moments about any two points on a given line are always same. Show that if the resultant is not zero, it must be parallel to the given line.* [KU(H) 2003]

Proof. Let CD be the given line whose equation is $ax + by + c = 0$.

Let $A(x_1, y_1)$ and $B(x_2, y_2)$ be two points on CD such that the sum of the moments of coplanar forces about them are always same.

Let G_1 and G_2 be the algebraic sum of moments of all the coplanar forces about A and B. Then

$$G_1 = G + y_1X - x_1Y$$
$$\text{and } G_2 = G + y_2X - x_2Y.$$

Given

$$G_1 = G_2 \Rightarrow G + y_1X - x_1Y = G + y_2X - x_2Y$$
$$\Rightarrow X(y_1 - y_2) = Y(x_1 - x_2)$$
$$\Rightarrow X/Y = (x_1 - x_2)/(y_1 - y_2). \tag{1}$$

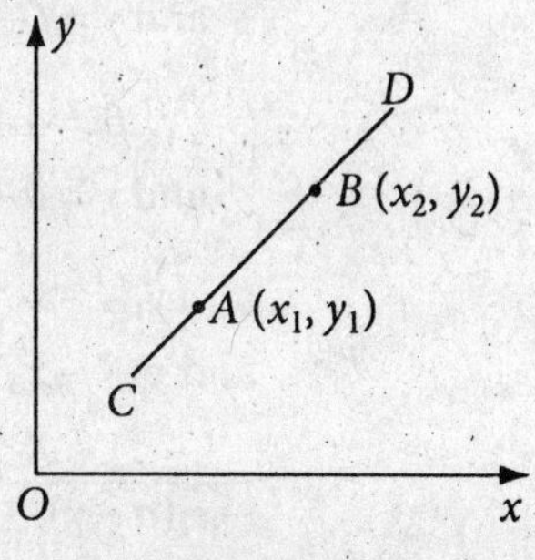

Fig. 2.20

Since (x_1, y_1) and (x_2, y_2) lie on (1),

$$\therefore\ ax_1 + by_1 + c = 0$$
$$\text{and } ax_2 + by_2 + c = 0.$$

By cross-multiplication,

$$\frac{a}{y_1 - y_2} = \frac{b}{x_2 - x_1} = \frac{c}{x_1y_2 - x_2y_1} \Rightarrow \frac{a}{b} = \frac{y_1 - y_2}{x_2 - x_1} \Rightarrow -\frac{a}{b} = \frac{y_1 - y_2}{x_1 - x_2}. \quad (2)$$

Now equation of line of action of the resultant of the given system of coplanar forces is

$$G + yX - xY = 0. \quad (3)$$

Slope m_1 of (3),

$$\begin{aligned}\frac{Y}{X} &= \frac{y_1 - y_2}{x_1 - x_2} \text{ [from (1)]}\\ &= -\frac{a}{b} \text{ [from (2)]}\\ &= \text{slope } m_2 \text{ of the straight line (1).}\end{aligned}$$

Hence, the line of action of the resultant is parallel to the given line.

Theorem 4. *A system of coplanar forces act on a rigid body and the moment of the system about three non-linear points in the plane of the forces are α, β, γ. Prove that* (i) *if α, β, γ are not equal and not zero, the system is equivalent to a single force;* (ii) *if $\alpha = \beta = \gamma = k (k \neq 0)$, the system is equivalent to a couple;* (iii) *if $\alpha = \beta = \gamma = 0$, the system is in equilibrium.*

[CU(H) 1992]

Proof. Let O, A, B be any three non-collinear points in the plane of the forces. Let the coordinates of A and B be (x_1, y_1) and (x_2, y_2) respectively.

Let X, Y be the algebraic sum of resolved parts of the system of forces parallel to the axes. Let G be the algebraic sum of moments of all the forces about O.

Then given $\alpha = G$,

$$\beta = G + y_1X - x_1Y$$
$$\text{and } \gamma = G + y_2X - x_2Y.$$

$$\therefore\ \beta = \alpha + y_1X - x_1Y$$
$$\text{and } \gamma = \alpha + y_2X - x_2Y.$$

$$y_1X - x_1Y + (\alpha - \beta) = 0$$
$$\text{and } y_2X - x_2Y + (\alpha - \beta) = 0.$$

Fig. 2.21

$$\Rightarrow \frac{X}{-x_1(\alpha - \gamma) + x_2(\alpha - \beta)} = \frac{Y}{y_2(\alpha - \beta) - y_1(\alpha - \gamma)} = \frac{1}{-y_1x_2 + x_1y_2}$$

$$\Rightarrow X = \frac{x_2(\alpha - \beta) - x_1(\alpha - \gamma)}{x_1y_2 - x_2y_1},\ Y = \frac{y_2(\alpha - \beta) - y_1(\alpha - \gamma)}{x_1y_2 - x_2y_1}.$$

Since O, A, B are non-collinear,

$$\therefore\ x_1(y_2-0)+x_2(0-y_2)\neq 0 \Rightarrow x_1y_2-x_2y_1\neq 0.$$

1. If α, β, γ are not equal and not zero, then $X\neq 0$, $Y\neq 0$. Thus, we get a non-zero resultant R. Hence, the system is equivalent to a single force.
2. If $\alpha=\beta=\gamma\neq 0$, then $X=0$ and $Y=0$. So $R=0$ and $G\neq 0$. Hence, the system is equivalent to a couple.
3. If $\alpha=\beta=\gamma=0$, then $X=0$, $Y=0 \Rightarrow R=0$ and $G=0$. Hence, the system is in equilibrium.

WORKED-OUT EXAMPLES

1. The algebraic sums of the moments of a system of forces about points whose coordinates are (x_1, y_1), (x_2, y_2), (x_3, y_3) and (x_4, y_4) referred to OX and OY as axes are G_1, G_2, G_3 and G_4 respectively. Prove that

$$\begin{vmatrix} 1 & x_1 & y_1 & G_1 \\ 1 & x_2 & y_2 & G_2 \\ 1 & x_3 & y_3 & G_3 \\ 1 & x_4 & y_4 & G_4 \end{vmatrix}=0.$$ [PU(H) 2001]

Solution: Let G be the moment of the system about the origin. If G' be the moment of the system about any point (h, k), then we know that

$$G'=G-hY+kX,$$
$$\text{i.e., } G-hY+kX-G'=0,$$

where X, Y are the resolved parts of the resultant along OX and OY.
Hence, taking moment of the system about the four given points, we have

$$G-x_1Y+y_1X-G_1=0 \quad (1)$$
$$G-x_2Y+y_2X-G_2=0 \quad (2)$$
$$G-x_3Y+y_3X-G_3=0 \quad (3)$$
$$G-x_4Y+y_4X-G_4=0. \quad (4)$$

Eliminating G, Y and X from (1), (2), (3) and (4), we get required result:

$$\begin{vmatrix} 1 & x_1 & y_1 & G_1 \\ 1 & x_2 & y_2 & G_2 \\ 1 & x_3 & y_3 & G_3 \\ 1 & x_4 & y_4 & G_4 \end{vmatrix}=0 \text{ (Proved).}$$

2. Three forces P, Q, R act along the sides of a triangle formed by the lines $x+y=1$, $y-x=1$, $y=2$. Find the equation to the line of action of resultant.

[PU(H) 2008, LNMU(H) 1997, MU(H) 2009, BVABU(H) 2009, TMBU(H) 2011, VU(H) 2007, CU(H) 1996, KU(H) 2002]

Solution: *First Method.* Let the three forces. P, Q, R act along the sides AB, BC and CA of $\triangle ABC$.

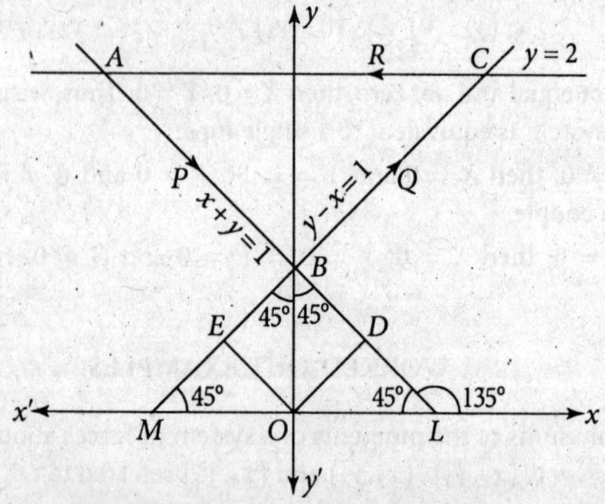

Fig. 2.22

Equation of AB is $x+y=1$.

So $OL=1, OB=1$.

Equation of BC is $y-x=1$, i.e., $-x+y=1$.

$\therefore\ OM=-1, OB=1$.

Since equation of AC is $y=2$.

$\therefore\ OF=2$.

From O, we draw OD and OE perpendicular to BL and BM respectively.

In $\triangle OBE$,

$$\sin 45^\circ=\frac{OE}{OB}=\frac{OE}{1}\Rightarrow OE=\frac{1}{\sqrt{2}}.$$

In $\triangle OBD$,

$$\sin 45^\circ=\frac{OD}{OB}=\frac{OD}{1}\Rightarrow OD=\frac{1}{\sqrt{2}}.$$

Now,

$$\begin{aligned}X &= \text{algebraic sum of resolved part of all the forces along } x\text{-axis}\\ &= P\cos 45^\circ+Q\cos 45^\circ+R\cos 180^\circ\\ &= \frac{P+Q-\sqrt{2}R}{\sqrt{2}}.\end{aligned}$$

$$\begin{aligned}Y &= \text{algebraic sum of resolved part of all the forces along } y\text{-axis}\\ &= -P\sin 45^\circ+Q\sin 45^\circ+R\sin 180^\circ\\ &= -\frac{P+Q}{\sqrt{2}}=\frac{Q-P}{\sqrt{2}}.\end{aligned}$$

$$\begin{aligned} G &= \text{algebraic sum of moments of all the forces about } O \\ &= -P\cdot OD - Q\cdot OE + R\cdot OF \\ &= -\frac{P}{\sqrt{2}} - \frac{Q}{\sqrt{2}} + 2R = \frac{2\sqrt{2}R - P - Q}{\sqrt{2}}. \end{aligned}$$

Hence, the equation of line of action of the resultant is

$$\begin{aligned} xY - yX = G \Rightarrow \frac{x(Q-P)}{\sqrt{2}} - \frac{y\left(P+Q-\sqrt{2}R\right)}{\sqrt{2}} &= \frac{2\sqrt{2}R-P-Q}{\sqrt{2}} \\ \Rightarrow x(Q-P) - y\left(P+Q-\sqrt{2}R\right) &= 2\sqrt{2}R - P - Q \\ \Rightarrow x(P-Q) + y\left(P+Q-\sqrt{2}R\right) &= P+Q-2\sqrt{2}R. \end{aligned}$$

Second Method. (Vector Method) (See figure of first method). Let $\vec{i}$ and $\vec{j}$ be the unit vectors along the x-axis and the y-axis respectively.

Then, we have

$$\overrightarrow{BL} = \overrightarrow{BO} + \overrightarrow{OL} = -\vec{j} + \vec{i}.$$

So unit vector along

$$\overrightarrow{BL} = \frac{\vec{i}-\vec{j}}{\sqrt{1^2+(-1)^2}} = \frac{\vec{i}-\vec{j}}{\sqrt{2}}. \qquad \therefore \vec{P} = \frac{P}{\sqrt{2}}\left(\vec{i}-\vec{j}\right).$$

Also,

$$\overrightarrow{MB} = \overrightarrow{MO} + \overrightarrow{OB} = \vec{i} + \vec{j}.$$

So unit vector along

$$\overrightarrow{MB} = \frac{\vec{i}+\vec{j}}{\sqrt{2}}. \qquad \therefore \vec{Q} = \frac{Q}{\sqrt{2}}\left(\vec{i}+\vec{j}\right) \text{ and } \vec{R} = -R\vec{i}.$$

Let $\vec{F}$ be the resultant of the forces $\vec{P}, \vec{Q}$ and $\vec{R}$.

$$\begin{aligned} \therefore \vec{F} = \vec{P} + \vec{Q} + \vec{R} \Rightarrow \vec{F} &= \frac{P}{\sqrt{2}}\left(\vec{i}-\vec{j}\right) + \frac{Q}{\sqrt{2}}\left(\vec{i}+\vec{j}\right) - R\vec{i} \\ \Rightarrow X\vec{i} + Y\vec{j} &= \frac{\vec{i}}{\sqrt{2}}\left(P+Q-\sqrt{2}R\right) + \frac{\vec{j}}{\sqrt{2}}(QP) \\ \Rightarrow X &= \frac{1}{\sqrt{2}}\left(P+Q-\sqrt{2}R\right) \text{ and } Y = \frac{1}{\sqrt{2}}(Q-P). \end{aligned}$$

Also

$$\begin{aligned} \vec{G} &= \text{algebraic sum of moments of all the forces about } O \\ &= \overrightarrow{OL} \times \frac{P}{\sqrt{2}}\left(\vec{i}-\vec{j}\right) + \overrightarrow{OM} \times \frac{Q}{\sqrt{2}}\left(\vec{i}+\vec{j}\right) + \overrightarrow{OF} \times \left(-R\vec{i}\right) \\ &= \vec{i} \times \frac{P}{\sqrt{2}}\left(\vec{i}-\vec{j}\right) + \left(-\vec{i}\right) \times \frac{Q}{\sqrt{2}}\left(\vec{i}+\vec{j}\right) + 2\vec{j} \times \left(-R\vec{i}\right) \\ &= \frac{1}{\sqrt{2}}\left[-P\vec{k} - Q\vec{k}\right] + 2R\vec{k} \end{aligned}$$

$$\Rightarrow \vec{G} = \frac{\vec{k}}{2}\left(-P - Q + 2\sqrt{2}R\right)$$
$$\Rightarrow G = \frac{1}{\sqrt{2}}\left(-P - Q + 2\sqrt{2}R\right).$$

Now proceed as in first method.

3. Forces P, Q, R act along the lines $x = 0$, $y = 0$ and $x \cos\alpha + y \sin\alpha = p$. Find the magnitude of the resultant and the equation of its line of action.

[LNMU(H) 2011, MU(H) 2010, PU(H) 2007, TMBU(H) 2010, BVABU(H) 2011]

Solution: *First Method.* Let the equation of the line AB be

$$x \cos\alpha + y \sin\alpha = p.$$

Let the forces P, Q and R act along OB, OA and AB respectively.

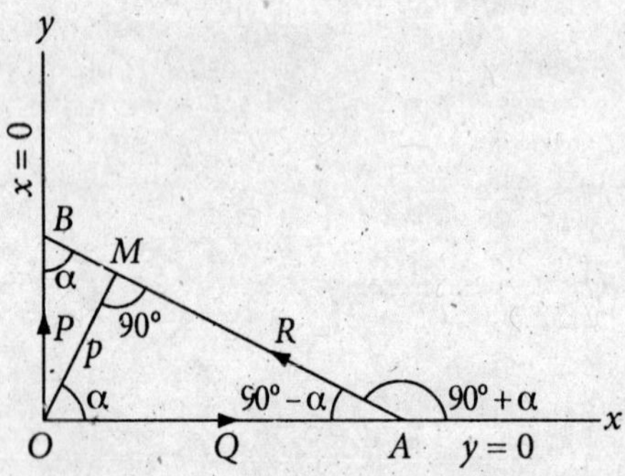

Fig. 2.23

If X, Y be the resolved parts of the forces parallel to the axes and G the sum of their moments about O, then

$$X = Q + R\cos(90 + \alpha) = Q - R\sin\alpha$$
$$Y = P + R\cos\alpha.$$

Then the magnitude of the resultant is given by

$$\sqrt{X^2 + Y^2} = \sqrt{(Q - R\sin\alpha)^2 + (P + R\cos\alpha)^2}$$
$$= \sqrt{Q^2 + R^2\sin^2\alpha - 2QR\sin\alpha + P^2 + R^2\cos^2\alpha + 2PR\cos\alpha}$$
$$= \sqrt{P^2 + Q^2 + R^2 - 2QR\sin\alpha + 2PR\cos\alpha}.$$

Also G = the algebraic sum of the moments of all the forces about $O = Rp$.

The equation to the line of action of resultant is

$$G + yX - xY = 0,$$
$$\Rightarrow Rp + y(Q - R\sin\alpha) - x(P + R\cos\alpha) = 0$$
$$\Rightarrow x(P + R\cos\alpha) - y(Q - R\sin\alpha) - Rp = 0.$$

Second Method. (Vector Method). See figure of first method.

Let $\vec{i}$ and $\vec{j}$ be the unit vectors along the x-axis and y-axis respectively.

In ΔOAM,

$$\cos\alpha = \frac{OM}{OA} = \frac{p}{OA} \Rightarrow OA = p\sec\alpha.$$

In ΔOBM,

$$\sin\alpha = \frac{OM}{OB} = \frac{p}{OB} \Rightarrow OB = p\,\text{cosec}\,\alpha.$$

Now

$$\overrightarrow{AB} = \overrightarrow{AO} + \overrightarrow{OB} = -OA\vec{i} + OB\,\vec{j} = -p\sec\alpha\vec{i} + p\,\text{cosec}\,\alpha\vec{j}.$$

So unit vector along $\overrightarrow{AB}$

$$= \frac{p\left(\text{cosec}\,\alpha\vec{j} - \sec\alpha\vec{i}\right)}{p\sqrt{\text{cosec}^2\alpha + \sec^2\alpha}} = \cos\alpha\vec{j} - \sin\alpha\vec{i}$$

$$\therefore\ \vec{R} = R\left(\cos\alpha\vec{j} - \sin\alpha\vec{i}\right).$$

Also $\vec{Q} = Q\vec{i}$ and $\vec{P} = P\vec{j}$.

Let $\vec{F}$ be the resultant of the given forces, then

$$\vec{F} = \vec{P} + \vec{Q} + \vec{R} = P\vec{j} + Q\vec{i} + R\left(\cos\alpha\vec{j} - \sin\alpha\vec{i}\right) = \vec{i}(Q - R\sin\alpha) + \vec{j}(P + R\cos\alpha)$$

$$\Rightarrow F = \sqrt{(Q - R\sin\alpha)^2 + (P + R\cos\alpha)^2}$$

$$= \sqrt{P^2 + Q^2 + R^2 - 2QR\sin\alpha + 2PR\cos\alpha}.$$

Also

$$\vec{G} = \text{algebraic sum of moments of all the forces about } O$$

$$= \overrightarrow{OA} \times \vec{R} = p\sec\alpha\vec{i} \times R\left(\cos\alpha\vec{j} - \sin\alpha\vec{i}\right)$$

$$= pR\left(\vec{i} \times \vec{j}\right) = pR\vec{k}$$

$$\Rightarrow G = pR.$$

So equation of line of action of the resultant is

$$G + yX - xY = 0$$

$$\Rightarrow pR + y(Q - R\sin\alpha) - x(P + R\cos\alpha) = 0$$

$$\Rightarrow x(P + R\cos\alpha) - y(Q - R\sin\alpha) - pR = 0.$$

4. A system of forces in one plane reduces to a couple whose moment is G. When each force is turned round its point of application through a right angle, it reduces to a couple of moment H. Show that when each force is turned through an angle α, the system is equivalent to a couple whose moment is $G\cos\alpha + H\sin\alpha$. For what value of α, the moment of the new couple be equal to the moment of the old couple.

[BVABU(H) 1999, RU(H) 1998, CU(H) 1994, KU(H) 2001, VU(H) 2000]

Solution: Let P_1 be one of the forces of the system acting at a point $A_1(x_1, y_1)$ of the rigid body and making an angle θ_1 with the x-axis.

Let the components of P_1 parallel to axes be X_1, Y_1, so for the other forces.

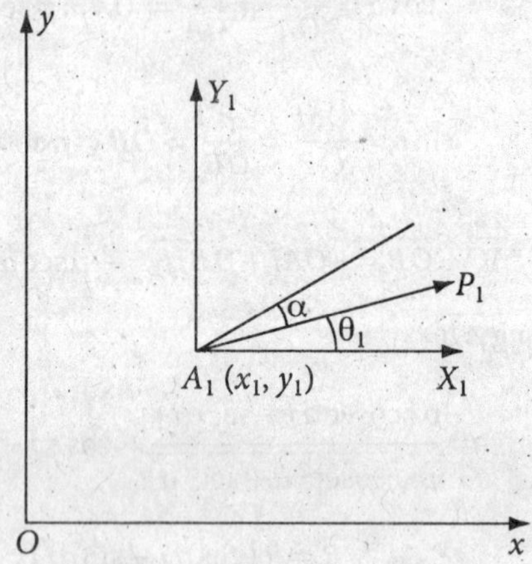

Fig. 2.24

Also if X, Y be the components of the system along the axes and G, the algebraic sum of moments of the forces about the origin, then

$$X = \Sigma X_1 = \Sigma P_1 \cos\theta_1$$

$$Y = \Sigma Y_1 = \Sigma P_1 \sin\theta_1$$

and $G = \Sigma(x_1 Y_1 - y_1 X_1) = \Sigma(x_1 P_1 \sin\theta_1 - y_1 P_1 \cos\theta_1) \Rightarrow G = \Sigma P_1 (x_1 \sin\theta_1 - y_1 \cos\theta_1)$

Since the system reduces to a couple G, therefore, we have $X = 0$, $Y = 0$.

$$\Sigma P_1 \cos\theta_1 = 0 \tag{1}$$

$$\Sigma P_1 \sin\theta_1 = 0 \tag{2}$$

and

$$G = \Sigma P_1 (x_1 \sin\theta_1 - y_1 \cos\theta_1). \tag{3}$$

When each force is turned through a right angle, then putting $\frac{\pi}{2} + \theta_1$ for θ_1 in (1), (2) and (3), we get

$$\Sigma P_1 \cos\left(\frac{\pi}{2} + \theta_1\right) = -\Sigma P_1 \sin\theta_1 = 0 \text{ [from (2)]}$$

$$\Sigma P_1 \sin\left(\frac{\pi}{2} + \theta_1\right) = \Sigma P_1 \cos\theta_1 = 0 \text{ [from (3)].}$$

So the resultant is not a single force, sum of the moment of the forces about the origin, that is

$$H = \Sigma P_1 \left[x_1 \sin\left(\frac{\pi}{2} + \theta_1\right) - y_1 \cos\left(\frac{\pi}{2} + \theta_1\right)\right] \Rightarrow H = \Sigma P_1 (x_1 \cos\theta_1 + y_1 \sin\theta_1). \tag{4}$$

Hence, on account of turning the forces through an angle $\pi/2$, the system reduces only to a couple whose moment H is given in (4).

When each force is turned through an angle α, then putting $\theta_1+\alpha$ for θ, we have

$$\begin{aligned} X' &= \Sigma P_1\cos(\theta_1+\alpha) = \Sigma P_1(\cos\theta_1\cos\alpha - \sin\theta_1\sin\alpha) \\ &= \cos\alpha(\Sigma P_1\cos\theta_1) - \sin\alpha(\Sigma P_1\sin\theta_1) \\ &= \cos\alpha\cdot 0 - \sin\alpha\cdot 0 \text{ [from (1) and (2)].} \end{aligned}$$

$$\begin{aligned} Y' &= \Sigma P_1\sin(\theta_1+\alpha) = \Sigma P_1(\sin\theta_1\cos\alpha + \cos\theta_1\sin\alpha) \\ &= \cos\alpha(\Sigma P_1\sin\theta_1) + \sin\alpha(\Sigma P_1\cos\theta_1) \\ &= \cos\alpha\cdot 0 + \sin\alpha\cdot 0 \text{ [from (1) and (2)]} \\ &= 0. \end{aligned}$$

So the resultant is not a single force.
Also

$$\begin{aligned} G' &= \Sigma P_1\left[x_1\sin(\theta_1+\alpha) - y_1\cos(\theta_1+\alpha)\right] \\ &= \Sigma P_1\left[x_1(\sin\theta_1\cos\alpha + \cos\theta_1\sin\alpha) - y_1(\cos\theta_1\cos\alpha - \sin\theta_1\sin\alpha)\right] \\ &= \Sigma P_1\left[\cos\alpha\,(x_1\sin\theta_1 - y_1\cos\theta_1) + \sin\alpha\,(x_1\cos\theta_1 + y_1\sin\theta_1)\right] \\ &= \cos\alpha\left[\Sigma P_1\,(x_1\sin\theta_1 - y_1\cos\theta_1)\right] + \sin\alpha\left[\Sigma P_1\,(x_1\cos\theta_1 + y_1\sin\theta_1)\right] \\ &= G\cos\alpha + H\sin\alpha \text{ [from (3)].} \end{aligned}$$

Hence, on account of turning of each force through an angle α, the system reduces only to a couple whose moment is $G\cos\alpha + H\sin\alpha$.

For the second part

$$\begin{aligned} \because \text{ Given: } G\cos\alpha + H\sin\alpha = G \Rightarrow G(1-\cos\alpha) &= H\sin\alpha \\ &\Rightarrow 2G\sin^2\frac{\alpha}{2} - 2H\sin\frac{\alpha}{2}\cos\frac{\alpha}{2} = 0 \\ &\Rightarrow 2\sin\frac{\alpha}{2}\left(G\sin\frac{\alpha}{2} - H\cos\frac{\alpha}{2}\right) = 0 \\ &\Rightarrow G\sin\frac{\alpha}{2} - H\cos\frac{\alpha}{2} = 0 \text{ as } \sin\frac{\alpha}{2} \neq 0 \\ &\Rightarrow \tan\frac{\alpha}{2} = \frac{H}{G} \Rightarrow \alpha = 2\tan^{-1}\left(\frac{H}{G}\right). \end{aligned}$$

5. A uniform beam of length $2a$ rests in equilibrium with one end resting against a smooth vertical wall and with a point of its length resting upon a smooth horizontal rod which is parallel to the wall and at a distance b from it. Show that inclination of the beam to the vertical is $\sin^{-1}(b/a)^{1/3}$. [MU(H) 2001, TMBU(H) 2009, BVABU(H) 2008]

Solution: Let the reaction R perpendicular to the wall and the reaction S perpendicular to the rod AB acting through C meet at O, then for the sake of equilibrium the vertical through G along which acts weight W of the rod will also pass through O.
Here $AB = 2a$, $AG = GB = a$, $CL = b$.

$$\angle AOG = \angle ACO = \angle ALC = \pi/2 = \angle LAO$$
$$\angle LAG = \angle AGO = \theta,\ \angle CAO = \pi/2 - \theta.$$

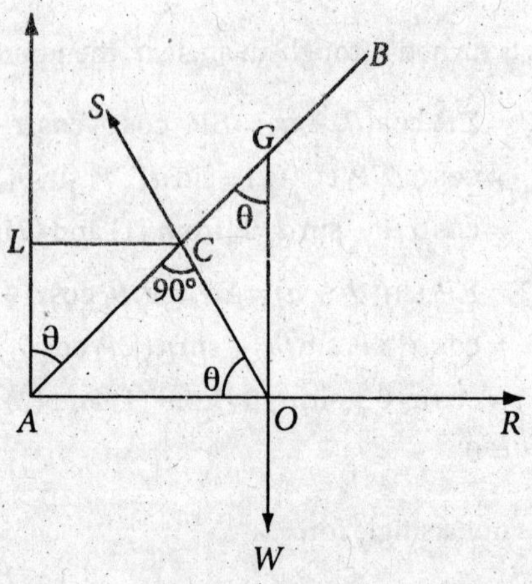

Fig. 2.25

From ΔACL,

$$\sin\theta = CL/AC. \tag{1}$$

From ΔCAO,

$$\sin\theta = AC/OA. \tag{2}$$

From ΔAOG,

$$\sin\theta = OA/AG. \tag{3}$$

Multiplying (1), (2) and (3), we get

$$\sin^3\theta = \frac{CL}{AC}\cdot\frac{AC}{OA}\cdot\frac{OA}{AG} = \frac{CL}{AG} = \frac{b}{a};\ \sin\theta = \left(\frac{b}{a}\right)^{1/3}.$$

Hence,

$$\theta = \sin^{-1}(b/a)^{1/3}.$$

6. A heavy uniform rod of length $2a$ rests partly within and partly without a fixed hemispherical bowl of radius r, the rim of the bowl is horizontal. One end of the rod is in contact with the rim and the other is on the surface. If θ be the inclination of the rod to the horizontal, show that $2r\cos 2\theta = a\cos\theta$.

Hence, show that greatest inclination to the horizontal which a heavy uniform rod can rest partly within and partly without a fixed hemispherical bowl is $\sin^{-1}\sqrt{1/3}$.

Solution: Let $AB = 2a =$ length of the rod, $OD = r =$ radius of the hemispherical bowl.

The forces acting on the rod are:

(i) The reaction R through the centre O;

(ii) The reaction S at D, at right angle to AB;

(iii) The weight W of the rod through its CG G, vertically downwards.

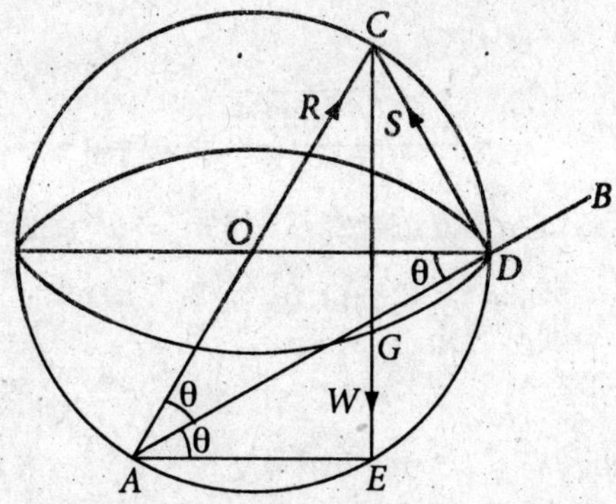

Fig. 2.26

The forces R and S meet on the circle at C. Hence, for equilibrium, the third force W must also pass through C and as such CG is vertical.

Also $\angle AEC = 90°$ since angle at the semicircle is a right angle, therefore, AE must be horizontal. Given $\angle BAE = \theta$.

$\therefore \angle OAD = \angle ADO = \angle BAE = \theta$.

In $\triangle ACE$,

$$\cos CAE = AE/AC \text{ or, } \cos 2\theta = AE/2r.$$

$$\therefore AE = 2r\cos 2\theta. \tag{1}$$

In $\triangle GAE$,

$$\cos\theta = AE/AG = AE/a.$$

$$\therefore AE = a\cos\theta \left[\because AG = GB = \tfrac{1}{2}AB = a\right]. \tag{2}$$

From (1) and (2),

$$a\cos\theta = 2r\cos 2\theta \text{ or, } a\cos\theta = 2r\left(2\cos^2\theta - 1\right) \text{ or, } 4r\cos^2\theta - a\cos\theta - 2r = 0.$$

$$\therefore \cos\theta = \frac{a \pm \sqrt{a^2 + 32r^2}}{8r}.$$

Taking positive sign as $\cos\theta$ is positive, we get

$$\cos\theta = \frac{a + \sqrt{a^2 + 32r^2}}{8r}. \tag{3}$$

From (3), we find that as a decreases, $\cos\theta$ decreases and hence, θ increases.

The angle is, therefore, greatest when the rod has the least length which is AD. It can be easily seen that

$$AD = 2a = 2r\cos\theta \text{ or, } \cos\theta = a/r. \tag{4}$$

Hence, from (3), we get

$$\frac{a}{r} = \frac{a+\sqrt{a^2+32r^2}}{8r} \Rightarrow 8a = a+\sqrt{a^2+32r^2}$$
$$\Rightarrow 7a = \sqrt{a^2+32r^2} \Rightarrow 49a^2 = a^2+32r^2$$
$$\Rightarrow 48a^2 = 32r^2 \text{ or, } 3a^2 = 2r^2,\ a/r = \sqrt{2/3}.$$

So from (4), we get

$$\cos\theta = \sqrt{2/3} \text{ or, } \sin\theta = \sqrt{1-\cos^2\theta} = \sqrt{1-\tfrac{2}{3}} = 1/\sqrt{3}.$$

$$\therefore\ \theta = \sin^{-1} 1/\sqrt{3}.$$

7. A square of side $2a$ is placed with its plane vertical between two smooth pegs which are in the same horizontal line at a distance c apart. Show that it will be in equilibrium when the inclination of one of the edges to the horizontal is either $\frac{\pi}{4}$ or, $\frac{1}{2}\sin^{-1}\left(\frac{a^2-c^2}{c^2}\right)$.

[CU(H) 2005, PU(H) 2003, MU(H) 1994]

Solution: Let the sides AB and AD of the square $ABCD$ be taken as axes of coordinates and AB be inclined at an angle θ to the horizontal. Let E and F be the pegs in the horizontal line on which the sides AB and AD rest respectively, and let the reactions at E and F meet at O, through which must also pass the vertical through G. Hence, GO is perpendicular to EF.

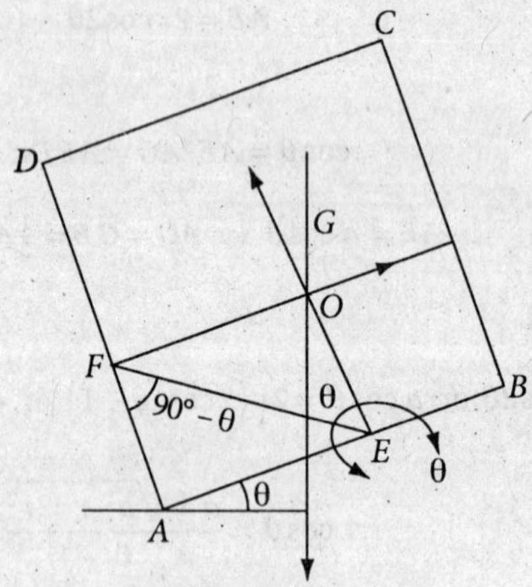

Fig. 2.27

Let m_1 be the slope of GO and m_2 be the slope of EF.

We shall now find the coordinates of G, O, E and F referred to AB and AD as axes. $G = (a, a)$.

In ΔAEF,

$$\cos\theta = AE/EF = AE/c \Rightarrow AE = c\cos\theta.$$

Also

$$\sin\theta = AF/EF = AF/c \Rightarrow AF = c\sin\theta.$$

Thus, coordinates of O, E and F are

$$(c\cos\theta, c\sin\theta),\ (c\cos\theta, 0),\ (0, c\sin\theta).$$

$$m_1 = \frac{c\sin\theta - a}{c\cos\theta - a} \text{ and } m_2 = \frac{c\sin\theta - 0}{0 - c\cos\theta} = \tan\theta.$$

Since GO is perpendicular to EF,

$$m_1 m_2 = -1$$
$$\Rightarrow \frac{c\sin\theta - a}{c\cos\theta - a} \times \frac{-\sin\theta}{\cos\theta} = -1$$
$$\Rightarrow c\sin^2\theta - a\sin\theta = c\cos^2\theta - a\cos\theta$$
$$\Rightarrow c(\cos^2\theta - \sin^2\theta) = a(\cos\theta - \sin\theta)$$
$$\Rightarrow c(\cos\theta - \sin\theta)(\cos\theta + \sin\theta) - a(\cos\theta - \sin\theta) = 0$$
$$\Rightarrow (\cos\theta - \sin\theta)\{c(\cos\theta + \sin\theta) - a\} = 0.$$

If $\cos\theta - \sin\theta = 0$, i.e., $\tan\theta = 1, \theta = \pi/4$.
If $c(\cos\theta + \sin\theta) = a$, squaring, we get

$$c^2(\cos\theta + \sin\theta)^2 = a^2$$
$$\Rightarrow c^2(\cos^2\theta + \sin^2\theta + 2\sin\theta\cos\theta) = a^2$$
$$\Rightarrow c^2(1 + \sin 2\theta) = a^2$$
$$\Rightarrow 1 + \sin 2\theta = \frac{a^2}{c^2} \Rightarrow \sin 2\theta = \frac{a^2}{c^2} - 1 = \frac{a^2 - c^2}{c^2} \Rightarrow \theta = \frac{1}{2}\sin^{-1}\left(\frac{a^2 - c^2}{c^2}\right).$$

8. A triangular lamina ABC, right angle at A, rests with its plane vertical and with sides AB and AC supported by smooth fixed pegs, D, E, in horizontal line. Prove that the inclination θ of AC to the horizontal is given by $AC\cos\theta - AB\sin\theta = 3DE\cos 2\theta$.

Solution: Let $AC = b$ and $AB = c$. Then take these lines as axes of coordinates, A being the origin. Thus, coordinates of B and C are $(0, c)$ and $(b, 0)$. Also the coordinates of centroid G are $(b/3, c/3)$.

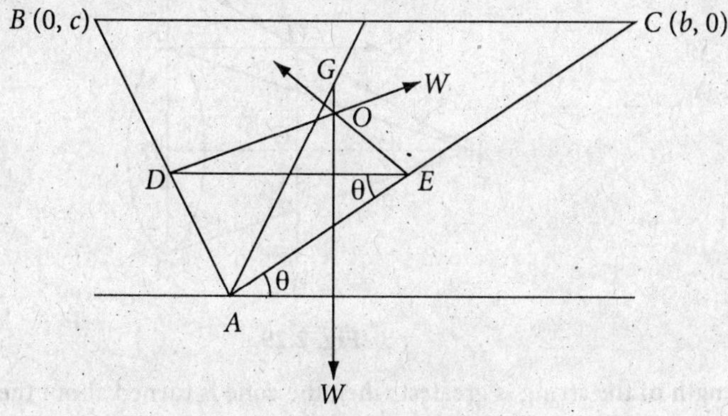

Fig. 2.28

Let $DE = d$.
In ΔADE,

$$\cos\theta = AE/DE = AE/d$$
$$\Rightarrow AE = d\cos\theta \text{ and } \sin\theta = AD/DE = AD/d$$
$$AD = d\sin\theta.$$

Thus, coordinates of O are $(d\cos\theta, d\sin\theta)$. Also coordinates of E and D are $(d\cos\theta, 0)$ and $(0, d\sin\theta)$

Now GO is perpendicular to DE.

Let m_1 and m_2 be the slopes of GO and DE.

Then

$$m_1 = \frac{d\sin\theta - \frac{c}{3}}{d\cos\theta - \frac{b}{3}} = \frac{3d\sin\theta - c}{3d\cos\theta - b}, \; m_2 = \frac{0 - d\sin\theta}{d\cos\theta - 0} = -\tan\theta.$$

Since GO is perpendicular to DE,

$$\therefore m_1 m_2 = -1$$
$$\Rightarrow \frac{3d\sin\theta - c}{3d\cos\theta - b} \times \frac{-\sin\theta}{\cos\theta} = -1$$
$$\Rightarrow 3d\sin^2\theta - c\sin\theta = 3d\cos^2\theta - b\cos\theta,$$
$$\Rightarrow b\cos\theta - c\sin\theta = 3d(\cos^2\theta - \sin^2\theta),$$
$$\Rightarrow AC\cos\theta - AB\sin\theta = 3DE\cos 2\theta \text{ (Proved).}$$

9. A solid cone of height h and semi-vertical angle α is placed with its base against a smooth vertical wall and is supported by a string attached to its vertex and to a point in the wall. Show that the greatest possible length of the string is $h\sqrt{1 + \frac{16}{9}\tan^2\alpha}$.

[RU(H) 1996, MU(H) 2011, TMBU(H) 2007]

Solution: Let A be the vertex of the cone and BC its base. Let AD be the axis of the cone and G its CG, then $GD = h/4$.

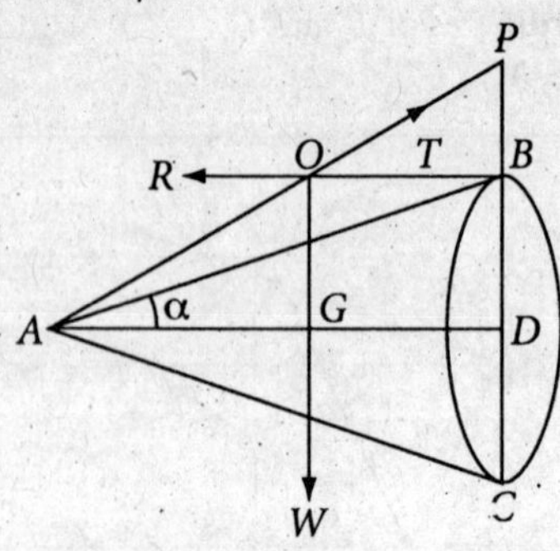

Fig. 2.29

The length of the string is greatest when the cone is turned about the point B. Thus, the reaction of the wall passes through B. Let it meet the vertical through G in O. Then AO gives the direction of the string which is tied to the wall at P.

In ΔABD,

$$\tan\alpha = BD/AD = BD/h$$
$$OG = BD = h\tan\alpha,$$
$$AG = AD - GD = h - \frac{h}{4} = \frac{3}{4}h.$$

In ΔOAG,

$$OA^2 = OG^2 + AG^2 = h^2\tan^2\alpha + \frac{9}{16}h^2 = \frac{9}{16}h^2\left(1 + \frac{16}{9}\tan^2\alpha\right)$$

$$\Rightarrow OA = \frac{3}{4}h\sqrt{1 + \frac{16}{9}\tan^2\alpha}. \quad (1)$$

Also ΔAOG and ΔAPD are similar;

$$\therefore \frac{AO}{AP} = \frac{AG}{AD} = \frac{\frac{3}{4}h}{h} = \frac{3}{4}.$$

$$\therefore AP = \frac{4}{3}OA = \frac{4}{3}\cdot\frac{3}{4}h\sqrt{1 + \frac{16}{9}\tan^2\alpha} = h\sqrt{1 + \frac{16}{9}\tan^2\alpha} \text{ [from (1)]}.$$

10. Two uniform rods AB, BC rigidly jointed at B so that angle ABC is a right angle, hang freely in equilibrium from a fixed point A. The lengths of the rods are a and b and their weights are wa and wb. Prove that if AB makes an angle θ with the vertical, then $\tan\theta = \frac{b^2}{a^2+2ab}$.

Solution: We consider the equilibrium of the two rods combined, their mutual actions at B neutralise and the forces left are then

(i) weight aw acting through D, the middle point of AB,
(ii) weight bw acting through E, the middle point of BC,
(iii) the action at the fixed point A.

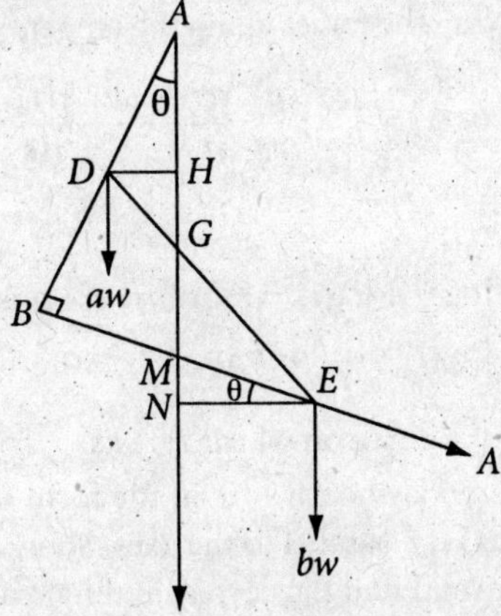

Fig. 2.30

Here the three forces are in equilibrium, therefore, common CG of aw and bw the point G is vertically below A, hence AG is vertical. Taking moments about the point A, we have

$$aw\cdot DH - bw\cdot EN = 0 \Rightarrow awAD\sin\theta = bwEM\cos\theta$$
$$\Rightarrow aw\cdot\frac{a}{2}\sin\theta = bw(BE-BM)\cos\theta$$
$$\Rightarrow \frac{a^2}{2}\sin\theta = b\left(\frac{b}{2}-a\tan\theta\right)\cos\theta$$
$$\Rightarrow a^2\sin\theta = b(b\cos\theta - 2a\sin\theta)$$
$$\Rightarrow \left(a^2+2ab\right)\sin\theta = b^2\cos\theta$$
$$\Rightarrow \tan\theta = \frac{b^2}{a^2+2ab}\ \textbf{(Proved).}$$

11. A particle P is acted on by forces μ/r_1 and μ/r_2 respectively, towards two fixed points O_1 and O_2. Show that it will rest at any point of a smooth groove whose equation is $r_1r_2 =$ constant, where $O_1P = r_1$ and $O_2P = r_2$. [TMBU(H) 1998]

Solution: We suppose that O_1P and O_2P make angles ϕ_1 and ϕ_2 respectively, with the tangent TP at P of the curve, where $O_1P = r_1$, $O_2P = r_2$.

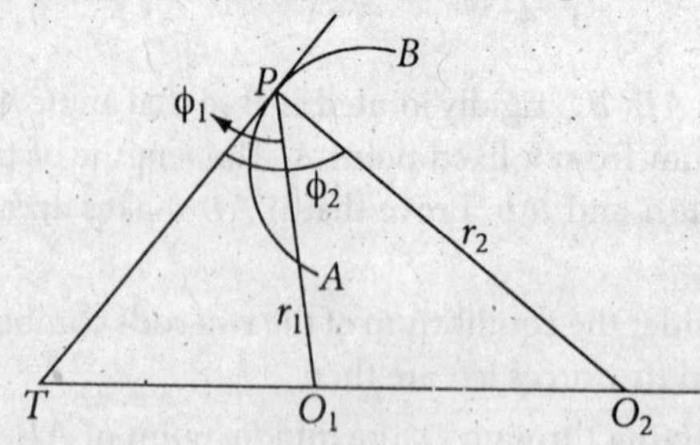

Fig. 2.31

Let arc $AP = s$, where A is a fixed point on the curve.

For equilibrium, resolving the forces along the tangent PT, we get

$$\frac{\mu}{r_1}\cos\phi_1 + \frac{\mu}{r_2}\cos\phi_2 = 0 \text{ or, } \frac{\mu}{r_1}\cdot\frac{dr_1}{ds} + \frac{\mu}{r_2}\cdot\frac{dr_2}{ds} = 0 \text{ or, } \frac{dr_1}{r_1} + \frac{dr_2}{r_2} = 0.$$

Integrating, we get

$$\log r_1 + \log r_2 = \log a, \text{ where } a \text{ is a constant}$$
$$\text{or, } \log(r_1r_2) = \log a \text{ or, } r_1r_2 = a.$$

This is the equation of the curve on which P rests.

12. A bead is free to move on a smooth wire in the form of a circle $x^2+y^2=a^2$ and is acted upon by forces μx, μy parallel to the axes. Show that it can rest in equilibrium at any point on the wire and find the corresponding reaction.

Solution: The circular wire is

$$x^2+y^2=a^2. \tag{1}$$

We take a point (x,y) on it, then the components of the forces parallel to the axes are given by

$$X=\mu x,\ Y=\mu y. \tag{2}$$

Differentiating (1) w.r.t. x, we get

$$2x+2y\frac{dy}{dx}=0 \text{ or, } \frac{dy}{dx}=-\frac{x}{y}. \tag{3}$$

Now the position of equilibrium is given by

$$X+Y\frac{dy}{dx}=0$$

$$X-Y\frac{x}{y}=0 \text{ [from (3)].}$$

From (2),

$$\mu x-\mu y\frac{x}{y}=0, \text{ i.e., } 0=0.$$

This is true for all positions of the bead on the circle (1). Hence, the bead can rest in equilibrium at any point on the wire. Also the reaction R is given by

$$R=\sqrt{X^2+Y^2}=\sqrt{\mu^2\left(x^2+y^2\right)}=\mu a \text{ [from (1) and (2)].}$$

13. A hemisphere of radius a and weight w is placed with its curved surface on a smooth table and a string of length l $(< a)$ is attached to a point on its rim and a point on the table. Find the position of equilibrium and prove that the tension of the string is $\frac{3W}{8}\cdot\frac{a-l}{\sqrt{2al-l^2}}$. [MU(H) 1993]

Solution: Let OC be the central radius of the hemisphere. Then G, the CG of the hemisphere lies on OC such that $OG=3a/8$.

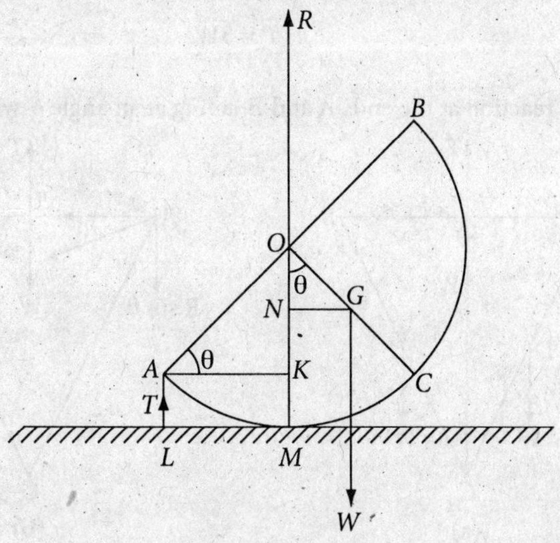

Fig. 2.32

Let M be the point of the hemisphere in contact with the smooth table.

The string is $AL = l$. Let T be the tension in the string. The weight W of the hemisphere acts vertically downwards through G. The reaction R at M acts vertically upwards and passes through the centre O.

From figure,

$$MO = MK + KO = AL + KO = l + OA\cdot\sin\theta = l + a\sin\theta$$
$$\Rightarrow a = l + a\sin\theta \Rightarrow a - l = a\sin\theta \Rightarrow \sin\theta = (a-l)/a.$$

$$\therefore \tan\theta = \frac{a-l}{\sqrt{2al - a^2}}, \qquad (1)$$

which gives the position of equilibrium.

Also taking moments about O, we have

$$T\cdot AK = W\cdot NG \Rightarrow T\cdot OA\cos\theta = W\cdot OG\sin\theta$$
$$\Rightarrow Ta\cos\theta = W\frac{3a}{8}\sin\theta \Rightarrow T = \frac{3W}{8}\tan\theta = \frac{3W}{8}\frac{a-l}{\sqrt{2al-l^2}} \text{ (Proved).}$$

14. Three equal uniform rods each of weight W are smoothly jointed so as to form an equilateral triangle. If the system be supported at the middle point of one of the rods, show that the action at the lowest angle is $\frac{\sqrt{3}}{6}W$ and that at each of the other is $W\sqrt{\frac{13}{12}}$.

[LNMU(H) 1996]

Solution: Let three given rods be AB, BC and CA. The system is supported at the middle point E of the rod AB. Let T be the reaction at E.

Resolving the forces, acting on the whole system in Fig. 2.33(a), vertically, we get

$$T = 3W. \qquad (1)$$

Let R be the reaction at the ends A and B acting at an angle θ with AB.

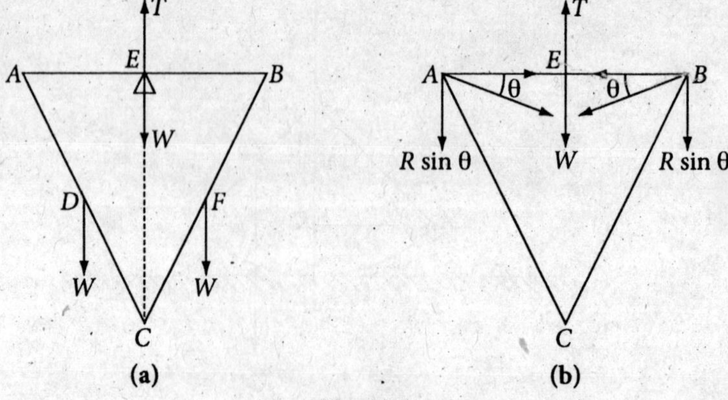

Fig. 2.33

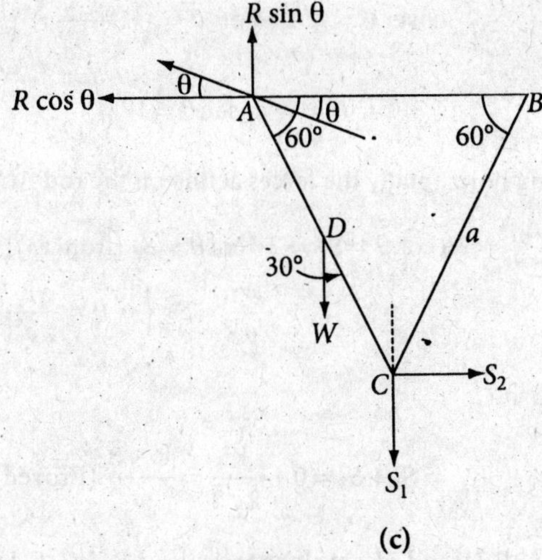

(c)

Fig. 2.33

We consider the forces acting on the rod AB, and resolving these forces vertically in Fig. 2.33(b), we get

$$T = W + 2R\sin\theta. \tag{2}$$

From (1) and (2), we get

$$3W = W + 2R\sin\theta \Rightarrow 2W = 2R\sin\theta \Rightarrow W = R\sin\theta \Rightarrow R = W\,\text{cosec}\,\theta. \tag{3}$$

We consider the forces acting on the rod AC. These are the reaction R acting at A. The direction of R acting on the rod AC in Fig. 2.33(c) will be in the direction reverse to that in case of Fig. 2.33(b), where we considered the force acting on the rod AB, the weight W acting at the mid-point D of the rod AC and the horizontal and vertical components S_2 and S_1 of the reaction at C.
Resolving vertically the forces acting on the rod AC, we get

$$R\sin\theta = W + S_1 \Rightarrow W = W + S_1 \text{ [from (3)]}$$
$$\Rightarrow S_1 = 0,$$

i.e., the reaction at C is horizontal.
Taking moment of the forces acting on the rod AC about C, we get

$$Wa\sin 30° - R\sin\theta 2a\sin 30° + R\cos\theta \cdot 2a\sin 60° = 0$$
$$\Rightarrow \frac{W}{2} - R\sin\theta + 2R\cos\theta\frac{\sqrt{3}}{2} = 0$$
$$\Rightarrow \frac{W}{2} - W + W\sqrt{3}\cot\theta = 0 \text{ [from (3)]}$$
$$\Rightarrow \cot\theta = \frac{1}{2\sqrt{3}}.$$

$$\therefore\ \operatorname{cosec}\theta = \sqrt{1+\cot^2\theta} = \sqrt{1+\frac{1}{12}} = \sqrt{\frac{13}{12}}.$$

$$\therefore\ R = W\operatorname{cosec}\theta = W\sqrt{\frac{13}{12}}.$$

Again, resolving horizontally the forces acting on the rod AC, we get

$$R\cos\theta = S_2 \Rightarrow W\cot\theta = S_2 \text{ [from (3)]}$$
$$\Rightarrow \frac{W}{2\sqrt{3}} = S_2 \left(\because\ \theta = \frac{1}{2\sqrt{3}}\right).$$

So the reaction at C

$$= S_1 + S_2 = 0 + \frac{W}{2\sqrt{3}} = \frac{W\sqrt{3}}{6} \text{ (Proved).}$$

15. Forces equal to $3P$, $7P$ and $5P$ act along the sides AB, BC, CA of an equilateral triangle ABC. Find the magnitude, direction and line of action of the resultant.

[LNMU(H) 1999, TMBU(H) 1990, PU(H) 2013]

Solution: Let $\hat{a}$, $\hat{b}$, $\hat{c}$ be the unit vectors along the sides AB, BC and CA of $\triangle ABC$. Let $\vec{F}$ be the resultant of these forces. Then

$$\vec{F} = P\left(3\hat{a}+7\hat{b}+5\hat{c}\right) \Rightarrow F^2 = P^2\left[9+49+25+42\left(\hat{a}\cdot\hat{b}\right)+70\left(\hat{b}\cdot\hat{c}\right)+30(\hat{a}\cdot\hat{c})\right]$$
$$\Rightarrow F^2 = P^2[83+142\cos 120°] = P^2[83-71] = 12P^2$$
$$\Rightarrow F = 2\sqrt{3}P.$$

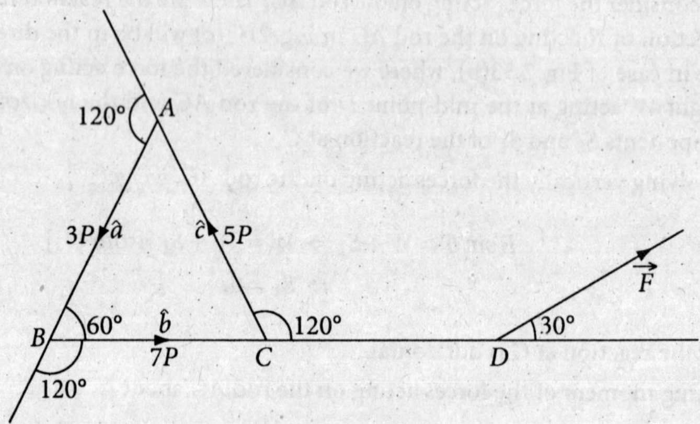

Fig. 2.34

If $\vec{F}$ makes angle θ with $\hat{b}$ then

$$\vec{F}\cdot\hat{b} = P\left[3\left(\hat{a}\cdot\hat{b}\right)+7+5\left(\hat{b}\cdot\hat{c}\right)\right] \Rightarrow F\cos\theta = P[8\cos 120°+7] = P\left(-\tfrac{8}{2}+7\right)$$
$$\Rightarrow 2\sqrt{3}\,P\cos\theta = 3P \Rightarrow \cos\theta = \sqrt{3}/2$$
$$\Rightarrow \theta = 30°.$$

So the resultant makes an angle 30° with BC.

If $\vec{F}$ act at D, then taking moment of all the forces about D, we get

$$\overrightarrow{DB}\times 3P\hat{a}+\overrightarrow{DC}\times 5P\hat{c}=0\Rightarrow \left|\overrightarrow{DB}\times 3\hat{a}\right|=\left|\overrightarrow{CD}\times 5\hat{c}\right|$$
$$\Rightarrow 3BD\sin 120=5CD\sin 120$$
$$\Rightarrow 3BC=5CD\Rightarrow\ CD=\tfrac{3}{2}BC.$$

So the resultant acts at a point on D, which lies on the line BC produced such that $CD=\frac{3}{2}BC$.

16. A frustum of a uniform right circular cone whose semi-vertical angle is α, is made by cutting $1/n$th of the axis. Prove that the frustum will rest with a slant side on a horizontal plane if $\tan^2\alpha<\frac{1}{4}\frac{3n^4-4n^3+1}{n^3-1}$. [PU(H) 1991]

Solution: The figure is the section of the cone by vertical plane through its CG.

Let $AD=h$ be the axis of the cone and BCC_1B_1, the frustum. Let the centres of gravity of the cone ABC, cone AB_1C_1 and the frustum BCC_1B_1 be G_1, G_2 and G respectively.

Then

$$AG_1=\frac{3}{4}h,\ AG_2=\frac{3}{4}\frac{h}{n}.$$

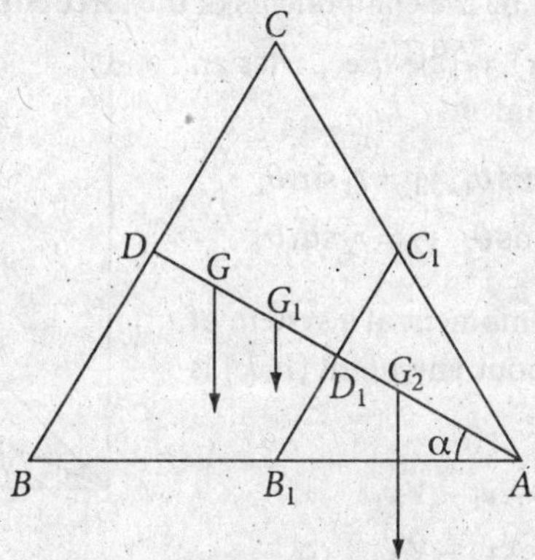

Fig. 2.35

If the weight of the cone ABC be W, then the weight of the cone AB_1C_1 is W/n^3. Hence, taking moments about A, we get

$$W\cdot\frac{3h}{4}-\frac{W}{n^3}\cdot\frac{3h}{4n}=\left(1-\frac{1}{n^3}\right)W\cdot AG. \tag{1}$$

The condition for equilibrium is that the line of action of the weights through G must pass through a point between B and B_1 and so

$$AG\cos\alpha>AB_1=AD_1\sec\alpha=\frac{h}{n}\sec\alpha.$$

Using (1), we have

$$\frac{\frac{3h}{4}\left(1-\frac{1}{n^4}\right)}{1-\frac{1}{n^3}}\cos\alpha > \frac{h}{n}\sec\alpha \Rightarrow \frac{3}{4}\left(1-\frac{1}{n^4}\right) > \sec^2\alpha\left(\frac{1}{n}-\frac{1}{n^4}\right)$$

$$\Rightarrow \frac{3}{4}-\frac{3}{4n^4} > \left(\frac{1}{n}-\frac{1}{n^4}\right)(1+\tan^2\alpha)$$

$$\Rightarrow \frac{3}{4}-\frac{3}{4n^4}-\frac{1}{n}+\frac{1}{n^4} > \left(\frac{1}{n}-\frac{1}{n^4}\right)\tan^2\alpha$$

$$\Rightarrow \frac{3n^4-3-4n^3+4}{4n^4} > \frac{(n^3-1)\tan^2\alpha}{n^4}$$

$$\Rightarrow \frac{3n^4-4n^3+1}{4} > (n^3-1)\tan^2\alpha$$

$$\Rightarrow \tan^2\alpha < \frac{3n^4-4n^3+1}{4(n^3-1)} \quad \textbf{(Proved).}$$

17. Moments of the resultant R of a system of coplanar forces about the three points O, A and B lying in the plane of the forces are G, $G+J_1$ and $G+J_2$ respectively. If referred to O as the origin, the polar coordinates of A and B be (r_1, θ_1) and (r_2, θ_2), then show that $R^2\sin(\theta_1-\theta_2) = \frac{J_1^2}{r_1^2}+\frac{J_2^2}{r_2^2}-\frac{2J_1J_2\cos(\theta_1-\theta_2)}{r_1r_2}$. [CU(H) 2002, KU(H) 2001]

Solution: Let X, Y be the components of the force R parallel to Ox and Oy.

Let (x_1, y_1) and (x_2, y_2) be the Cartesian coordinates of A and B.

$$\therefore\ x_1 = r_1\cos\theta_1,\ y_1 = r_1\sin\theta_1$$

$$\text{and } x_2 = r_2\cos\theta_2,\ y_2 = r_2\sin\theta_2$$

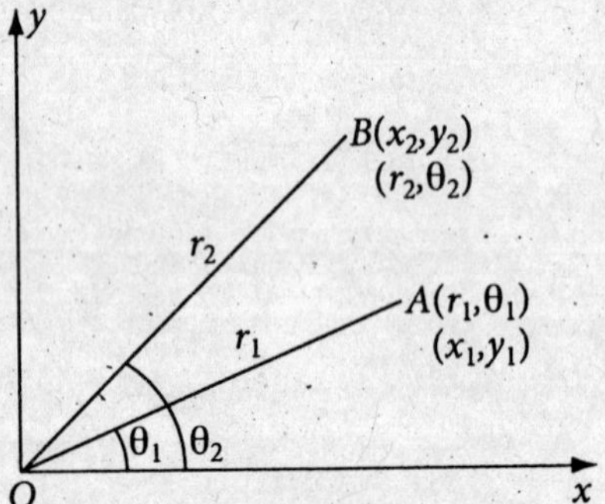

Fig. 2.36

We know that the moment of a system of coplanar forces about any point (h, k) is $G+Xk-Yh$.

$$\therefore\ G+J_1 = G+Xy_1-Yx_1$$

$$\text{and } G+J_2 = G+Xy_2-Yx_2.$$

$$G+J_1 = G+Xr_1\sin\theta_1-Yr_1\cos\theta_1$$

$$\text{and } G+J_2 = G+Xr_1\sin\theta_r-Yr_2\cos\theta_2.$$

$$J_1 = r_1\sin\theta_1 X - r_1\cos\theta_1 Y$$

$$J_1/r_1 = X\sin\theta_1 - Y\cos\theta_1$$

$$\text{and } J_2/r_2 = X\sin\theta_2 - Y\cos\theta_2.$$

$$X\sin\theta_1 - Y\cos\theta_1 - (J_1/r_1) = 0$$

$$\text{and } X\sin\theta_2 - Y\cos\theta_2 - (J_2/r_2) = 0.$$

$$\therefore \frac{X}{(J_2/r_2)\cos\theta_1-(J_1/r_1)\cos\theta_2}=\frac{Y}{-(J_2/r_2)\sin\theta_1+(J_1/r_1)\sin\theta_2}$$
$$=\frac{1}{-\sin\theta_1\cos\theta_2+\sin\theta_2\cos\theta_1}$$
$$=\frac{1}{-\sin(\theta_1-\theta_2)}.$$

$$X=\frac{(J_1/r_1)\cos\theta_2-(J_1/r_1)\cos\theta_1}{\sin(\theta_1-\theta_2)}$$
$$Y=\frac{(J_2/r_2)\sin\theta_1-(J_1/r_2)\sin\theta_2}{\sin(\theta_1-\theta_2)}.$$

Now

$$R^2=X^2+Y^2=\frac{1}{\sin^2(\theta_1-\theta_2)}\left[\left(\frac{J_1}{r_1}\cos\theta_2-\frac{J_2}{r_2}\cos\theta_1\right)^2+\left(\frac{J_2}{r_2}\sin\theta_1-\frac{J_1}{r_1}\sin\theta_2\right)^2\right].$$

$$\Rightarrow R^2\sin^2(\theta_1-\theta_2)=\frac{J_1^2}{r_1^2}\cos^2\theta_2+\frac{J_2^2}{r_2^2}\cos^2\theta_1-\frac{2J_1J_2}{r_1r_2}\cos\theta_1\cos\theta_2$$
$$+\frac{J_2^2}{r_2^2}\sin^2\theta_1+\frac{J_1^2}{r_1^2}\sin^2\theta_2-\frac{2J_1J_2}{r_1r_2}\sin\theta_1\sin\theta_2.$$

$$\Rightarrow R^2\sin^2(\theta_1-\theta_2)=\frac{J_1^2}{r_1^2}\left(\cos^2\theta_2+\sin^2\theta_2\right)+\frac{J_2^2}{r_2^2}\left(\cos^2\theta_1+\sin^2\theta_1\right)$$
$$-\frac{2J_1J_2}{r_1r_2}(\cos\theta_1\cos\theta_2+\sin\theta_1\sin\theta_2).$$

$$\Rightarrow R^2\sin^2(\theta_1-\theta_2)=\frac{J_1^2}{r_1^2}+\frac{J_2^2}{r_2^2}-\frac{2J_1J_2}{r_1r_2}\cos(\theta_1-\theta_2)\ \textbf{(Proved)}.$$

18. M_1, M_2, M_3 are the moments of a system of forces acting in the xy-plane about any three non-collinear points (x_1,y_1), (x_2,y_2), (x_3,y_3) respectively. If the resultant of the system is a single force at the origin, then show that

$$M_1(x_2y_3-x_3y_2)+M_2(x_3y_1-x_1y_3)+M_3(x_1y_2-x_2y_1)=0.$$

Find the magnitude and direction of the resultant. **[CU(H) 1999, KU(H) 2004]**

Solution: Let A, B, C be the three non-collinear given points.

Let X, Y be the sum of components of a system of forces along coordinate axes Ox and Oy. Let G be the moment of the system about O.

$\therefore$ $M_1=G-x_1Y+y_1X$, $M_2=G-x_2Y+y_2X$ and $M_3=G-x_3Y+y_3X$.

If the system reduces to a single force at $O(0,0)$, then $G=0$.

$$\therefore M_1+x_1Y-y_1X=0 \quad (1)$$
$$M_2+x_2Y-y_3X=0 \quad (2)$$
$$\text{and } M_3+x_3Y-y_3X=0. \quad (3)$$

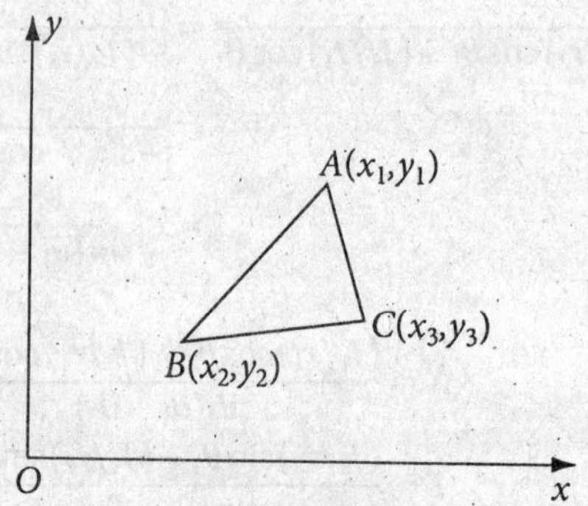

Fig. 2.37

Eliminating X, Y from (1), (2), (3), we get

$$\begin{vmatrix} M_1 & x_1 & y_1 \\ M_2 & x_2 & y_2 \\ M_3 & x_3 & y_3 \end{vmatrix} = 0 \Rightarrow M_1(x_2y_3 - x_3y_2) - M_2(x_1y_3 - x_3y_1) + M_3(x_1y_2 - x_2y_1) = 0$$

$$\Rightarrow M_1(x_2y_3 - x_3y_2) + M_2(x_3y_1 - x_1y_3) + M_3(x_1y_2 - x_2y_1) = 0.$$

Again, from (1) and (2), we have

$$\frac{1}{-x_1y_2 + x_2y_1} = \frac{Y}{-M_1y_2 + M_2y_1} = \frac{X}{M_1x_2 - M_2x_1}$$

$$\Rightarrow X = \frac{M_1x_2 - M_2x_1}{x_2y_1 - x_1y_2}, \quad Y = \frac{M_2y_1 - M_1y_2}{x_2y_1 - x_1y_2}.$$

Thus, the magnitude of the resultant force is

$$R = \sqrt{X^2 + Y^2} = \frac{\sqrt{(M_1x_2 - M_2x_1)^2 + (M_2y_1 - M_1y_2)^2}}{(x_2y_1 - x_1y_2)}.$$

If the resultant makes angle θ with the positive direction of the x-axis, then

$$\tan\theta = \frac{Y}{X} = \frac{M_1y_2 - M_2y_1}{M_1x_2 - M_2x_1} \Rightarrow \theta = \tan^{-1}\left(\frac{M_1y_2 - M_2y_1}{M_1x_2 - M_2x_1}\right).$$

19. The moments of a system of coplanar forces acting in the xy-plane about $(0,0)$, $(a,0)$ and $(0,a)$ are aW, $2aW$ and $3aW$ respectively. Find the components parallel to the coordinate axes and the line of action of the single force to which the system is equivalent.

Solution: We know that the system of forces can be reduced to components X and Y along any two rectangular axes Ox and Oy and a couple of moment G.

Also we know that the moment G' of the system of forces about (h,k) is $G' = G + Xk - Yh$.

Now, taking moment about $(0,0)$, $(a,0)$ and $(0,a)$, we have

$$aW = G, \; 2aW = G - Ya, \; 3aW = G + aX$$

$$\Rightarrow 2aW = aW - Ya, \; 3aW = aW + aX$$

$$\Rightarrow aW = -Ya, \; 2aW = aX$$

$$\Rightarrow Y = -W, \; X = 2W.$$

Also the equation of line of action of the resultant is

$$G+yX-xY=0 \Rightarrow aW+y(2W)-x(-W)=0$$
$$\Rightarrow aW+2yW+xW=0$$
$$\Rightarrow x+2y+a=0.$$

20. A system of coplanar forces has the total moments H, $2H$ respectively about points whose coordinates are $(2a,0)$, $(0,a)$ referred to fixed rectangular axes. The total resolved parts of the forces along the line $y=x$ vanishes. Find the points at which the line of action of the resultant meets the coordinate axes.

Solution: We know that the system of forces can be reduced to components X and Y along any two rectangular axes Ox and Oy and a couple of moment G about O.

Also the moment G' of the system of forces about any point (x_1, y_1) is given by $G' = G + y_1 X - x_1 Y$.

Thus the moments of the system of forces about $A(2a,0)$ and $B(0,a)$ is given by

$$H = G - 2aY \qquad (1)$$
$$2H = G + aX. \qquad (2)$$

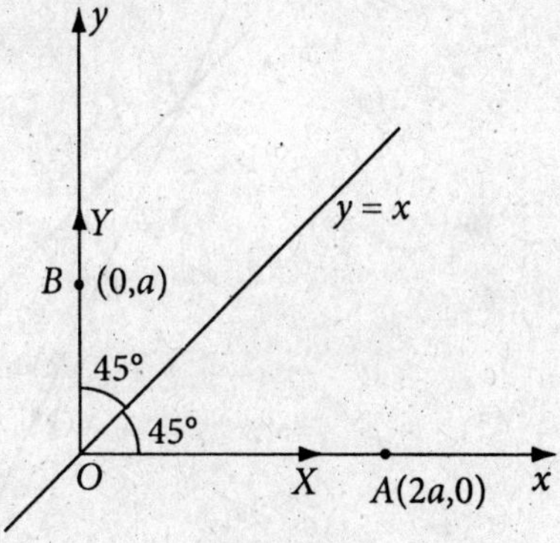

Fig. 2.38

Also given that the sum of the resolved part of the forces along the line $y=x$ vanishes. So, we have

$$X\cos 45° + Y\sin 45° = 0 \Rightarrow X+Y=0. \qquad (3)$$

Subtracting (1) from (2), we get

$$H = aX + 2aY \Rightarrow H = aX - 2aX \text{ [from (3)]}$$
$$\Rightarrow H = -aX \Rightarrow X = -H/a.$$

$\therefore\ Y = -X = H/a.$

From (2),

$$2H = G - H \Rightarrow G = 3H.$$

Now, the equation of the line of action of the resultant is

$$G - xY + yX = 0 \Rightarrow 3H - x\frac{H}{a} + y\left(-\frac{H}{a}\right) = 0 \Rightarrow 3 - \frac{x}{a} - \frac{y}{a} = 0$$
$$\Rightarrow x+y=3a \Rightarrow \frac{x}{3a} + \frac{y}{3a} = 1,$$

which shows that the line of action of the resultant meets the coordinate axes at $(3a,0)$ and $(0,3a)$.

21. Three forces P, Q, R act along the sides of a triangle formed by the lines $x+y=3$, $2x+y=1$ and $x-y+1=0$. Find the equation of the line of action of the resultant.

[VU(H) 2005]

Solution: *First Method.* Let (h, k) be a point on the line of action of the resultant. Then sum of the moments of P, Q, R about (h, k) is zero.

$$\therefore\ -\frac{P(h+k-3)}{\sqrt{2}}+\frac{Q}{\sqrt{5}}(2h+k-1)-\frac{R}{\sqrt{2}}(h-k+1)=0$$

$$\frac{P}{\sqrt{2}}(h+k-3)-\frac{Q}{\sqrt{5}}(2h+k-1)+\frac{R}{\sqrt{2}}(h-k+1)=0.$$

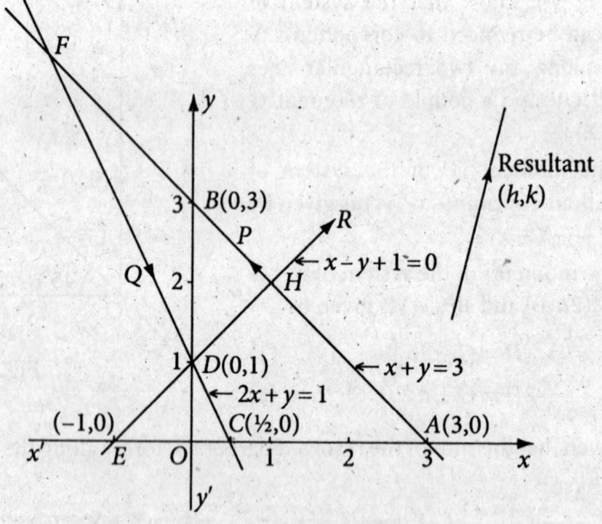

Fig. 2.39

Generalising (h, k), we have the required equation as

$$\frac{P}{\sqrt{2}}(x+y-3)-\frac{Q}{\sqrt{5}}(2x+y-1)+\frac{R}{\sqrt{2}}(x-y+1)=0$$

$$\Rightarrow P\sqrt{5}(x+y-3)-Q\sqrt{2}(2x+y-1)+R\sqrt{5}(x-y+1)=0$$

$$\Rightarrow x\left(\sqrt{5}P-2\sqrt{2}Q+\sqrt{5}R\right)+y\left(\sqrt{5}P-Q\sqrt{2}-\sqrt{5}R\right)+\sqrt{2}Q-3\sqrt{5}P+\sqrt{5}R=0.$$

Second Method. (Vector Method). See figure of method I.

Let forces P, Q, R act along the sides BF, FD and DH of $\triangle FDH$ respectively.

Let $\vec{i}$ and $\vec{j}$ be the unit vectors along the x-axis and the y-axis respectively.

We have $\overrightarrow{AB}=\overrightarrow{AO}+\overrightarrow{OB}=-3\vec{i}+3\vec{j}=3\left(\vec{j}-\vec{i}\right)$.

So unit vector along

$$\overrightarrow{AB}=\frac{3\left(-\vec{i}+\vec{j}\right)}{\sqrt{(-3)^2+3^2}}=\frac{\vec{j}-\vec{i}}{\sqrt{2}}.\qquad \therefore\ \vec{P}=\frac{P\left(\vec{j}-\vec{i}\right)}{\sqrt{2}}.$$

Also, $\overrightarrow{DC}=\overrightarrow{DO}+\overrightarrow{OC}=-\vec{j}+\frac{1}{2}\vec{i}$.

So unit vector along $\overrightarrow{DC}$

$$=\frac{\frac{1}{2}\vec{i}-\vec{j}}{\sqrt{\left(\frac{1}{2}\right)^2+(-1)^2}}=\frac{\vec{i}-2\vec{j}}{\sqrt{5}}. \qquad \therefore \vec{Q}=\frac{Q}{\sqrt{5}}\left(\vec{i}-2\vec{j}\right).$$

Again, $\overrightarrow{ED}=\overrightarrow{EO}+\overrightarrow{OD}=\vec{i}+\vec{j}$.

So unit vector along

$$\overrightarrow{ED}=\frac{\vec{i}+\vec{j}}{\sqrt{1^2+1^2}}=\frac{\vec{i}+\vec{j}}{\sqrt{2}}.$$

Let $\vec{F}$ be the resultant of the forces $\vec{P},\vec{Q}$ and $\vec{R}$.

$$\therefore \vec{F}=\vec{P}+\vec{Q}+\vec{R}$$

$$\Rightarrow \vec{F}=\frac{P}{\sqrt{2}}\left(\vec{j}-\vec{i}\right)+\frac{Q}{\sqrt{5}}\left(\vec{i}-2\vec{j}\right)+\frac{R}{\sqrt{2}}\left(\vec{i}+\vec{j}\right)$$

$$\Rightarrow X\vec{i}+Y\vec{j}=\vec{i}\left(-\frac{P}{\sqrt{2}}+\frac{Q}{\sqrt{5}}+\frac{R}{\sqrt{2}}\right)+\vec{j}\left(\frac{P}{\sqrt{2}}-\frac{2Q}{\sqrt{5}}+\frac{R}{\sqrt{2}}\right)$$

$$\Rightarrow X=\frac{R-P}{\sqrt{2}}+\frac{Q}{\sqrt{5}} \text{ and } Y=\frac{P+R}{\sqrt{2}}-\frac{2Q}{\sqrt{5}}.$$

Also,

$$\vec{G}=\text{algebraic sum of moments of all the forces about } O$$

$$=\overrightarrow{OA}\times\vec{P}+\overrightarrow{OC}\times\vec{Q}+\overrightarrow{OD}\times\vec{R}$$

$$=3\vec{i}\times\frac{P}{\sqrt{2}}\left(\vec{j}-\vec{i}\right)+\frac{\vec{i}}{2}\times\frac{Q}{\sqrt{5}}\left(\vec{i}-2\vec{j}\right)+\vec{j}\times\frac{R}{\sqrt{2}}\left(\vec{i}+\vec{j}\right)$$

$$=\frac{3P}{\sqrt{2}}\vec{k}-\frac{Q}{\sqrt{5}}\vec{k}-\frac{R}{\sqrt{2}}\vec{k}$$

$$\Rightarrow \vec{G}=\left(\frac{3P-R}{\sqrt{2}}-\frac{Q}{\sqrt{5}}\right)\vec{k}$$

$$\Rightarrow G=\frac{3P-R}{\sqrt{2}}-\frac{Q}{\sqrt{5}}.$$

Hence, the equation of the resultants

$$xY-yX=G$$

$$\Rightarrow x\left(\frac{P+R}{\sqrt{2}}-\frac{2Q}{\sqrt{5}}\right)-y\left(\frac{R-P}{\sqrt{2}}+\frac{Q}{\sqrt{5}}\right)=\frac{3P-R}{\sqrt{2}}-\frac{Q}{\sqrt{5}}$$

$$\Rightarrow x\left[\sqrt{5}(P+R)-2\sqrt{2}Q\right]-y\left[\sqrt{5}(R-P)+\sqrt{2}Q\right]=\sqrt{5}(3P-R)-Q\sqrt{2}.$$

22. A system of coplanar forces acting in the plane of rectangular axes. Ox, Oy is equivalent to forces (X, Y) along the axes and a couple G at the origin. Show how to reduce the same system w.r.t. to another point O' whose coordinates w.r.t. O are (α, β). If the

couple component after reduction w.r.t. O' is G, then show that the couple component when the system is reduced w.r.t. the middle point OO' is $\frac{1}{2}(G+G')$. [CU(H) 2004]

Solution: For the first part, write the theorem 2.2.

We know that the system of forces can be reduced to components X and Y along any two rectangular axes Ox and Oy and a couple of moment G about O.

Then

$$G' = G + \beta X - \alpha Y. \qquad (1)$$

Let O'' be the mid-point of OO'. So coordinates of O'' are $(\alpha/2, \beta/2)$. Let G'' be the moment of the system of forces about the point O''.

$$\therefore\ G'' = G + \frac{\beta}{2}X - \frac{\alpha}{2}Y \Rightarrow 2G'' = 2G + \beta X - \alpha Y$$

$$\Rightarrow 2G'' = 2G + G' - G \text{ [from (1)]}$$

$$\Rightarrow 2G'' = G + G'$$

$$\Rightarrow G'' = \frac{1}{2}(G+G').$$

23. Show that three coplanar forces P, Q, R acting at the points A, B, C are in a static if equilibrium, if they meet at a point on the circumcircle of the triangle ABC and of $P:Q:R = a:b:c$, where a, b, c are the sides of the triangle ABC. [CU(H) 2002]

Solution: Let the forces P, Q, R acting at A, B, C meet at O, on the circumcircle of $\triangle ABC$.

According to the question,

$$P/a = Q/b = R/c.$$

But in $\triangle ABC$,

$$\frac{a}{\sin A} = \frac{b}{\sin B} = \frac{c}{\sin C}.$$

$$\therefore\ \frac{P}{\sin A} = \frac{Q}{\sin B} = \frac{R}{\sin C}. \qquad (1)$$

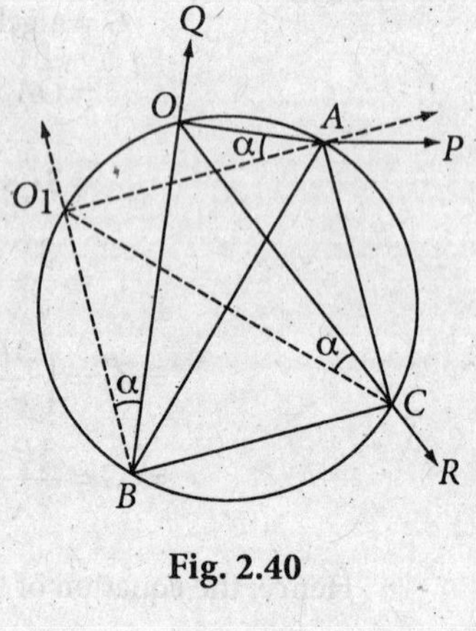

From Fig. 2.40, it is clear that

$$\angle BAC = \angle A = \angle BOC$$

$$\angle ABC = \angle B = \angle COA$$

$$\angle ACB = \angle C = 180° - \angle AOB.$$

Fig. 2.40

So from (1), we have

$$\frac{P}{\sin BOC} = \frac{Q}{\sin COA} = \frac{R}{\sin AOB}. \qquad (2)$$

Hence, the forces are in equilibrium. Now, we turn the line of action of P through an angle α about A and let it meet the circle in O_1. We join BO_1 and CO_1.

Clearly, $\angle OAO_1 = \angle OBO_1 = \angle OCO_1 = \alpha$, as they are on the same arc OO_1 and on the same side of it.

Hence, the forces again meet at O_1 on the circumcircle of $\triangle ABC$, if each force is turned through the same angle α.

From figure, we find that

$$\angle BOC = \angle BO_1C$$
$$\angle COA = \angle CO_1A$$
$$\angle AOB = \angle AO_1B.$$

Therefore, from (1), we get

$$\frac{P}{\sin BO_1C} = \frac{Q}{\sin CO_1A} = \frac{R}{\sin AO_1B}.$$

Hence, in this case also the forces are in equilibrium, i.e., the equilibrium is a static.

24. A plane system of forces is equivalent to a couple of moment M, and if the forces are turned through a right angle about the respective points of application in the same sense, then they are equivalent to a couple N. Prove that when each force is turned about its point of application through an angle α in the same sense the system will be in equilibrium, if $\tan\alpha = -M/M$. Show that if the forces be turned through an angle $2\tan^{-1}(N/M)$, then the system is equivalent to a couple of moment M. [KU(H) 2007]

Solution: Let P_1 be one of the forces of the system acting at a point $A_1(x_1, y_1)$ and making an angle θ_1 with the x-axis. Let the components of P_1 parallel to the axes be X_1, Y_1; so for the other forces.

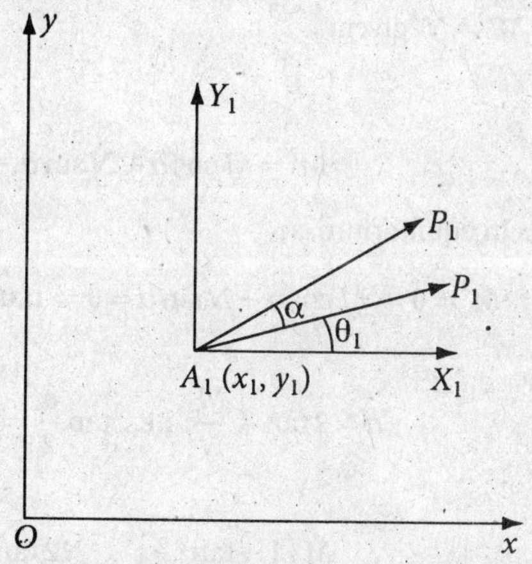

Fig. 2.41

We know that the system of coplanar forces is equivalent to X, Y and M, where

$$X = \Sigma X_1 = \Sigma P_1 \cos\theta_1$$
$$Y = \Sigma Y_1 = \Sigma P_1 \sin\theta_1$$

and

$$M = \Sigma(Y_1x_1 - X_1y_1) = \Sigma P_1(x_1 \sin\theta_1 - y_1 \cos_1). \tag{1}$$

Since the system reduces to a couple, therefore,

$$X=\Sigma X_1=\Sigma P_1\cos\theta_1=0 \tag{2}$$
$$Y=\Sigma Y_1=\Sigma P_1\sin\theta_1=0. \tag{3}$$

When each force is turned through an angle α, then

$$\begin{aligned}X'&=\Sigma P_1\cos(\theta_1+\alpha)=\Sigma P_1(\cos\theta_1\cos\alpha-\sin\theta_1\sin\alpha)\\&=\cos\alpha(\Sigma P_1\cos\theta_1)-\sin\alpha(\Sigma P_1\sin\theta_1)\\&=\cos\alpha\cdot 0-\sin\alpha\cdot 0\text{ [from (2) and (3)]}\\&=0.\end{aligned}$$

$$\begin{aligned}Y'&=\Sigma P_1\sin(\theta_1+\alpha)=\Sigma P_1(\sin\theta_1\cos\alpha+\cos\theta_1\sin\alpha)\\&=\cos\alpha(\Sigma P_1\sin\theta_1)+\sin\alpha(\Sigma P_1\cos\theta_1)\\&=\cos\alpha\cdot 0+\sin\alpha\cdot 0\text{ [from (2) and (3)]}\\&=0.\end{aligned}$$

$$\begin{aligned}\therefore\ M'&=\Sigma\left[x_1P_1\sin(\theta_1+\alpha)-y_1P_1\cos(\theta_1+\alpha)\right]\\&=\Sigma P_1\left[x_1(\sin\theta_1\cos\alpha+\cos\theta_1\sin\alpha)-y_1(\cos\theta_1\cos\alpha-\sin\theta_1\sin\alpha)\right]\\&=\cos\alpha\left[\Sigma P_1(x_1\sin\theta_1-y_1\cos\theta_1)\right]+\sin\alpha\left[\Sigma P_1(x_1\cos\theta_1+y_1\sin\theta_1)\right]\\&=M\cos\alpha+V\sin\alpha\text{ [from (1)]},\end{aligned}$$

where $V=\Sigma P_1(x_1\cos\theta_1+y_1\sin\theta_1)$.

When $\alpha=90°$, $M'=N$, given.

$\therefore\ N=V$.

Thus,

$$M'=M\cos\alpha+N\sin\alpha.$$

As the system is in equilibrium, so

$$M'=0\Rightarrow M\cos\alpha+N\sin\alpha=0\Rightarrow\tan\alpha=-M/N.$$

When

$$\alpha=2\tan^{-1}\frac{M}{N},\text{ i.e., }\tan\frac{\alpha}{2}=\frac{N}{M},$$

then

$$M'=M\cos\alpha+N\sin\alpha=\frac{M\left(1-\tan^2\frac{\alpha}{2}\right)}{1+\tan^2\frac{\alpha}{2}}+\frac{N2\tan\frac{\alpha}{2}}{1+\tan^2\frac{\alpha}{2}}=\frac{M\left(1-\frac{N^2}{M^2}\right)+2N\frac{N}{M}}{1+\frac{N^2}{M^2}}$$
$$=\frac{(M^2-N^2)+2N^2}{M}\times\frac{M^2}{M^2+N^2}=\frac{M(M^2+N^2)}{M^2+N^2}=M.$$

25. $OABC$ is a rectangle whose vertices are $O(0,0)$, $A(0,0)$, $B(0,7)$, $C(0,7)$. Forces of magnitudes P, Q, R, S act along OA, OC, BA, BC respectively and reduce to a resultant along the line $4x+3y-12=0$. The moment of the system about O is 12 units in the sense of rotation from Ox and Oy. If the first force is a magnitude 9 units, find Q, R, S.

Solution: Let X, Y be the algebraic sum of resolved parts of the forces along x-axis and y-axis, respectively.

$\therefore X = P - S$ and $Y = Q - R$.

The sum of the moments of the forces about O is $G = 7S - 9R = 12$, given.

The equation of line of action of the resultant is $xY - yX - G = 0$.

Fig. 2.42

$$x(Q-R) - y(P-S) - 12 = 0. \quad (1)$$

But the equation of line of action of the resultant is

$$4x + 3y - 12 = 0. \qquad (2)$$

Comparing (1) and (2), we get

$$\frac{Q-R}{4} = \frac{S-P}{3} = 1 \Rightarrow Q - R = 4,\ S - P = 3.$$

Given, $P = 9$. $\therefore S = P + 3 = 12$.

Also

$$7S - 9R = 12 \Rightarrow 7 \times 12 - 12 = 9R \Rightarrow 9R = 72 \Rightarrow R = 8.$$

$$\therefore Q = R + 4 = 8 + 4 = 12.$$

$$Q = 12,\ R = 8,\ S = 12.$$

26. Two equal uniform heavy rods are connected at one extremity by a thin string and the system is placed symmetrically in a vertical plane with rods resting on two smooth pegs in the same horizontal line. If a be the length of each rod, b the distance between the pegs and $l(< b)$ is the length of the connecting string, then show that in equilibrium position the inclination of each rod to the horizon is $\cos^{-1}\left(\sqrt[3]{\frac{b-l}{a}}\right)$. [CU(H) 2000]

Solution: Let AB, CD be the two rods, each of length l. AC is the string. Let T be the tension in the string.

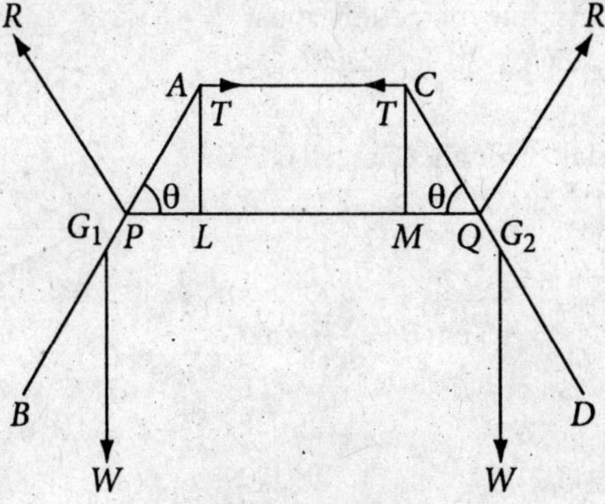

Fig. 2.43

Let P and Q be the pegs and R, the normal reaction there.

Given: $AB = CD = a$, $PQ = b$, $AC = l$, $AG_1 = a/2$.

We draw AL and CM perpendicular to PQ.

Let θ be the inclination of each rod to the horizon.

In $\triangle APL$,

$$\cos\theta = \frac{PL}{AP} \Rightarrow PL = AP\cos\theta.$$

$$\therefore\ MQ = PL = AP\cos\theta.$$

$$\therefore\ PQ = PL + LM + MQ \Rightarrow b = 2PL + AC \Rightarrow b = 2AP\cos\theta + l$$

$$\Rightarrow AP = \frac{b-l}{2\cos\theta}. \qquad (1)$$

For the equilibrium of rod AB, resolving vertically, we get

$$W = R\sin(90+\theta) \Rightarrow R = W/\cos\theta. \qquad (2)$$

Taking moment of all the forces about A, we get

$$W\cdot AG_1\cos\theta = R\cdot AP \Rightarrow W\frac{a}{2}\cos\theta = \frac{W}{\cos\theta}\cdot\frac{b-l}{2\cos\theta} \text{ [from (1) and (2)]}$$

$$\Rightarrow \cos^2\theta = \frac{b-l}{a} \Rightarrow \cos\theta = \sqrt[3]{\frac{b-l}{a}} \Rightarrow \theta = \cos^{-1}\left(\sqrt[3]{\frac{b-l}{a}}\right).$$

27. The altitude of a cone is h and the radius of the base is r. A string is fastened to the vertex and to a point on the circumference of the circular base and is then put over a smooth peg. Show that if the cone rests with its axis horizontally, the length of the string must be $\sqrt{h^2+4r^2}$. **[CU(H) 2004]**

Solution: Let ABC be the given cone whose weight is W.

Since the string passes over a peg, the tension in two parts of string is the same. And since the weight balances the two equal tensions, its line of action must bisect the angle AOC. So let $\angle AOG = \angle COG = \theta$.

Let the horizontal through C meet OG in K.

In $\triangle OAG$,

$$\cot\theta = \frac{OG}{AG} \Rightarrow OG = AG\cot\theta = \frac{3}{4}h\cot\theta. \qquad (1)$$

Also in $\triangle OKC$,

$$\cot\theta = \frac{OK}{KC} = \frac{OK}{GD} \Rightarrow OK = GD\cot\theta = \frac{h}{4}\cot\theta.$$

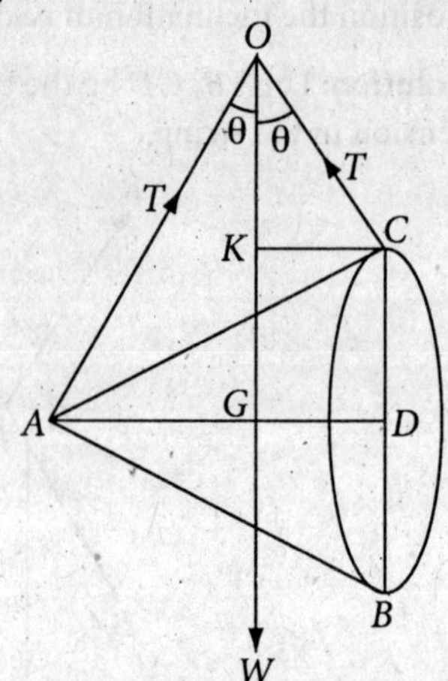

Fig. 2.44

Now,

$$\begin{aligned}
OG &= OK + KG = \frac{h}{4}\cot\theta + CD = \frac{h}{4}\cot\theta + r \\
\Rightarrow OG &= r + \frac{h}{4}\cot\theta \Rightarrow \frac{3}{4}h\cot\theta = r + \frac{h}{4}\cot\theta \text{ [from (1)]} \\
\Rightarrow \frac{2h}{4}\cot\theta &= r \Rightarrow \cot\theta = \frac{2r}{h}. \qquad (2)
\end{aligned}$$

In $\triangle OKC$,

$$\sin\theta = \frac{KC}{OC} \Rightarrow OC = KC\operatorname{cosec}\theta = GD\operatorname{cosec}\theta = \frac{h}{4}\operatorname{cosec}\theta.$$

Also, in $\triangle OAG$,

$$\sin\theta = \frac{AG}{OA} \Rightarrow OA = AG\operatorname{cosec}\theta = \frac{3}{4}h\operatorname{cosec}\theta.$$

$\therefore$ length of the string

$$\begin{aligned}
AO + OC &= \frac{3}{4}h\operatorname{cosec}\theta + \frac{h}{4}\operatorname{cosec}\theta = h\operatorname{cosec}\theta = h\sqrt{1+\cot^2\theta} \\
&= h\sqrt{1+\frac{4r^2}{h^2}} = \sqrt{h^2+4r^2}.
\end{aligned}$$

28. A square frame $ABCD$ of four equal jointed rods hangs from A, the shape being maintained by a string joining mid-points of AB and BC. Prove that the ratio of the tension of the string to the reaction at C is $8/\sqrt{5}$. [VU(H) 2005, CU(H) 2003]

Solution: Let W be the weight of each rod and $2a$ be the length of each side of the square.

Let T be the tension of the string. Let X and Y be the resolved part of the action at C of rod BC on CD and CD on BC.

We consider the equilibrium of the rod BC. Taking moments about B, we get

$$\begin{aligned}
&W\cdot a\sin 45° - T\cdot a\sin 45° + X\cdot 2a\sin 45° + Y\cdot 2a\cos 45° = 0 \\
&\Rightarrow W - T + 2X + 2Y = 0 \\
&\Rightarrow T = W + 2X + 2Y. \qquad (1)
\end{aligned}$$

Now, we consider the equilibrium of the rod CD. Taking moments about D, we get

$$\begin{aligned}
&-Wa\cos 45° - X2a\sin 45° + Y2a\cos 45° = 0 \\
&\Rightarrow -W - 2X + 2Y = 0 \\
&\Rightarrow 2Y = W + 2X. \qquad (2)
\end{aligned}$$

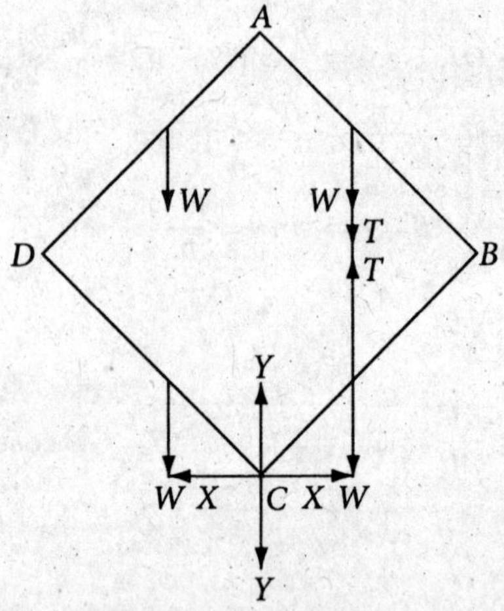

Fig. 2.45

Finally, we consider the equilibrium of the rods AB and BC together jointed at B. Taking moments about A, we get

$$X4a\cos 45° - 2Wa\sin 45° + Ta\sin 45° - Ta\sin 45° = 0$$
$$\Rightarrow 4X - 2W = 0$$
$$\Rightarrow X = W/2.$$

From (2),

$$2Y = W + W = 2W \Rightarrow Y = W.$$

From (1) and (2), we get

$$T = 4Y = 4W.$$

Then action R at C is

$$\sqrt{X^2 + Y^2} = \sqrt{\frac{W^2}{4} + W^2} = \frac{\sqrt{5}}{2}W.$$

Hence, the required ratio

$$\frac{T}{R} = \frac{4W}{\frac{\sqrt{5}}{2}W} = \frac{8}{\sqrt{5}}.$$

29. Two uniform similar rods of same material PQ and QT of length $2a$ and $2b$ respectively are rigidly united at Q and suspended freely from P. If they are inclines at an angle α and β respectively to the vertical, prove that $(a^2 + 2ab)\sin\alpha = b^2 \sin\beta$. [CU(H) 2004]

Solution: Let W_1 and W_2 be the weights of the rods PQ and QT. Let G_1 and G_2 be the mid-points of the rods PQ and QT, where their weights W_1 and W_2 act. Let w be the weight per unit length of the rod. $\therefore$ $W_1 = 2aw$ and $W_2 = 2bw$.

We consider the equilibrium of the two rods combined, their mutual actions at B neutralise and the forces left are the

(i) weight W_1 acting through G_1, the mid-point of PQ,
(ii) weight W_2 acting through G_2, the mid-point of QT,
(iii) the action at the fixed point A.

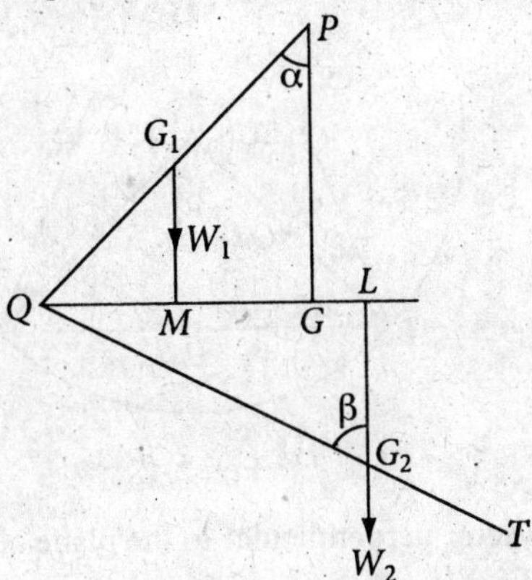

Fig. 2.46

Here, the three forces are in equilibrium, therefore, common centre of gravity of G_1 and G_2, i.e., the point G is vertically P; hence PG is vertical.

Taking moments about the point P, we have

$$\begin{aligned} W_1 \cdot GM - W_2 \cdot GL = 0 &\Rightarrow W_1 \cdot PG \sin\alpha = W_2(QL - QG) \\ &\Rightarrow W_1 \cdot a \sin\alpha = W_2\left(QG_2 \sin\beta - QP \sin\alpha\right) \\ &\Rightarrow W_1 \cdot a \sin\alpha = W_2(b \sin\beta - 2a \sin\alpha) \\ &\Rightarrow 2aw \cdot a \sin\alpha = 2bw(b \sin\beta - 2a \sin\alpha) \\ &\Rightarrow a^2 \sin\alpha = b(b \sin\beta - 2a \sin\alpha) \\ &\Rightarrow \left(a^2 + 2ab\right) \sin\alpha = b^2 \sin\beta \text{ (Proved).} \end{aligned}$$

30. Forces P, Q, R act along the sides BC, CA and AB of the triangle ABC and forces P', Q', R' act along OA, OB, OC, where O is the circum-centre in the senses indicated by the order of the letter. If the sin forces are in equilibrium, show that $P\cos A + Q\cos B + R\cos C = 0$ and $\frac{PP'}{a} + \frac{QQ'}{b} + \frac{RR'}{c} = 0$. [CU(H) 2001]

Solution: Since the system of sin forces P, Q, R, P', Q', R' is in equilibrium, so the algebraic sum of their moments about any point in their plane is zero.

Let r be the circum-radius of $\triangle ABC$. $\therefore\ OA = OB = OC = r$.

Also, $\angle BOC = 2\angle A$.

$\because\ OA = OB \Rightarrow \angle OAB = \angle OBA = \pi/2 - C$.

Let $\hat{a}, \hat{b}, \hat{c}, \hat{d}, \hat{e}, \hat{f}$ be the unit vectors along the forces P, Q, R, P', Q', R' respectively.

Taking moments of all the forces about O, we have

$$\overrightarrow{OB} \times P\hat{a} + \overrightarrow{OC} \times Q\hat{b} + \overrightarrow{OA} \times R\hat{c} = \vec{0}$$

$$\Rightarrow \left[OBP\sin\left(\frac{\pi}{2} - A\right) + OCQ\sin\left(\frac{\pi}{2} - B\right) + OAR\sin\left(\frac{\pi}{2} - C\right)\right] r = 0,$$

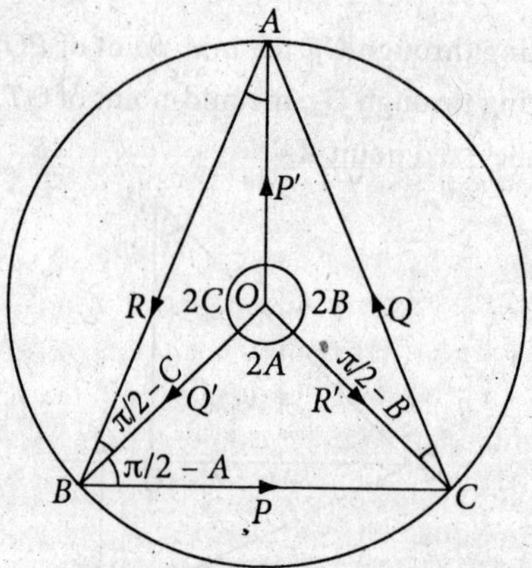

Fig. 2.47

where $\hat{n}$ is the unit vector perpendicular to the plane of $\triangle ABC$.

$$\Rightarrow P\cdot r\cos A+Q\cdot r\cos B+R\cdot r\cos C=0$$

$$\Rightarrow P\cos A+Q\cos B+R\cos C=0 \text{ (Proved).}$$

Now taking moments of all the forces about A, we get

$$\overrightarrow{AB}\times P\hat{a}+\overrightarrow{AB}\times Q'\hat{e}+\overrightarrow{AC}\times R'\hat{f}=\vec{0}$$

$$\Rightarrow\overrightarrow{AB}\times P\hat{a}+\overrightarrow{AB}\times Q'\hat{e}+\overrightarrow{AC}\times R'\hat{f}=\vec{0}$$

$$\Rightarrow P\cdot AB\sin B-Q'c\sin(90°-C)+R'b\sin(90°-B)=0$$

$$\Rightarrow P\cdot c\sin B-Q'c\cos C+R'b\cos B=0$$

$$\Rightarrow P\cdot c\frac{2\triangle}{ac}-Q'c\cos C+R'b\cos B=0\ \left[\because \triangle=\tfrac{1}{2}ac\sin B\right]$$

$$\Rightarrow\frac{P\cdot 2\triangle}{a}-Q'c\cos C+R'b\cos B=0. \tag{1}$$

Similarly, taking moments about B and C, we get

$$Q\cdot\frac{2\triangle}{c}-R'a\cos A+P'c\cos C=0 \tag{2}$$

$$\text{and } R\cdot\frac{2\triangle}{c}-P'b\cos B+Q'a\cos A=0. \tag{3}$$

Multiplying (1) by P', (2) by Q' and (3) by R', we get

$$\frac{PP'}{a}\cdot 2\triangle-Q'P'c\cos C+R'P'b\cos B=0 \tag{4}$$

$$\frac{QQ'}{b}\cdot 2\triangle-Q'R'a\cos A+Q'P'c\cos C=0 \tag{5}$$

and

$$\frac{RR'}{c}\cdot 2\triangle-PR'b\cos B+Q'R'a\cos A=0. \tag{6}$$

Adding (4), (5) and (6), we get

$$\left(\frac{PP'}{a}+\frac{QQ'}{b}+\frac{RR'}{c}\right)\cdot 2\Delta=0 \Rightarrow \frac{PP'}{a}+\frac{QQ'}{b}+\frac{RR'}{c}=0 \text{ (Proved).}$$

31. A step ladder in the form of the letter A, with each of its legs inclined at an angle α to the vertical, is placed on a horizontal floor and is held by a cord connecting the middle points of its legs, there being no friction anywhere. Show that when a weight W is placed on one of the steps at a height from the floor equal to $1/n$ of the height of the ladder, the increase in the tension of the cord is $\frac{1}{n}W\tan\alpha$.

Solution: Let AB and AC be the two legs of the ladder held up by the string DE, where D and E are mid-points of AB and AC.

Let W be the weight of each leg and T the tension in the string DE.

Let R and S be the normal reactions at B and C respectively, when a weight W is placed at F such that

$$FN=\frac{1}{n}AK. \qquad (1)$$

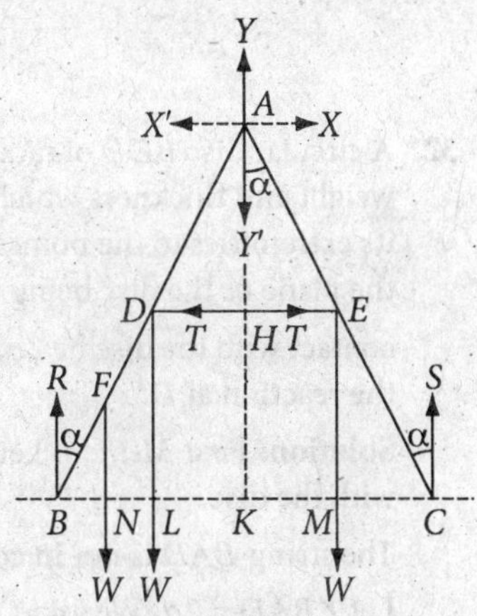

Fig. 2.48

Taking moment about B for the whole system, we get

$$\begin{aligned} S\cdot BC &= W\cdot BL+W\cdot BM+W\cdot BN \\ &= W(MC+BM)+W\cdot BN \\ &= W\cdot BC+W\cdot BN. \qquad (2) \end{aligned}$$

Now in similar triangles BNF and BKA.

$$\frac{BN}{BK}=\frac{FN}{AK}=\frac{1}{n} \text{ [from (1)]}$$

$$\Rightarrow BN=\frac{1}{n}BK=\frac{1}{2n}BC.$$

So from (2),

$$S\cdot BC=W\cdot BC+W\cdot\frac{1}{2n}BC \Rightarrow S=W+\frac{W}{2n}. \qquad (3)$$

Now, taking moments of the forces acting on the leg AC about A, we have

$$\begin{aligned} &-T\cdot EH-W\cdot KM+S\cdot KC=0 \\ &\Rightarrow T\cdot EM+W\cdot KM=S\cdot KC \\ &\Rightarrow T\cdot AE\cos\alpha+W\cdot AE\sin\alpha=S\cdot AC\sin\alpha \\ &\Rightarrow T\cdot\frac{AC}{2}\cos\alpha+W\cdot\frac{AC}{2}\sin\alpha=S\cdot AC\sin\alpha \end{aligned}$$

$$\Rightarrow T\cos\alpha + W\sin\alpha = 2S\sin\alpha$$

$$\Rightarrow T = (2S - W)\tan\alpha$$

$$\Rightarrow T = \left[2\left(W + \frac{W}{2n}\right) - W\right]\tan\alpha \text{ [from (3)]}$$

$$\Rightarrow T = \left[W + \frac{W}{n}\right]\tan\alpha.$$

Let T_0 be the tension when weight W is not attached at F. $\therefore$ $T_0 = W\tan\alpha$.

Hence, increase in the tension

$$T - T_0 = \left[W + \frac{W}{n} - W\right]\tan\alpha = \frac{W}{n}\tan\alpha.$$

32. A circular disc BCD of radius a and weight W is supported by a band of inappreciable weight and thickness which surrounds the disc along the arc BCD and is fastened at its extremities to the point A in a vertical wall, the portion AD touching the wall and the plane of the disc being at right angles to the wall. If the length of the band not in contact with the disc be $2b$, show that the tension of the band is $\frac{W}{2}\cdot\frac{(a^2+b^2)}{b^2}$. Also find the reaction at D.

Solution: *First Method.* Let $ABCD$ be the string, the portion BCD being in contact with the disc.

The string BAD is not in contact with the disc. String $BAD = 2b$. $\therefore$ $AB = AD = b$.

Let $\angle BAD = 2\alpha$. We join OA. Then $\angle BAO = \angle DAO = \alpha$.

Let T be the tension along BA and along DA each.

Resultant of these two tensions $= 2T\cos\alpha$ along OA.

Thus, T along BA and T along DA have been reduced to be a single force $2T\cos\alpha$ along OA.

The forces acting on the disc are:

1. Weight W acting vertically downwards through O.
2. Reactor R at D perpendicular to the wall.
3. Force $2T\cos\alpha$ along OA.

By Lami's theorem to forces at O, we have

$$\frac{2T\cos\alpha}{\sin 90^\circ} = \frac{R}{\sin(180^\circ - \alpha)} = \frac{W}{\sin(90^\circ + \alpha)}$$

$$\Rightarrow \frac{2T\cos\alpha}{1} = \frac{R}{\sin\alpha} = \frac{W}{\cos\alpha}$$

$$\Rightarrow T = \frac{W}{2}\sec^2\alpha,\ R = W\tan\alpha.$$

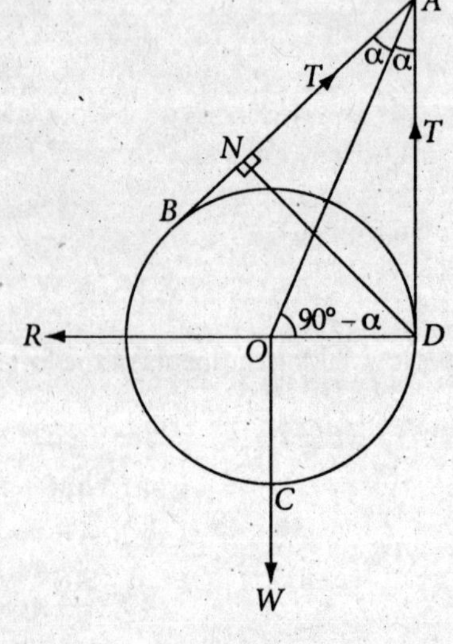

Fig. 2.49

In ΔOAD,

$$\tan\alpha = \frac{OD}{AD} = \frac{a}{b}.$$

$$\therefore\ T = \frac{W}{2}\sec^2\alpha = \frac{W}{2}\left(1+\tan^2\alpha\right) = \frac{W}{2}\left(1+\frac{a^2}{b^2}\right) = \frac{W}{2}\cdot\frac{(a^2+b^2)}{b^2}$$

and

$$R = W\tan\alpha = W\frac{a}{b}.$$

Second Method. Draw the figure of the first method. From D, we draw DN **perpendicular** to AB.

Taking moment of all the forces about D, we get

$$-T\cdot DN + W\cdot OD = 0 \Rightarrow T\cdot DN = W\cdot OD \Rightarrow T\cdot AD\sin 2\alpha = W\cdot a$$

$$\Rightarrow Tb\frac{2\tan\alpha}{1+\tan^2\alpha} = Wa \Rightarrow \frac{Tb2\frac{a}{b}}{1+\frac{a^2}{b^2}} = Wa$$

$$\Rightarrow \frac{T\cdot 2ab^2}{a^2+b^2} = Wa \Rightarrow T = \frac{W(a^2+b^2)}{2b^2}.$$

Now taking moment of all the forces about A, we get

$$-R\cdot AD + W\cdot OD = 0 \Rightarrow R\cdot AD = W\cdot OD \Rightarrow R\cdot b = W\cdot a \Rightarrow R = \frac{Wa}{b}.$$

33. Four rods are smoothly jointed at their extremities to form a quadrilateral $ABCD$ and the opposite corners A, C and B, D are joined by light strings. If T, T' be the tensions in AC and BD respectively, show that $T\left(\frac{1}{AO}+\frac{1}{OC}\right) = T'\left(\frac{1}{BO}+\frac{1}{OD}\right)$, where O is the point intersection of the diagonals. **[VU(H) 2007]**

Solution: Let R_1, R_2, R_3, R_4 be the reactions of the rods AB, BC, CD, DA respectively as shown in the figure.

Therefore, each corner A, B, C, D are in equilibrium under action of T and reactions R_1, R_4; action of T' and reactions R_1, R_2; action of T and reactions R_2, R_3; action of T' and reactions of R_3, R_4 respectively.

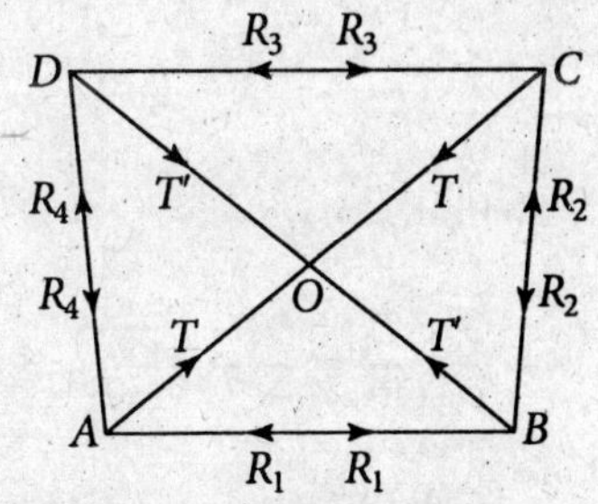

Fig. 2.50

Taking moments about D, for the rod AD, we have

$$T\cdot AD\sin OAD - R_1\cdot AD\sin BAD = 0$$

as system is in equilibrium

$\Rightarrow T\cdot AD\sin OAD = R_1\cdot AD\sin BAD.$ (1)

Again, taking moment of all the forces about C, we have

$$T'\cdot BC\sin CBD = R_1\cdot BC\sin ABC \qquad (2)$$

as system is in equilibrium.

Dividing (1) by (2), we get

$$\frac{T\cdot AD\sin OAD}{T'\cdot BC\sin CBD}=\frac{R_1\cdot AD\sin BAD}{R_1\cdot BC\sin ABC}\Rightarrow\frac{T\cdot AD\sin OAD}{T'\cdot BC\sin CBD}=\frac{AD\sin BAD}{BC\sin ABC}. \quad (3)$$

In $\triangle AOD$,

$$\frac{OD}{\sin OAD}=\frac{AD}{\sin AOD}\Rightarrow AD\cdot\sin OAD=OD\cdot\sin AOD. \quad (4)$$

In $\triangle ABD$,

$$\frac{DB}{\sin DAB}=\frac{AD}{\sin ABD}\Rightarrow AD\sin DAB=DB\sin ABD. \quad (5)$$

In $\triangle BOC$,

$$\frac{OC}{\sin OBC}=\frac{CB}{\sin BOC}\Rightarrow CB\sin OBC=OC\sin BOC. \quad (6)$$

In $\triangle ACB$,

$$\frac{AC}{\sin ABC}=\frac{BC}{\sin CAB}\Rightarrow BC\sin ABC=AC\sin CAB. \quad (7)$$

From (3), (5) and (7), we get

$$\frac{T\cdot AD\sin OAD}{T'\cdot BC\sin CBD}=\frac{DB\sin ABD}{AC\sin CAB}\Rightarrow\frac{T\cdot AD\sin OAD}{T'\cdot BC\sin CBD}=\frac{DB\cdot\sin ABO}{AC\cdot\sin OAD}. \quad (8)$$

For $\triangle AOB$,

$$\triangle AOB=\frac{1}{2}AO\cdot AB\sin OAB\Rightarrow\sin OAB=\sin CAB=\frac{2\triangle AOB}{AO\cdot AB}.$$

Also for $\triangle AOB$,

$$\triangle AOB=\frac{1}{2}BO\cdot AB\sin OBA\Rightarrow\sin OBA=\sin DBA=\frac{2\triangle AOB}{BO\cdot AB}.$$

So from (8), we get

$$\frac{T\cdot OD}{T'\cdot OC}=\frac{DB}{AC}\cdot\frac{\frac{2\triangle AOB}{BO\cdot AB}}{\frac{2\triangle AOB}{AO\cdot AB}}\Rightarrow\frac{T\cdot OD}{T'\cdot OC}=\frac{DB}{AC}\cdot\frac{AO}{BO}\Rightarrow\frac{T\cdot AC}{OC\cdot AO}=\frac{T'\cdot DB}{OD\cdot BO}$$

$$\Rightarrow\frac{T(AO+OC)}{OC\cdot AO}=\frac{T'\cdot(DO+OB)}{OD\cdot BO}\Rightarrow T\left(\frac{1}{OA}+\frac{1}{OC}\right)=T'\cdot\left(\frac{1}{BO}+\frac{1}{OD}\right) \text{ (Proved).}$$

34. A ladder of weight 10 kg rests on a smooth horizontal ground leaning against a smooth vertical wall at an inclination $\tan^{-1}2$ with the horizon and is prevented from slipping by a string attached at its lower end, and to the junction of the wall and the floor. A boy of weight 30 kg begins to ascend the ladder. If the string can bear a tension of 10 kg, how far along the ladder can the boy rise with safety? **[IAS 2008]**

Solution: Let AB be the ladder which is in contact with the floor at A and rests at B against the vertical wall. The weight 10 kg of the ladder acts through G, the mid-point of AB. A is tied by a string OA, whose tension is T. Let the man of weight 30 kg be standing at M, where $AM = x$.

Let $AB = 2a$ and $\angle OAB = \theta$.

Resolving horizontally and vertically, we get

$$R' = T \quad (1)$$

and $R = (10+30)\text{kg} = 40$ kg.

Now, taking moment of all the forces about A, we get

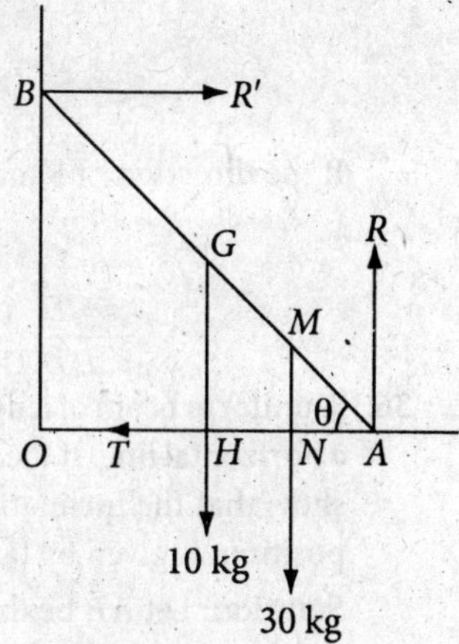

Fig. 2.51

$$R' \cdot BO + 30 \cdot AN + 10 \cdot AM = 0 \Rightarrow -T \cdot 2a \sin\theta + 30 \cdot AM \cos\theta + 10 \cdot AG \cos\theta = 0$$

$$\Rightarrow -T \cdot 2a \sin\theta + 30 \cdot x \cos\theta + 10a \cos\theta = 0$$

$$\Rightarrow T \cdot 2a \sin\theta = 10\cos\theta(a+3x)$$

$$\Rightarrow T \cdot 2a \tan\theta = 10(a+3x) \Rightarrow T \cdot 2a \cdot 2 = 10(a+3x)$$

$$\Rightarrow 2Ta = 5a + 15x \Rightarrow 15x = (2T-5)a. \quad (2)$$

So from (2), we get $15x = 15a \Rightarrow x = a = \frac{1}{2}(2a)$.

Thus, the body can rise a distance half of the ladder with safety.

35. A uniform rod AB is a movable about a hinge at A and rests with one end in contact with a smooth vertical wall. If the rod is inclined at an angle of 30° with the horizontal, find the reaction at the hinge in magnitude and direction. [IAS 2009]

Solution: Let W be the weight of the rod acting vertically downwards through G, the mid-point of AB.

Let R be the normal reaction of the wall.

Let the lines of action of W and R meet at O. Then S, in the reaction of the hinge at A, must also pass through O and act along AO.

Let $\angle ABO = \alpha$, $\angle AOG = \theta$.

By 'm-n theorem' in ΔAOB,

$$(a+a)\cot 60° = a\cot\theta - a\cot 90°,$$

where $AB = 2a$.

$$\Rightarrow \frac{2a}{\sqrt{3}} = a\cot\theta \Rightarrow \cot\theta = \frac{2}{\sqrt{3}}. \quad (1)$$

By Lami's theorem to forces at O, we have

$$\frac{S}{\sin 90} = \frac{W}{\sin(90° + \theta)}.$$

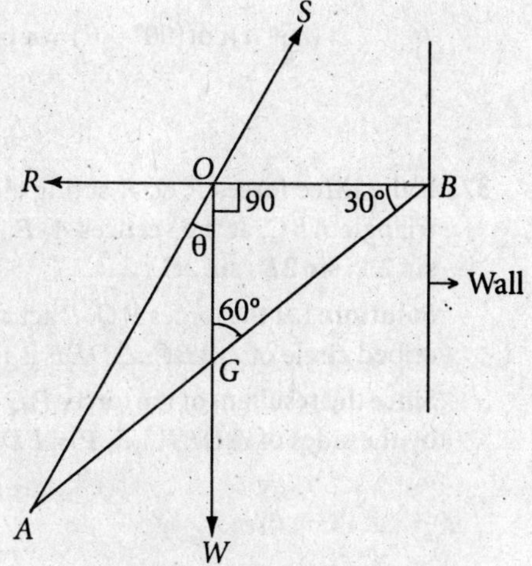

Fig. 2.52

$$\Rightarrow S = W\sec\theta = W\sqrt{1+\tan^2\theta} = W\sqrt{1+\frac{3}{4}} = \frac{W}{2}\sqrt{7}.$$

If the direction of S makes an angle θ with the vertical, then

$$\tan\theta = \frac{\sqrt{3}}{2} \Rightarrow \theta = \tan^{-1}\left(\frac{\sqrt{3}}{2}\right).$$

36. A uniform beam of length l rests with its ends on two smooth plane which intersects in a horizontal line. If the inclinations of the planes to the horizontal are α and β ($\beta > \alpha$), show that the inclination θ of the beam to the horizontal, in one of the equilibrium position, is given by $\tan\theta = \frac{1}{2}(\cot\alpha - \cot\beta)$. **[IAS 2007]**

Solution: Let AB be the rod with its ends on the planes CA and CB respectively. The reaction R at A is perpendicular to CA and the reaction S at B is perpendicular to CB. Let the lines of action of R and S meet at O. Then the third force, the weight W of the rod must pass through O, so that OG is vertical, where G is the mid-point of AB.

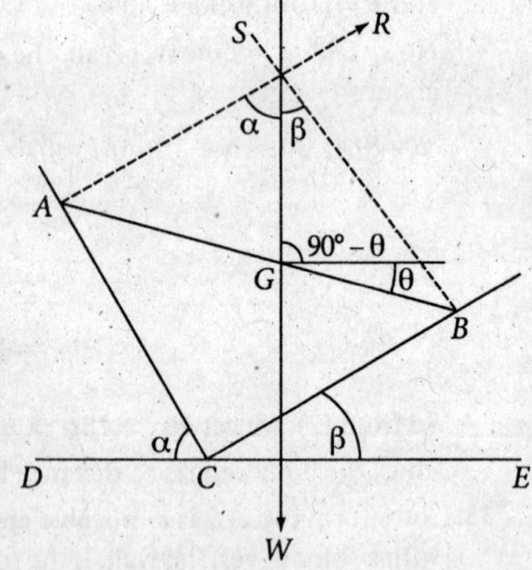

Fig. 2.53

Since angle between two lines is equal to the angle between their perpendiculars.

$\therefore$ angle between AO and OG = angle between CA and $CD = \alpha$.

Similarly, $\angle BOG = \beta$.

Let $AG = GB = a$.

Then by 'm-n theorem' in ΔAOB, we have

$$(a+a)\cot(90° - \theta) = a\cot\alpha - a\cot\beta \Rightarrow 2\tan\theta = (\cot\alpha - \cot\beta)$$

$$\Rightarrow \tan\theta = \frac{1}{2}(\cot\alpha - \cot\beta)$$

37. If the three forces P, Q, R acting along the tangent to the circumscribed circle of the triangle ABC, at the vertices A, B, C are equivalent to a couple, show that $P:Q:R = \sin 2A : \sin 2B : \sin 2C$. **[MU(H) 2007]**

Solution: Let the forces P, Q, R act along the tangents. FD, DE and EF to the circumscribed circle of ΔABC. ΔDEF is the triangle formed by these three tangents.

Since the resultant of the forces P, Q and R is a couple, so they are completely represented by the sides of ΔDEF, i.e., $P = FD$, $Q = DE$ and $R = EF$.

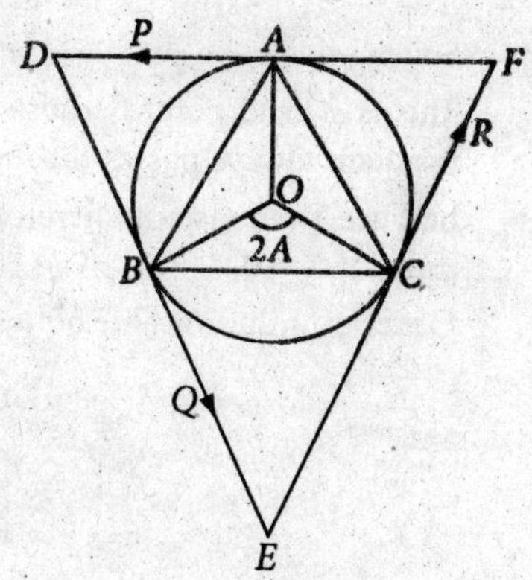

Fig. 2.54

In ΔDEF, by sine formula

$$\frac{FD}{\sin DEF}=\frac{DE}{\sin DFE}=\frac{EF}{\sin FDE}$$

$$\Rightarrow\frac{P}{\sin DEF}=\frac{Q}{\sin DFE}=\frac{Q}{\sin FDE}. \quad (1)$$

Now

$$\angle DEF+\angle BOC=\pi$$
$$\Rightarrow \angle DEF+2A=\pi$$
$$\Rightarrow \angle DEF=\pi-2A.$$

Similarly,

$$\angle DEF=\pi-2B, \angle FDE=\pi-2C.$$

So from (1), we get

$$\frac{P}{\sin(\pi-2A)}=\frac{Q}{\sin(\pi-2B)}=\frac{R}{\sin(\pi-2C)}\Rightarrow\frac{P}{\sin 2A}=\frac{Q}{\sin 2B}=\frac{R}{\sin 2C},$$

i.e., $P:Q:R=\sin 2A:\sin 2B:\sin 2C$ **(Proved).**

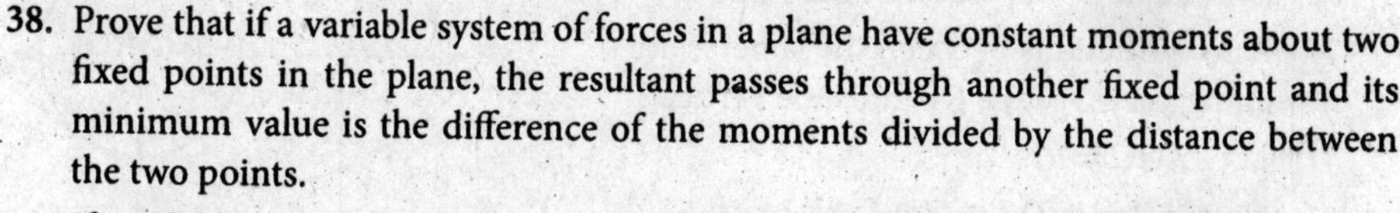

38. Prove that if a variable system of forces in a plane have constant moments about two fixed points in the plane, the resultant passes through another fixed point and its minimum value is the difference of the moments divided by the distance between the two points.

If available system of coplanar forces have constant moments G, G' about two fixed points O, A in the plane of the forces, show that the resultant passes through another fixed point and its least value is $\frac{G-G'}{OA}$. **[CU(H) 2007, KU(H) 2006]**

Solution: Let the two fixed points be O and A. Take O as origin and OA as x-axis and coordinates of A as $(a,0)$. Let G be the moment of the system about the origin O. If G' is the moment about the point A, then

$$G'=G+0\cdot y-aY=G-aY. \quad (1)$$

O

A(a,0)

Fig. 2.55

Since G and G' are given to be constants, vis-à-vis since

$\therefore$ Y = constant.

Equation of line of action of the resultant is

$$xY-yX-G=0\Rightarrow(xY-G)-yX=0.$$

This equation for all values of X represents line passing through the intersection of $xy-G=0$ and $y=0$.

Solving these, we get $x = G/Y$ and $y = 0$, that is the point $(G/Y, 0)$.

This is a fixed point because G and Y are constant. Hence, for all values of X the resultant always passes through a fixed point whose coordinates are $(G/Y, 0)$. Now, because Y is constant, therefore, the resultant which is always $\sqrt{X^2+Y^2}$ will be minimum, if $X = 0$.

Hence, minimum value of the resultant

$$= Y = \frac{G-G'}{a} \text{ [from (1)]}$$

$$= \frac{\text{difference between the moments}}{\text{difference between the points}}.$$

39. Two smooth spheres, each of radius a and weight W, lie on a smooth spherical bowl of radius $na(n > 2)$. Show that the mutual pressure between the sphere is $\frac{W}{\sqrt{n^2-2n}}$.

[BVABU(H) 2005]

Solution: As the two spheres are equal and rest in equilibrium, the reactions between the bowl and spheres at D and E are equal, say, each ZR and both passes is through O, the centre of the bowl.

Also the actions between the spheres are equal, each S (say), OC must be vertical.

Let $\angle AOC = \theta$.

Now, $AC = a$, $OA = OD - AD = na - a$.

In ΔOAC,

$$\sin\theta = \frac{AC}{OA} = \frac{a}{na-a} = \frac{1}{n-1}.$$

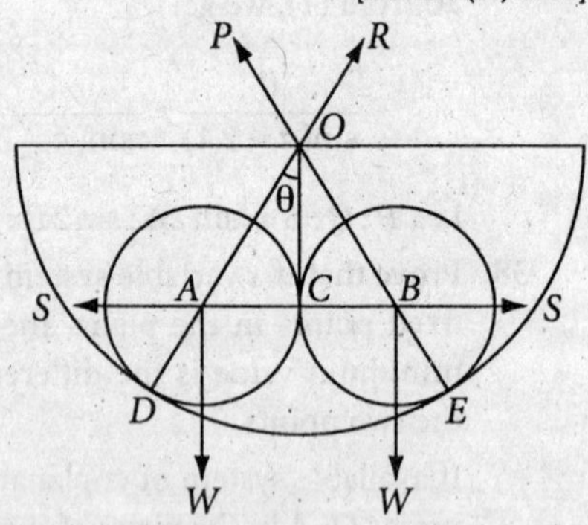

Fig. 2.56

For the equilibrium of sphere with centre A, by Lami's theorem, we have

$$\frac{S}{\sin(180^\circ-\theta)} = \frac{W}{\sin(90+\theta)} \Rightarrow \frac{S}{\sin\theta} = \frac{W}{\cos\theta} \Rightarrow S = \frac{W\sin\theta}{\cos\theta}.$$

$$\Rightarrow S = \frac{W\sin\theta}{\sqrt{1-\sin^2\theta}} = \frac{W\cdot\frac{1}{n-1}}{\sqrt{1-\frac{1}{(n-1)^2}}} = \frac{W}{\sqrt{(n-1)^2-1}} = \frac{W}{\sqrt{n^2-2n}}.$$

Second Method. Taking moment about O, we get

$$W\cdot AC - S\cdot OC = 0 \Rightarrow W\cdot a = S\cdot a\cot\theta$$

$$\Rightarrow W = S\cot\theta \Rightarrow S = W\tan\theta = W/\sqrt{n^2-2n}.$$

40. A small bead P slides on a smooth elliptic wire. It is attached towards the foci S and H by forces proportional to SPW and HPW respectively. Find the position of equilibrium.

[CU(H) 2006]

Solution: The bead is in equilibrium at $P(a\cos\phi, b\sin\phi)$ under three forces; viz., the attractions towards foci and the normal reaction of the elliptic wire. Hence, the components of attraction along tangent at P are equal and opposite. By a well-known property of ellipses, PS and PH are equally inclined to the tangent at P, say, at angles α each, so $PS^m \cos\alpha = PH^m \cos\alpha \Rightarrow PS = PH$.

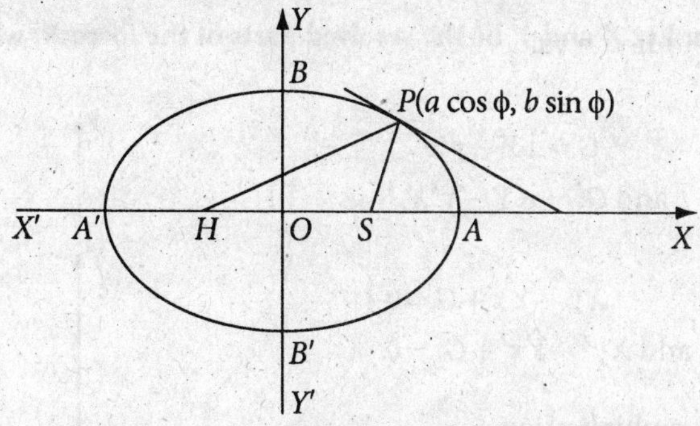

Fig. 2.57

Hence, the bead will be in equilibrium at the extremities of the minor axes.

41. Forces P, Q, R act along the sides BC, CA, AB of a triangle ABC. Find the condition that their resultant should be parallel to BC, and determine its magnitude. [MU(H) 2004]

Solution: Let $\hat{a}$, $\hat{b}$, $\hat{c}$ be the unit vectors along the sides BC, CA, AB of the $\triangle ABC$. Let $\vec{F}$ be the resultant of the forces.

Then

This is the required condition.

$$\vec{F} = P\hat{a} + Q\hat{b} + R\hat{c}$$

$$\Rightarrow \vec{F}\times\hat{a} = P\left(\hat{a}\times\hat{b}\right) + Q\left(\hat{b}\times\hat{a}\right) + R(\hat{c}\times\hat{a})$$

$$\Rightarrow \vec{0} = -Q\left(\hat{a}\times\hat{b}\right) + R(\hat{c}\times\hat{a})$$

$$\Rightarrow Q\left|\hat{a}\times\hat{b}\right| = R\left|\hat{c}\times\hat{a}\right|$$

$$\Rightarrow Q\sin(\pi - C) = R\sin(\pi - B)$$

$$\Rightarrow Q\sin C = R\sin B$$

$$\Rightarrow \frac{Q}{R} = \frac{\sin B}{\sin C} = \frac{b}{c} \Rightarrow Qc = Rb. \qquad (1)$$

Fig. 2.58

Again, from (1),

$$\vec{F}\cdot\hat{a} = P(\hat{a}\cdot\hat{a}) + Q\left(\hat{b}\cdot\hat{a}\right) + R(\hat{c}\cdot\hat{a}) \Rightarrow F\cos 0 = P + Q\cos(\pi - C) + R\cos(\pi - B)$$

$$\Rightarrow F = P - Q\cos C - R\cos B$$

$$\Rightarrow F = P - Q\cos C - \frac{Qc}{b}\cos B \text{ [from (1)]}$$

$$\Rightarrow F = P - \frac{Q}{b}(b\cos C + c\cos B)$$

$$\Rightarrow F = P - \frac{Qa}{b} \quad (\because\ a = b\cos C + c\cos B).$$

This gives the magnitude of the resultant $\vec{F}$.

42. **OX, OY are two straight lines at right angle and a force acting in their plane at O has moments G and G' about the two points whose coordinates are (x, y) and (x', y') respectively w.r.t. the lines OX and OY as axes of coordinates. If $xy' - yx'$ is not zero, prove that the magnitude R of the force and angle θ between its line of action and OX are given by $R^2 = \frac{(xG'-x'G)^2+(yG'-y'G)^2}{(xy'-yx')^2}$ and $\tan\theta = \frac{yG'-y'G}{xG'-x'G}$.** **[KU(H) 2001]**

Solution: Let X and Y be the resolved parts of the force R, which makes an angle θ with OX.

Then

$$G = xY - yX$$

$$\text{and } G' = x'Y - y'X.$$

$$Xy - Yx + G = 0$$

$$\text{and } Xy' - Yx' + G' = 0.$$

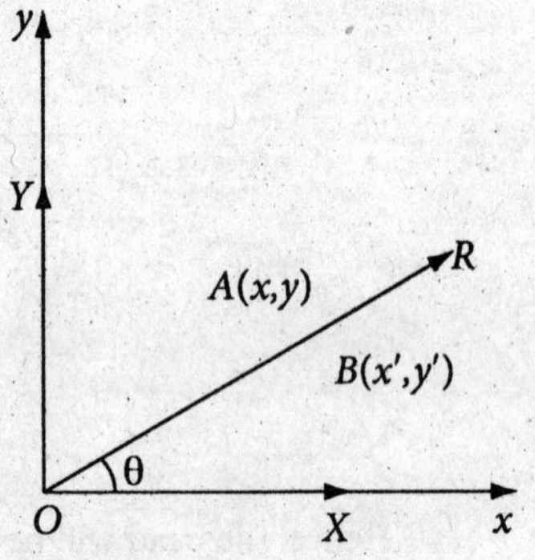

By cross-multiplication

$$\frac{X}{-xG' + x'G} = \frac{Y}{Gy' - G'y} = \frac{1}{-yx' + xy'}$$

$$\Rightarrow X = \frac{x'G - xG'}{xy' - yx'},\ Y = \frac{Gy' - G'y}{xy' - yx'}.$$

Fig. 2.59

Now

$$R^2 = X^2 + Y^2 = \frac{(x'G - xG')^2 + (yG - G'y)^2}{(xy' - yx')^2}$$

$$\text{and } \tan\theta = \frac{Y}{X} = \frac{y'G - yG'}{x'G - xG'} = \frac{yG' - y'G}{xG' - x'G}.$$

43. Three forces P, Q, R act along the sides BC, CA, AB of an equilateral triangle ABC. If their resultant is a force parallel to BC through the centroid of the triangle, prove that $Q = R = \frac{1}{2}P$. **[JPU(H) 2010]**

Solution: Let $\hat{a}, \hat{b}, \hat{c}$ be the unit vectors along the sides BC, CA and AB of $\triangle ABC$. Let $\vec{F}$ be the resultant of the forces $\vec{P}, \vec{Q}, \vec{R}$. Then

$$\vec{F} = P\hat{a} - Q\hat{b} - R\hat{c}. \qquad (1)$$

Let G be the centroid of $\triangle ABC$, then from (1)

$$\vec{F}\hat{a} = P(\hat{a} \times \hat{a}) - Q(\hat{b} \times \hat{a}) - R(\hat{c} \times \hat{a}) \Rightarrow \vec{0} = 0 + Q(\hat{a} \times \hat{b}) - R(\hat{c} \times \hat{a})$$

$$\Rightarrow R(\hat{c} \times \hat{a}) = Q(\hat{a} \times \hat{b})$$

$$\Rightarrow R|\hat{c} \times \hat{a}| = Q|\hat{a} \times \hat{b}|$$

$$\Rightarrow R\sin 120^\circ = Q\sin 120^\circ \Rightarrow R = Q. \qquad (2)$$

Since G is the centroid of the equilateral $\triangle ABC$, so AD, BE and CF are perpendiculars to BC, CA and AB respectively.

Also

$$GD = GE = GF.$$

Now, taking moment of all the forces about G, we get

$$\overrightarrow{GD} \times P\hat{a} - \overrightarrow{GE} \times Q\hat{b} - \overrightarrow{GF} \times R\hat{c} = \vec{0}$$
$$\Rightarrow (P \cdot GD \sin 90^\circ - Q \cdot GE \sin 90^\circ - R \cdot GF \sin 90^\circ)\,\hat{n} = \vec{0},$$

where $\hat{n}$ is the unit vector for perpendicular to the plane of $\triangle ABC$

$$\Rightarrow P - Q - R = 0 \ (\because GD = GE = GF)$$
$$\Rightarrow P = 2Q \Rightarrow \tfrac{1}{2}P = Q = R \text{ [from (2)] } \textbf{(Proved).}$$

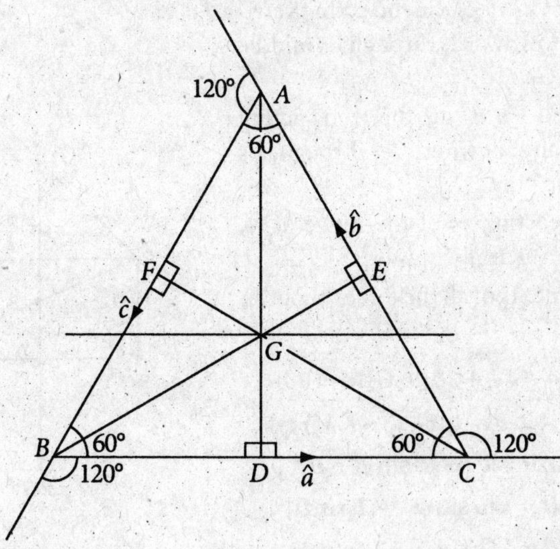

Fig. 2.60

44. The moments of a system of forces (not in equilibrium) acting on a rigid body in one plane about three collinear points A, B, C in the plane are G_1, G_2, G_3 respectively. Prove that $G_1 \cdot BC + G_2 \cdot CA + G_3 \cdot AB = 0$. **[PU(H) 2011]**

Solution: Let the line joining the points A, B and C be taken as axis of x and O be any point on it taken as origin. Let the coordinates of A, B, C be $(x_1, 0)$, $(x_2, 0)$, $(x_2, 0)$ respectively.

$G_1 =$ algebraic sum of the moments of the forces about $A = G - Yx_1$.

Similarly, $G_2 = G - Yx_2$ and $G_3 = G - Yx_3$.

Fig. 2.61

Now,

$$\begin{aligned} &G_1 \cdot BC + G_2 \cdot CA + G_3 \cdot AB \\ &= (G - Yx_1)(x_2 - x_3) + (G - Yx_2)(x_3 - x_1) + (G - Yx_3)(x_1 - x_2) \\ &= G(x_2 - x_3 + x_3 - x_1 + x_1 - x_2) \\ &\qquad - Y[x_1x_2 - x_3x_1 + x_2x_3 - x_1x_2 + x_3x_1 - x_2x_3] \\ &= 0 \textbf{ (Proved).} \end{aligned}$$

45. Two uniform rods AB and BC of lengths $2a$ and $2b$ respectively are rigidly jointed at B and are suspended freely from A. If they are inclined at angles θ and ϕ respectively to the vertical, prove that $\frac{\sin\theta}{\sin\phi} = \frac{b^2}{a^2+2ab}$. [PU(H) 2010]

Solution: Let G_1, G_2 be the mid-points of AB and BC respectively.

Let AM be the vertical through A. Then $\angle BAM = \theta, \angle BDA = \phi$.

Let w be the weight per unit length of the rod.

Then, weight of the rod $AB = 2aw$ and weight of the rod $BC = 2bw$.

We consider the equilibrium of the two rods combined, their mutual actions at B neutralise and the forces left are;

(i) Weight $2aw$ of the rod acting vertically downwards through its middle point G_1.

(ii) Weight $2bw$ of the rod acting vertically downwards through its middle point G_2.

(iii) The reaction R of the hinge at A, acting vertically upwards.

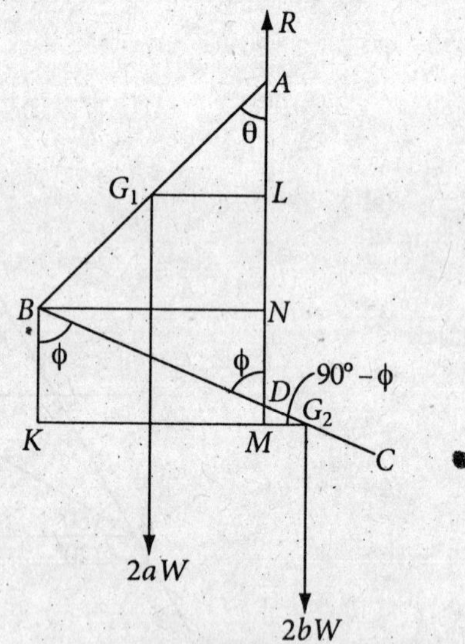

Fig. 2.62

Taking moments of all the forces about A, we get

$$-2aw \cdot G_1L + 2bw \cdot G_2M = 0$$

$$\Rightarrow a \cdot AG_1 \sin\theta = b(G_2K - KM)$$

$$\Rightarrow a \cdot a\sin\theta = b\,[G_2B\sin\phi - BN]$$

$$\Rightarrow a^2\sin\theta = b[b\sin\phi - AB\sin\theta]$$

$$\Rightarrow a^2\sin\theta = b(b\sin\phi - 2a\sin\theta)$$

$$\Rightarrow \left(a^2+2ab\right)\sin\theta = b^2\sin\phi$$

$$\Rightarrow \frac{\sin\theta}{\sin\phi} = \frac{b^2}{a^2+2ab} \text{ (Proved).}$$

46. Forces P, Q and R act along the sides of the triangle formed by the lines $x=0, y=0$ and $\frac{x}{a}+\frac{y}{a}=1$ respectively, axes are rectangular. Find the magnitude of the resultant and equation of its line of action. [PU(H) 2008]

Solution: *First Method.* Let $\frac{x}{a}+\frac{y}{b}=1$ be the equation of line AB.

From O, we draw ON perpendicular to AB.

Then

$$ON = \frac{1}{\sqrt{\frac{1}{a^2}+\frac{1}{b^2}}} = \frac{ab}{\sqrt{a^2+b^2}}.$$

Then $X =$ algebraic sum of resolved part of all the forces along x-axis $= Q + R\cos\theta$.

$Y =$ algebraic sum of resolved parts of all the forces along y-axis $= P + R\sin\theta$

and $G =$ algebraic sum of moments of all the forces about the origin O

$$= R \cdot OBN = \frac{Rab}{\sqrt{a^2+b^2}}.$$

Now slope of the line $\frac{x}{a}+\frac{y}{b}=1$ is

$$-\frac{1/a}{1/b}=-\frac{b}{a}, \text{ i.e., } \tan\theta=-\frac{b}{a}.$$

$$\therefore \cos\theta=-\frac{a}{\sqrt{a^2+b^2}}$$

$$\text{and } \sin\theta=\frac{b}{\sqrt{a^2+b^2}}.$$

$$\therefore X=Q-\frac{aR}{\sqrt{a^2+b^2}}$$

$$\text{and } Y=P+\frac{bR}{\sqrt{a^2+b^2}}.$$

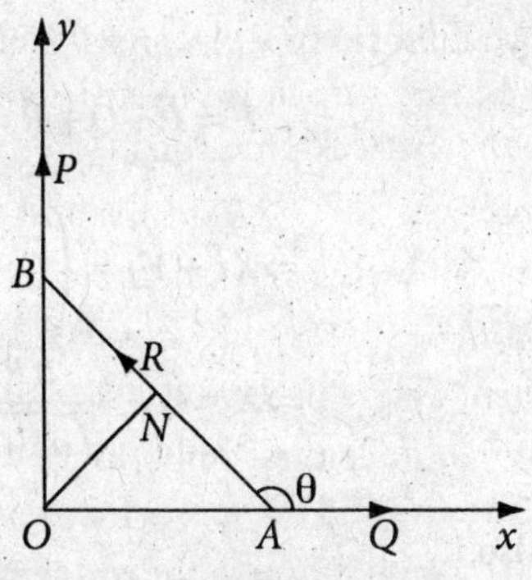

Fig. 2.63

So magnitude of the resultant

$$=\sqrt{X^2+Y^2}=\sqrt{\left(Q-\frac{aR}{\sqrt{a^2+b^2}}\right)^2+\left(P+\frac{bR}{\sqrt{a^2+b^2}}\right)^2}$$

$$=\sqrt{Q^2+P^2+\frac{R^2\left(a^2+b^2\right)}{a^2+b^2}-\frac{2aQR}{\sqrt{a^2+b^2}}+\frac{2bPR}{\sqrt{a^2+b^2}}}$$

$$=\sqrt{P^2+Q^2+R^2-\frac{2aQR}{\sqrt{a^2+b^2}}+\frac{2bPR}{\sqrt{a^2+b^2}}}.$$

Also equation of line of action of the resultant is

$$xY-yX=G\Rightarrow x\left(P+\frac{bR}{\sqrt{a^2+b^2}}\right)-y\left(Q-\frac{aR}{\sqrt{a^2+b^2}}\right)=R\cdot\frac{ab}{\sqrt{a^2+b^2}}$$

$$\Rightarrow x\left(P\sqrt{a^2+b^2}+bR\right)-y\left(Q\sqrt{a^2+b^2}-aR\right)=abR.$$

Second Method. (Vector Method). See the figure of the first method.

Let forces P, Q, R act along the sides OB, OA and AB of the $\triangle OAB$.

Let $\vec{i}$ and $\vec{j}$ be the the unit vectors along the x-axis and y-axis respectively.

$\therefore \vec{P}=P\vec{j}, \vec{Q}=Q\vec{i}.$

Also $\overrightarrow{AB}=\overrightarrow{AO}+\overrightarrow{OB}=-a\vec{i}+b\vec{j}.$

So unit vector along

$$\overrightarrow{AB}=\frac{b\vec{j}-a\vec{i}}{\sqrt{a^2+b^2}}. \quad \therefore \vec{R}=\frac{R\left(b\vec{j}-a\vec{i}\right)}{\sqrt{a^2+b^2}}.$$

Let $\vec{F}$ be the resultant of $\vec{P}, \vec{Q}, \vec{R}$.

So

$$\vec{F}=\vec{P}+\vec{Q}+\vec{R}=P\vec{j}+Q\vec{i}+\frac{R\left(b\vec{j}-a\vec{i}\right)}{\sqrt{a^2+b^2}}$$

$$\Rightarrow X\vec{i}+Y\vec{j}=\left(Q-\frac{aR}{\sqrt{a^2+b^2}}\right)\vec{i}+\left(P+\frac{bR}{\sqrt{a^2+b^2}}\right)\vec{j}$$

$$\Rightarrow X=Q-\frac{aR}{\sqrt{a^2+b^2}},\ Y=\frac{P+bR}{\sqrt{a^2+b^2}}.$$

Also,

$$\vec{G}=\text{algebraic sum of moments of all the forces about } O$$

$$=\overrightarrow{OA}\times\frac{R\left(b\vec{j}-a\vec{i}\right)}{\sqrt{a^2+b^2}}=\frac{a\vec{i}\times R\left(b\vec{j}-a\vec{i}\right)}{\sqrt{a^2+b^2}}$$

$$\Rightarrow\vec{G}=\frac{abR\vec{k}}{\sqrt{a^2+b^2}}\Rightarrow G=\frac{abR}{\sqrt{a^2+b^2}}.$$

Now magnitude of the resultant

$$=\sqrt{X^2+Y^2}=\sqrt{\left(Q-\frac{aR}{\sqrt{a^2+b^2}}\right)^2+\left(P+\frac{bR}{\sqrt{a^2+b^2}}\right)^2}$$

$$=\sqrt{Q^2+\frac{a^2R^2}{a^2+b^2}-\frac{2aQR}{\sqrt{a^2+b^2}}+P^2+\frac{b^2R^2}{a^2+b^2}+\frac{2bPR}{\sqrt{a^2+b^2}}}$$

$$=\sqrt{P^2+Q^2+R^2-\frac{2aQR}{\sqrt{a^2+b^2}}+\frac{2bPR}{\sqrt{a^2+b^2}}}$$

Also equation of line of action of the resultant is $xY-yX=G$

$$\Rightarrow x\left(P+\frac{bR}{\sqrt{a^2+b^2}}\right)-y\left(Q-\frac{aR}{\sqrt{a^2+b^2}}\right)=\frac{R\cdot ab}{\sqrt{a^2+b^2}}$$

$$\Rightarrow x\left(P\sqrt{a^2+b^2}+bR\right)-y\left(Q\sqrt{a^2+b^2}-aR\right)=Rab.$$

47. Show that the greatest inclination to the horizon at which a uniform rod can rest partly within and partly outside a fixed smooth hemispherical bowl is $\sin^{-1}\left(1/\sqrt{3}\right)$.

[PU(H) 2010]

Solution: Let the rod AB of length $2a$ and weight W, rest partially inside and partially outside a fixed hemispherical bowl whose centre is at O and radius is r.

Let the rod rest with the end A in contact the spherical surface and the the point C in contact with the rim of the bowl.

The rod is in equilibrium under the action of the following three forces only.

(i) R, the reaction of the spherical surface at A along the normal to the sphere at A and so passing through the centre O of the base of the bowl.

(ii) S, the reaction at C, acting at right angles to the rod.

(iii) W, the weight of the rod acting vertically downwards at the mid-point G of the rod.

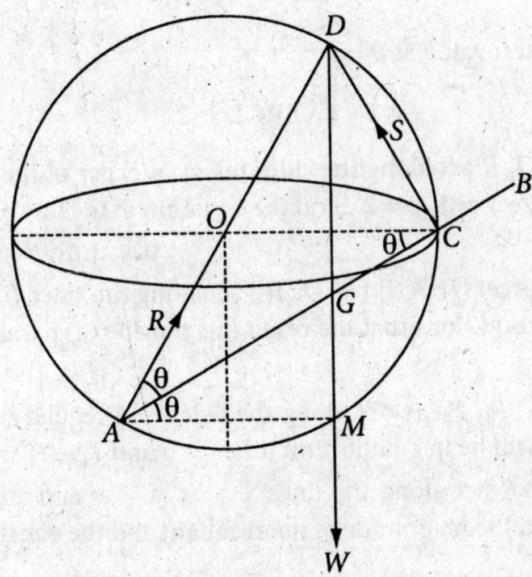

Fig. 2.64

Since the forces R and S meet at D, hence the line of action of the third force W must also pass through D as shown in the figure. Thus the line DG is vertical.

$\because \angle ACD = 90°$ and the line AD passes through the centre of the sphere of which the bowl is a part, therefore, the point D must be on this sphere, i.e., AD is the diameter of the sphere of which the bowl is a part. Suppose DG when produced meets the surface of the bowl at M. Join AM. Then $\angle AMD = 90°$, because the diameter AD of the sphere subtends a right angle at the point M on the surface of the sphere. Hence, the line AM is horizontal.

If the rod makes an angle θ to the horizontal, then $\angle CAM = \theta$. $\because$ OC and AM are parallel and AC meets them,

$\therefore \angle OCA = \angle CAM = \theta$.

But $OC = OA$ (being radii of the bowl). $\therefore \angle CAO = \angle OCA = \theta$.

$\therefore \angle DAM = 2\theta$.

Now from ΔDAM,

$$AM = AD\cos 2\theta = 2r\cos 2\theta \tag{1}$$

and from ΔGAM,

$$AM = AG\cos\theta = a\cos\theta. \tag{2}$$

From (1) and (2), we have

$$2r\cos 2\theta = a\cos\theta \Rightarrow a = \frac{2r\cos 2\theta}{\cos\theta}. \tag{3}$$

Since
$$AC \le AB \Rightarrow cAD\cos\theta \le 2a \Rightarrow 2r\cos\theta \le \frac{4r\cos 2\theta}{\cos\theta} \text{ [from (3)]}$$
$$\Rightarrow \cos^2\theta \le 2\left(1-2\sin^2\theta\right) \Rightarrow 1-\sin^2\theta \le 2-4\sin^2\theta$$
$$\Rightarrow 3\sin^2\theta \le 1 \Rightarrow \sin^2\theta \le 1/3$$
$$\Rightarrow -1/\sqrt{3} \le \sin\theta \le 1/\sqrt{3} \Rightarrow |\sin\theta| \le 1/\sqrt{3}.$$

Hence, the greatest value of $\theta = \sin^{-1}\frac{1}{\sqrt{3}}$.

EXERCISE 2

1. Three forces P, Q, R act along the sides taken in order of the triangle formed by the lines $x+y=1, x-y=1, y=2$. Find the equation of the line of action of the resultant. [LNMU(H) 2000, MU(H) 2003, RU(H) 1999]
2. Two system of forces $(P, Q, R), (P', Q', R')$ act along the sides BC, CA, AB of a triangle ABC. Find the conditions that the resultants may be (a) parallel, (b) at right angles. [PU(H) 1998, BVABU(H) 2007]
3. Forces $P_1, P_2, P_3, P_4, P_5, P_6$ act along the sides of a regular hexagon taken in order. Prove that they will be in equilibrium it $\Sigma P_1 = 0$ and $P_1 - P_4 = P_3 - P_6 = P_5 - P_2$.
4. Forces P, Q and R act along the lines $x = 0$, $y = 0$ and $x\cos 30° + y\sin 30° = p$ respectively. Find the magnitude of the resultant and the equation of its line of action. [BVABU(H) 1997]
5. The algebraic sums of moments of a system of coplanar forces about points whose coordinates are $(1,0)$, $(0,2)$ and $(2,3)$ referred to rectangular axes are G_1, G_2 and G_3 respectively. Find the tangent of the angle which the direction of the resultant force makes with the axis of x. [PU(H) 2002]
6. The straight line $4x+3y=5$ meets the rectangular axes Ox and Oy at the points A and B respectively. If the forces P, Q, R act along OB, OA and AB, then find the magnitude of the resultant and the equation of line of action.
7. A system of coplanar forces reduces to a couple whose moment is G and when each force is turned round its point of application in the plane of the forces through a right angle, the system reduces to a couple of moment $G\sqrt{3}$. Prove that when each forces is turned through 60° in the same plane, the system is also equivalent to a couple whose moment is $2G$. [CU(H) 1995]
8. A system of coplanar forces has the total moments H, $2H$ and $3H$ about the points $(0,0)$, $(0,1)$ and $(2,4)$ respectively. Find the magnitude and the equation of line of action of the resultant. [CU(H) 1982]
9. A uniform rod of length $2l$ rests with its lower end in contact with a smooth vertical wall. It is supported by a string of length a, one end of which is fastened to a point in the wall and the other end to a point in the rod at a distance b from its lower end. If the inclination of the string to the vertical be θ, then show that $\cos^2\theta = \frac{b^2(a^2-b^2)}{a^2 l(2b-l)}$.
10. AB and AC are similar uniform rods of length a, smoothly jointed at A. BD is a weightless bar of length b, smoothly jointed at B and fastened at D to a smooth ring sliding on AC. The system is hung on a small smooth pin at A. Show that the rod makes with the vertical an angle $\tan^{-1}\frac{b}{a+\sqrt{a^2-b^2}}$.

11. Three forces each equal to P act along the sides of a triangle ABC in order. Prove that the resultant R is given by $R = P\left(1-8\sin\frac{A}{2}\sin\frac{B}{2}\sin\frac{C}{2}\right)^{1/2}$ and find the distance of its line of action from A. Examine the case when the triangle is equilateral.

12. The algebraic sum of the moments of a system of coplanar forces about three non-collinear points A, B, C in their plane are L, M, N respectively. Prove that their resultant is given by $R^2 = \frac{\Sigma a^2(L-M)(L-N)}{4\pi^2}$, where a, b, c are the sides of the triangle ABC and Δ its area.

13. A system of forces in the plane of a rectangular plane $ABCD$ has moments G_1, G_2, G_3 about the points A, B, C respectively. Show that the magnitude of the resultant is $\sqrt{\left(\frac{G_1-G_2}{AB}\right)^2+\left(\frac{G_3-G_2}{BC}\right)^2}$ and find the distance from A of the point where its line of action intersects AB.

Discuss the case, when $G_1 = G_2 = G_3$.

14. A uniform rod whose weight is W is supported by two fine strings one attached to each end, which after passing over small fixed smooth pulleys, carry weights W_1 and W_2 respectively at the other ends. Show that the rod is inclined to the horizon at an angle $\sin^{-1}\frac{W_1^2-W_2^2}{W\sqrt{2(W_1^2+W_2^2)-W^2}}$.

15. A prism whose cross section is an equilateral triangle, rests with edges horizontal on smooth planes inclined at angles α and β to the horizon. It θ be the angle that the plane through these edges makes with the vertical, show that $\tan\theta = \frac{2\sqrt{3}\sin\alpha\sin\beta+\sin(\alpha+\beta)}{\sqrt{3}\sin(\alpha-\beta)}$.

16. A square lamina rests in a vertical plane on two smooth pegs at a distance c a part in the same horizontal plane. If θ be the angle the diagonal through the lowest corner makes with the vertical, x and y the distances of the CG from the sides which touch the pegs, prove that $(x+y)\sin\theta = (y-x)\cos\theta + \sqrt{2}c\sin 2\theta$.

17. When an elliptic lamina is acted upon at the extremities of a pair of conjugate diameters by forces in its own plane tending outwards and normal to the edge, show that there will and normal to the edge, show that there will and normal to the edge, show that there will be equilibrium, if the force at the end of each diameter is proportional to the conjugate diameter. [TMBU(H) 2000]

18. When a system of coplanar forces in equilibrium continues in equilibrium for all displacements in the plane of the forces, show that the a static centre of any number of them must be coincident with that of the remainder.

19. A rod rests wholly within a hemispherical bowl of radius r, its CG dividing the rod into two portions of length a and b. Show that if θ be the inclination of the rod to the horizon, in the position of equilibrium, then $\sin\theta = \frac{b-a}{2\sqrt{r^2-ab}}$.

Prove also that if 2α be the angle subtended by the rod at the centre of the sphere, then $\tan\theta = \frac{b-a}{b+a}\tan\alpha$.

20. A circular disc of weight w and radius a is suspended horizontally by three equal vertical strings of length b, attached symmetrically to the perimeter. Show that the magnitude of the horizontal couple required to keep it twisted through an angle θ is $\frac{wa^2\sin\theta}{\sqrt{b^2-4a^2\sin^2\frac{\theta}{2}}}$.

21. Two equal smooth spheres, each of weight w and radius r, are placed inside an smooth hollow cylinder open at both ends, which rests on a smooth horizontal plane with its axis vertical. If $a(< 2r)$ be the radius of the cylinder find the reaction between the spheres. Show that the condition necessary for the cylinder not to be upset or not to be overturned is

$$W' > 2W\left(1 - \frac{r}{a}\right),$$

where W' is the weight of the cylinder.

22. Two equal heavy rods, of weight W and length $2a$ are freely hinged to and placed. Symmetrically over a smooth fixed sphere of radius r. Show that the inclination of each rod to the horizon is given by $r(\tan^3\theta + \tan\theta) = a$.

23. A smooth hemispherical bowl of diameter a is placed so that its edge touches a smooth vertical wall; a uniform rod AB of weight W is in equilibrium; inclined at 60° to the horizon with one end A resting on the inner surface of the bowl and the other end B resting against the wall, show that the length of the rod must be $a + \frac{a}{\sqrt{13}}$. Find the reactions against the wall and the bowl.

24. One end of a uniform rod is attached to a huge and the other is supported by a string attached to the extremity of the rod. The rod and the string are inclined at the same angle θ to the horizontal. If W be the weight of the rod, show that the action at the hinge is $\frac{W}{4}\sqrt{9+\cot^2\theta}$. Find also the tension of the string.

25. Three uniform rods AB, BC and CD, whose weights are proportional to their lengths a, b and c, are jointed at B and C are in a horizontal position resting on two pegs P and Q, find the action at the joints B and C and show that the distance between the pegs must be $\frac{a^2}{2a+b} + \frac{c^2}{2c+b} + b$.

26. A circular disc of weight W and radius a is suspended horizontally by three equal vertical strings, of length attached symmetrically to its perimeter. Show that the magnitude of the horizontal couple required to keep it twisted through an angle θ is $Wa^2\frac{\sin\theta}{\sqrt{b^2-4a^2\sin^2\frac{\theta}{2}}}$.

Answers

1. $x(P-Q)+y\left(P+Q-\sqrt{2}R\right)=P-Q-2\sqrt{2}R$.
2. (a) $(QR'-Q'R)\sin A+(RP'-R'P)\sin B+(PQ'-P'Q)\sin C=0$;

   (b) $PP'+QQ'+RR'-(QR'+RQ')\cos A-(RP'+PR')\cos B-(PQ'+QP')\cos C=0$.
4. Magnitude $=\sqrt{P^2+Q^2+R^2+\sqrt{3}PR-QR}$. Equation of the line of action of the resultant is $x\left(P+\frac{\sqrt{3}}{2}R\right)-y\left(Q-\frac{R}{2}\right)=R\cdot p$.
5. $\tan\alpha=\frac{G_1-3G_2+2G_3}{2G_1-G_2-G_3}$.
6. Magnitude $=\sqrt{P^2+Q^2+R^2-\frac{6}{5}QR+\frac{8}{5}PR}$. Equation of the line of action of the resultant is $x\left(P+\frac{4}{5}R\right)-y\left(Q-\frac{3}{5}R\right)-R=0$.
8. As resultant $=\sqrt{2}H$. Equation of line of action of the resultant is $x-y-1=0$.
11. $\frac{2P\Delta}{aR}$, where Δ is the area of ΔABC and $a = BC$, when the triangle is equilateral resultant is zero.

13. $\frac{aG_1}{G_1-G_2}$, where $AB=a$, when $G_1=G_2=G_3$, the magnitude of the resultant vanishes.

21. Reaction $= W\sqrt{\frac{r}{2a-r}}$.

23. $\sqrt{\frac{13}{12}}W, \frac{W}{\sqrt{12}}$.

24. $\frac{W}{4}\operatorname{cosec}\theta$.

25. 0, $\frac{bw}{2}$, where w is the constant of proportionality.

Chapter 3

Virtual Work

3.1 Introduction

Before starting with the virtual work, we require the definition of real (actual) work. So we first define actual work.

Definition (Actual Work)

A force is said to do work on a body, when its points of application move in the direction of force. The amount of work done by a force is measured by the product of the magnitude of the force and the distance through which its points of application are moved in the direction of the force.

If a constant force $\vec{F}$ acts at O and displaces it through a distance d where $\vec{d} = \overrightarrow{OA}$ along $\vec{F}$, then the work done by

$$\vec{F} = Fd = |\vec{F}||\vec{d}|. \qquad (1)$$

Fig. 3.1

In general, if the displacement is not in the direction of the force, the work done by a force $\vec{F}$ when its point of application undergoes a small displacement, is defined to be the product of the magnitude of the force and the resolved part of displacement along $\vec{F}$.

If the force $\vec{F}$ acts on a particle at O and displaces it slightly through a distance d where $\overrightarrow{OA} = \vec{d}$, which is inclined at an angle θ with the direction of $\vec{F}$, then the displacement along $\vec{F}$ is $d\cos\theta$.

$\therefore$ Work done by the force $\vec{F}$

$$= F(d\cos\theta) = |\vec{F}||\vec{d}|\cos\theta = \vec{F}\cdot\vec{d}. \qquad (2)$$

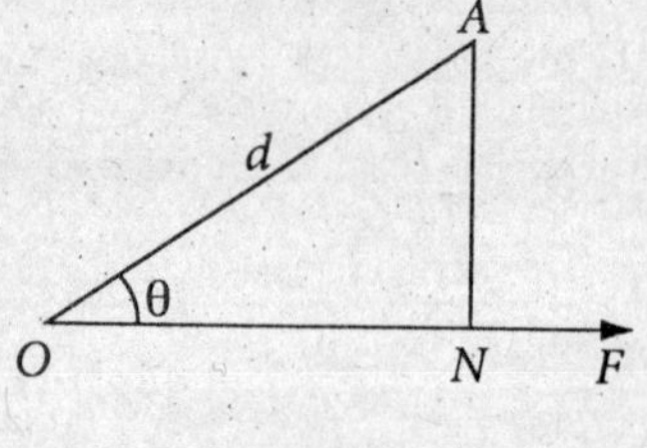

Fig. 3.2

It is clear that the work done $\gtreqless 0$ according as $\theta \lesseqgtr \pi/2$.

Convention of Signs

If the point of application of the force is displaced in the direction of force, the work done is positive and if in the opposite direction, the work done is negative.

Thus, if we lift a weight w through a height h, the work done is negative and is equal to $-wh$.

3.2 Work done by a System of Concurrent Forces

Theorem 1. *The algebraic sum of the works done by a system of forces acting on a particle is equal to the work done by the resultant.*

Classical Proof

Let $F_1, F_2, F_3, \ldots$ be a system of forces acting on a particle at O, which displaces it through a distance d where $OA = d$, making angle $\theta_1, \theta_2, \theta_3, \ldots$ respectively with the direction of OA. If F be the resultant of the system of forces, making an angle θ with OA, then we have algebraic sum of the resolved parts of the forces along OA = resolved part of the resultant along OA.

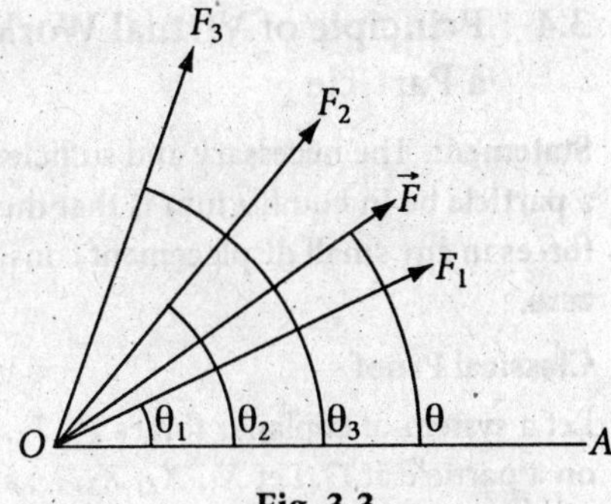

Fig. 3.3

$$\therefore\ \Sigma F_1 \cos\theta_1 = F_1 \cos\theta_1 + F_2 \cos\theta_2 + \ldots$$
$$= F\cos\theta. \qquad (1)$$

Hence, total work done by the system of forces

$$= F_1(d\cos\theta_1) + F_2(d\cos\theta_2) + \ldots$$
$$= d(F_1\cos\theta_1 + F_2\cos\theta_2 + \ldots)$$
$$= d(F\cos\theta)\ \text{[from (1)]}$$
$$= F(d\cos\theta)$$
$$= \text{work done by the resultant force } \vec{F}.$$

Vector Proof

Let $\vec{F}_1, \vec{F}_2, \vec{F}_3, \ldots$ be the forces of the system acting on a particle at O and it undergoes a small displacement $\overrightarrow{OA} = \vec{d}$. If $\vec{F}$ be the resultant of the system at O, then clearly

$$\vec{F} = \vec{F}_1 + \vec{F}_2 + \vec{F}_3 + \cdots = \Sigma F_1. \qquad (2)$$

Hence, total work done by the system of forces

$$= \vec{F}_1 \cdot \vec{d} + \vec{F}_2 \cdot \vec{d} + \ldots$$
$$= \left(\vec{F}_1 + \vec{F}_2 + \ldots\right)\cdot \vec{d}\ \text{(by distributive law)}$$
$$= \vec{F}\cdot\vec{d}\ \text{[from (2)]}$$
$$= \text{work done by their resultant } \vec{F}.$$

3.3 Virtual Work

If a system of forces acting on a particle (or rigid body) be in equilibrium and if the particle is imagined to undergo a slight displacement consistent with the geometrical conditions of the system, then the work done by the system during this imaginary displacement is called *virtual work.*

The imaginary displacement of the particle in equilibrium under the system of forces is called the *virtual displacement.*

Now since the work is after all a work, whether it is actual or virtual, therefore, all the theorems which are true for the actual work will also be true in case of the virtual work.

3.4 Principle of Virtual Work for a System of Coplanar Forces acting on a Particle

[MU(H) 2010]

Statement. The necessary and sufficient condition for a system of coplanar forces acting on a particle be in equilibrium is that the algebraic sum of virtual work done by the system of forces in any small displacement consistent with the geometrical conditions of the system is zero.

Classical Proof

Let a system of coplanar forces F_1, F_2, F_3 act on a particle at O. Let $X_1, X_2, X_3, \ldots$ and $Y_1, Y_2, Y_3, \ldots$ be the resolved part of these forces in two mutually perpendicular directions Ox and Oy respectively. Then the sum of the resolved parts in the directions Ox and Oy are

$$\left.\begin{aligned} \Sigma X_1 &= X_1 + X_2 + X_3 + \ldots \\ \text{and } \Sigma Y_1 &= Y_1 + Y_2 + Y_3 + \ldots \end{aligned}\right\}. \quad (1)$$

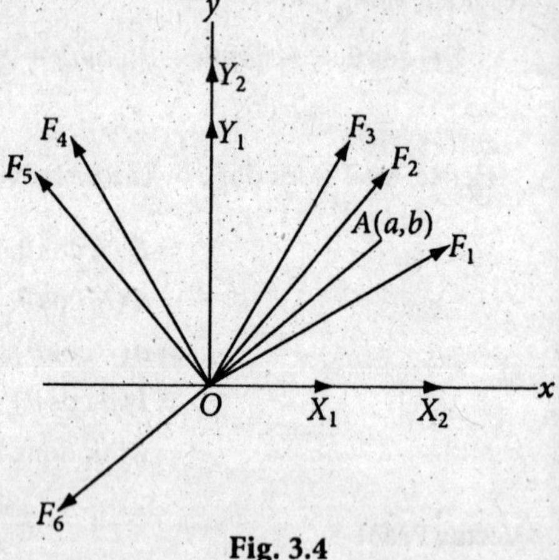

Fig. 3.4

Let OA represent an arbitrary small displacement of the particle from O where the coordinates of A are given by (a, b).

Then the virtual work done by the force F_1 = algebraic sum of works done by its resolved part X_1, Y_1 is $aX_1 + bY_1$.

Similarly, the virtual works done by the forces $\vec{F}_2$, $\vec{F}_3$, ... are

$$aX_2 + bY_2,\ aX_3 + bY_3,\ \ldots.$$

Hence, the algebraic sum of the virtual work done by the system of the forces through O

$$\begin{aligned} &= (aX_1 + bY_1) + (aX_2 + bY_2) + \ldots \\ &= a(X_1 + X_2 + \ldots) + b(Y_1 + Y_2 + \ldots) \\ &= a\Sigma X_1 + b\Sigma Y_1 \text{ [from (1)].} \end{aligned} \quad (2)$$

The Condition is necessary

Here we assume that the particle at O is in equilibrium. Then applying the condition of equilibrium, we have

$$X = \Sigma X_1 = 0,\ Y = \Sigma Y_1 = 0. \quad (3)$$

Using (3) in (2) we get the algebraic sum of virtual work done by the system $= a \cdot 0 + b \cdot 0 = 0$.

This proves the necessary condition.

The Condition is sufficient

Here we assume that the algebraic sum of the virtual work done by the system of coplanar forces acting on a particle is zero for any small displacement.

Then from (2),

$$a\Sigma X_1 + b\Sigma Y_1 = 0 \tag{4}$$

for all suitable small values of a and b. We have to show that system of forces are in equilibrium.

Let us give a displacement in which the particle is moved only through a distance a along x-axis. Hence, for this displacement we have $a \neq 0$, $b = 0$.

So from (4),

$$a\Sigma X_1 = 0. \therefore \Sigma X_1 = 0.$$

Next we choose a displacement in which the particle is moved only through a distance b along y-axis. Hence, for this displacement, we have $a = 0$, $b \neq 0$.

So from (4),

$$b\Sigma Y_1 = 0. \therefore \Sigma Y_1 = 0.$$

Hence, we find that $\Sigma X_1 = 0$, $\Sigma Y_1 = 0$.

Hence, the system of forces must be in equilibrium.

Vector Proof

We consider a system of coplanar forces $\vec{F}_1$, $\vec{F}_2$, $\vec{F}_3$, ... acting on a single particle at O and OA represents an arbitrary small displacement of the particle from O, where $\overrightarrow{OA} = \vec{d}$. If $\vec{F}$ be the resultant of the forces of the system at O, then clearly

$$\vec{F} = \vec{F}_1 + \vec{F}_2 + \vec{F}_3 + \cdots = \Sigma \vec{F}_1. \tag{1}$$

Hence, the algebraic sum of virtual work done by the system of forces

$$\begin{aligned} &= \vec{F}_1 \cdot \vec{d} + \vec{F}_2 \cdot \vec{d} + \ldots \\ &= \left(\vec{F}_1 + \vec{F}_2 + \ldots\right) \cdot \vec{d} \\ &= \vec{F} \cdot \vec{d} \text{ [from (1)].} \end{aligned} \tag{2}$$

The Condition is necessary

Here we assume that system of forces acting at O be in equilibrium, then $\vec{F} = 0$.

.. from (2), the algebraic sum of virtual work done by the system of forces $= \vec{F} \cdot \vec{d} = 0$.

The Condition is sufficient

Here we assume that algebraic sum of virtual work done by the system of forces acting at O be zero, for all small displacements of the particle.

$\therefore$ from (2), $\vec{F} \cdot \vec{d} = 0$ for a small displacement $\vec{d}$.

Hence, $\vec{F} = 0$.

This shows that the system of forces are in equilibrium.

Theorem 1. *Principle of virtual work for a system of coplanar forces acting at different points of a rigid body.*

[BVABU(H) 2011, LNMU(H) 2000, CU(H) 2008, MU(H) 2012, KU(H) 2007, PU(H) 2010, TMBU(H) 2011, RU(H) 1999]

Statement. If a system of coplanar forces acting at different points of a rigid body be in equilibrium and if the body be given a small virtual displacement consistent with the geometrical conditions of the system, then the algebraic sum of the virtual works done by the forces is zero. Conversely, if the algebraic sum of all the virtual works for any displacement is zero, the forces are in equilibrium.

Proof. Let OX and OY be the two rectangular axes in the plane of the forces. We suppose that any force F_1 of the system acts at the point P whose Cartesian coordinates are (x, y) and whose polar coordinates are (r, θ).

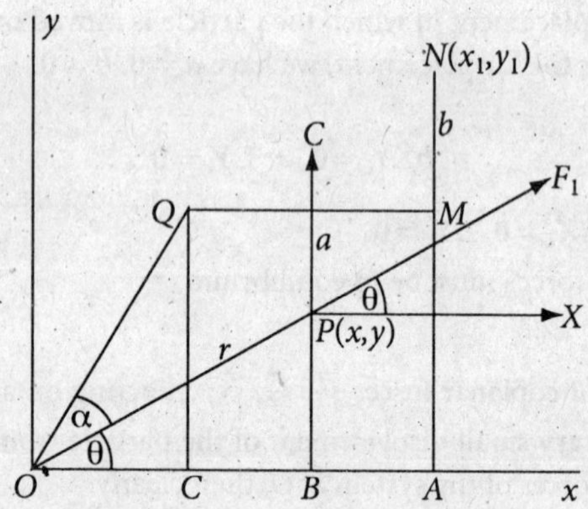

Fig. 3.5

Then $OP = r$, $\angle XOP = \theta$ and $x = r\cos\theta$, $y = r\sin\theta$.

Let the body undergoes a small displacement. We can do so by rotating the body through a suitable small angle α radians about O and then moving it through suitable distances a and b parallel to the axes.

By the above motion, the point P first moves to Q, then to M and finally to N.

If coordinates of N be (x_1, y_1), then $OA = x_1$, $NA = y_1$,

$$x_1 = OA = OC + CA = OC + a.$$

In ΔOQC,

$$\cos(\theta + \alpha) = \frac{OC}{OQ} = \frac{OC}{r}.$$

$$OC = r\cos(\theta + \alpha).$$

$$x_1 = a + r\cos(\theta + \alpha)$$

$$= a + r(\cos\theta\cos\alpha - \sin\theta\sin\alpha) \; [\because \alpha \text{ is small, } \cos\alpha = 1,\ \sin\alpha = \alpha]$$

$$= a + r(\cos\theta \cdot 1 - \sin\theta \cdot \alpha) = a + r\cos\theta - r\sin\theta \cdot \alpha$$

$$\Rightarrow x_1 = a + x - y\alpha \Rightarrow x_1 - x = a - y\alpha.$$

Also

$$y_1 = NA = NM + MA = b + QC.$$

In $\triangle OQC$,

$$\sin(\theta+\alpha) = \frac{QC}{OQ} = \frac{QC}{r}.$$

$$QC = r\sin(\theta+\alpha).$$

$$\therefore\ y_1 = b + r\sin(\theta+\alpha) = b + r(\sin\theta\cos\alpha + \cos\theta\sin\alpha)$$

$$= b + r(\sin\theta\cdot 1 + \cos\theta\cdot\alpha) = b + r\sin\theta + r\cos\theta\cdot\alpha.$$

$$y_1 = b + y + r\cos\theta\cdot\alpha,\ \therefore\ y_1 - y = b + x\alpha.$$

If X_1, Y_1 be the components of the force F_1 parallel to axes, then the virtual work done by the force F_1

$$= \text{algebraic sum of virtual work done by } X_1 \text{ and } Y_1$$

$$= X_1(a - \alpha y) + Y_1(b + \alpha x)$$

$$= aX_1 + bY_1 + \alpha(xY_1 - yX_1).$$

Similar expressions for the virtual work done by the other forces can be obtained, a, b and α being the same for each force.

The algebraic sum of the virtual work is, therefore,

$$a\Sigma X_1 + b\Sigma Y_1 + \alpha\Sigma(xY_1 - yX_1). \tag{1}$$

Now, if the forces of the system are in equilibrium, then applying the general conditions of equilibrium, we have

$$\left.\begin{array}{c}\Sigma X_1 = 0, \Sigma Y_1 = 0\\ \Sigma(xY_1 - yX_1) = 0\end{array}\right\}. \tag{2}$$

Separately using (2) in (1) we get algebraic sum of virtual works done by the forces of system = 0.

Conversely, we suppose that the algebraic sum of the virtual works done by the coplanar forces is zero.

[PU(H) 2011, TMBU(H) 2010, RU(H) 2000, BVABU(H) 1999, LNMU(H) 2011, MU(H) 2001, JPU(H) 2010]

So from (1),

$$a\Sigma X_1 + b\Sigma Y_1 + \alpha\Sigma(xY_1 - yX_1) = 0 \tag{3}$$

for all suitable values of a, b and α.

We show that the system will be in equilibrium.

Since a, b and α are arbitrary, we may choose them according to our convenience.

First, we choose a displacement such that the body is moved only through a distance a parallel to x-axis, then for this displacement, we have $a \neq 0$, $b = 0$ and $\alpha = 0$.

So from (3),

$$a\Sigma X_1 = 0;\ \therefore\ \Sigma X_1 = 0.$$

Next, we choose a displacement such that the body is moved only through a distance b parallel to y-axis, then for this displacement, we have $b \neq 0$, $a = 0$ and $\alpha = 0$.

Hence, from (3),

$$b\Sigma Y_1 = 0;\ \Sigma Y_1 = 0.$$

Finally, we choose a displacement such that the body is moved about O only through a small angle α, then for this displacement, we have $\alpha \neq 0$, $a = 0$ and $b = 0$.

So from (3),

$$\alpha\Sigma(xY_1 - yX_1) = 0$$
$$\Sigma(xY_1 - yX_1) = 0.$$

Thus, the conditions of equilibrium, namely $\Sigma X_1 = 0$, $\Sigma Y_1 = 0$ and $\Sigma(xY_1 - yX_1) = 0$ are satisfied. Hence, the forces are in equilibrium.

Vector Proof

We know that a system of forces can be reduced to a single force at an arbitrary point together with a couple. Now let $F_1, F_2, \ldots, F_n$ be the forces acting on a rigid body at points. $\vec{r}_1, \vec{r}_2, \ldots, \vec{r}_n$ and this system be equivalent to a single force $\vec{R}$ at O and a couple of moment $\vec{G}$. Hence,

$$\vec{R} = \vec{F}_1 + \vec{F}_2 + \cdots + \vec{F}_n \tag{1}$$

$$\text{and } \vec{G} = \vec{r}_1 \times \vec{F}_1 + \vec{r}_2 \times \vec{F}_2 + \cdots + \vec{r}_n \times \vec{F}_n. \tag{2}$$

Now if the rigid body undergoes a small virtual displacement which is the sum of constant translation $\vec{u}$ and a small rotation $\vec{e}$ about O, then if $\vec{r}_n$ is displaced to $\vec{r}_n + d\vec{r}_n$, we have

$$d\vec{r}_n = \vec{u} + \vec{e} \times \vec{r}_n,$$

where $n = 1, 2, 3, \ldots$.

Therefore, the algebraic sum of virtual work is

$$\begin{aligned}
&\vec{F}_1 \cdot d\vec{r}_1 + \vec{F}_2 \cdot d\vec{r}_2 + \cdots + \vec{F}_n \cdot d\vec{r}_n \\
&\quad = \vec{F}_1 \cdot (\vec{u} + \vec{e} \times \vec{r}_1) + \vec{F}_2 \cdot (\vec{u} + \vec{e} \times \vec{r}_2) + \cdots + \vec{F}_n \cdot (\vec{u} + \vec{e} \times \vec{r}_n) \\
&\quad = \vec{u} \cdot \left(\vec{F}_1 + \vec{F}_2 + \cdots + \vec{F}_n\right) + \vec{e} \cdot \left(\vec{r}_1 \times \vec{F}_1 + \vec{r}_2 \times \vec{F}_2 + \cdots + \vec{r}_n \times \vec{F}_n\right) \\
&\qquad\qquad\qquad\qquad \left[\because \vec{a}\left(\vec{b} \times \vec{c}\right) = \vec{b} \cdot (\vec{c} \times \vec{a})\right] \\
&\quad = \vec{u} \cdot \vec{R} + \vec{e} \cdot \vec{G} \text{ [from (1) and (2)].}
\end{aligned}$$

$$\therefore\ \Sigma\vec{F}_n \cdot d\vec{r}_n = \vec{u} \cdot \vec{R} + \vec{e} \cdot \vec{G}. \tag{3}$$

Since the system is in equilibrium

$$\therefore\ \vec{R} = 0, \vec{G} = 0 \Rightarrow \Sigma\vec{F}_n \cdot d\vec{r}_n = 0.$$

Hence, the algebraic sum of virtual work done is zero.

Conversely, we assume that the algebraic sum of virtual work done is zero.

$$\therefore\ \Sigma\vec{F} \cdot d\vec{r}_n = 0.$$

So from (3),

$$\vec{u}\cdot\vec{R}+\vec{e}\cdot\vec{G}=0. \tag{4}$$

Now let us choose a displacement in which the body has only translation $\vec{u}$.

Hence, by putting $\vec{u}=0$ in (4), we get

$$\vec{u}\cdot\vec{R}=0\Rightarrow\vec{R}=\vec{O}. \tag{5}$$

Again, we choose the displacement in which the body has only rotation $\vec{e}$, hence, by putting $\vec{u}=\vec{O}$ in (1), we get

$$\vec{e}\cdot\vec{G}=0\Rightarrow\vec{G}=\vec{O}. \tag{6}$$

By (5) and (6) we get

$\vec{R}=\vec{O}=\vec{G}$, the general conditions of equilibrium for a system of forces are satisfied, hence, the forces acting on the rigid body are in equilibrium.

3.5 Omission of the Forces

[PU(H) 2008, TMBU(H) 2008, RU(H) 1999, BVABU(H) 2007, LNMU(H) 1999, MU(H) 2005]

Forces which may be omitted in forming the equation of virtual work of a system of coplanar forces acting at different points of a rigid body.

There are certain forces which have no contribution to the virtual work, i.e., the virtual works done by them are zero. Hence, these forces can be excluded from consideration in forming the equation of virtual work.

These forces are as follows:

1. **Tension of an Inextensible String.** [MU(H) 2001]

Let AB be the string connecting two particles A and B whose positions vectors are $\vec{r}_1$ and $\vec{r}_2$.

Let $\vec{T}$ be the tension of the string. Let A' and B' be the displaced positions of A and B respectively. The position vectors of A' and B' are $\vec{r}_1+d\vec{r}_1$ and $\vec{r}_2+d\vec{r}_2$ respectively.

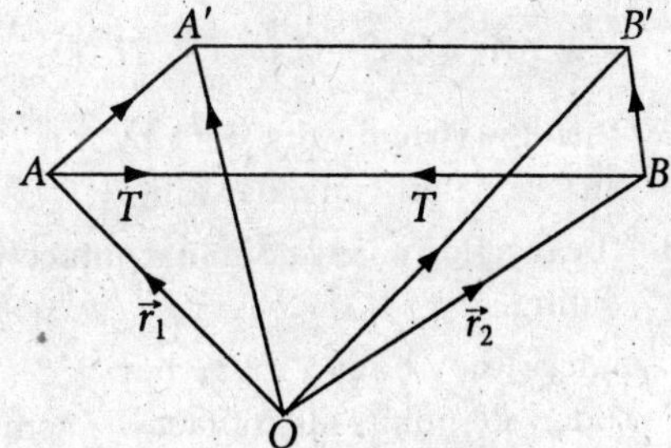

Fig. 3.6

$$\therefore\ \overrightarrow{OA}=\vec{r}_1;\ \overrightarrow{OB}=\vec{r}_2,$$

$$\overrightarrow{OA'}=\vec{r}_1+d\vec{r}_1$$

$$\overrightarrow{OB'}=\vec{r}_2+d\vec{r}_2$$

$$\overrightarrow{OA}+\overrightarrow{AA'}=\overrightarrow{OA'},$$

$$\overrightarrow{AA'}=\overrightarrow{OA'}-\overrightarrow{OA'}=d\vec{r}_1.$$

Also $\overrightarrow{OB}+\overrightarrow{BB'}=\overrightarrow{OB'}$.

$$\overrightarrow{BB'}=\vec{r}_2+d\vec{r}_2-\vec{r}_2=d\vec{r}_2.$$

The sum of the works done by the tensions $\vec{T}$ at A and $-\vec{T}$ at B is equal to

$$\vec{T}\cdot d\vec{r}_1+\left(-\vec{T}\cdot d\vec{r}_2\right)=\vec{T}\cdot(d\vec{r}_1-d\vec{r}_2).$$

Since the string is inextensible,

$AB = A'B'$, or a constant ($\neq 0$) for all virtual displacements.

Also
$$\overrightarrow{OA} + \overrightarrow{AB} = \overrightarrow{OB},\ \overrightarrow{AB} = \vec{r}_2 - \vec{r}_1,$$
$$AB = |\overrightarrow{AB}| = \text{a constant},$$
$$\Rightarrow \left(\overrightarrow{AB}\right)^2 = AB^2 = \text{a constant},$$
$$\Rightarrow (\vec{r}_2 - \vec{r}_1)^2 = \text{a constant},$$
$$\Rightarrow 2(\vec{r}_2 - \vec{r}_1)\cdot(d\vec{r}_2 - d\vec{r}_1) = 0,$$
$$\Rightarrow 2(\vec{r}_2 - \vec{r}_1)\cdot(d\vec{r}_2 - d\vec{r}_2) = 0,$$
$$\Rightarrow d\vec{r}_1 - d\vec{r}_2 = 0.$$

Since $\vec{r}_2 \neq \vec{r}_1 \Rightarrow \vec{T}(d\vec{r}_1 - d\vec{r}_2) = 0.$

Hence, the total virtual work done by the tensions $\vec{T}$ of A and $-\vec{T}$ at B is zero.

2. **The reaction of a smooth surface with which the body is in contact.**

Let A be the point of contact of the body with the smooth surface and $\vec{R}$, the reaction of the surface, normal to the surface.

Let the point A be displaced to point A'. Let $\vec{r}$ and $\vec{r} + d\vec{r}$ be the position vectors of A and A'.

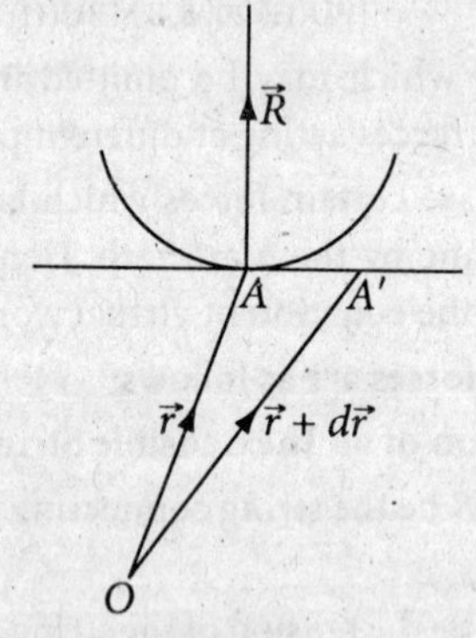

Fig. 3.7

So $\overrightarrow{OA} = \vec{r},\ \overrightarrow{OA'} = \vec{r} + d\vec{r},$

$\overrightarrow{OA} + \overrightarrow{AA'} = \overrightarrow{OA'}$

$\overrightarrow{AA'} = \overrightarrow{OA'} - \overrightarrow{OA} = \vec{r} + d\vec{r} - \vec{r} = d\vec{r}.$

Then the virtual work done by $\vec{R}$ is $\vec{R}\cdot d\vec{r}$ which is zero since $d\vec{r}$ being tangential to the surface is perpendicular to $\vec{R}$.

3. **The reaction at any point of contact with a fixed surface on which the body rolls without sliding.**

Since the body rolls but does not slide, so point of contact P comes to rest momentarily and, therefore, its displacement will be zero. Hence, the normal reaction and the friction at P have no displacement, so that the corresponding virtual work done by the normal reaction and the friction are all zero.

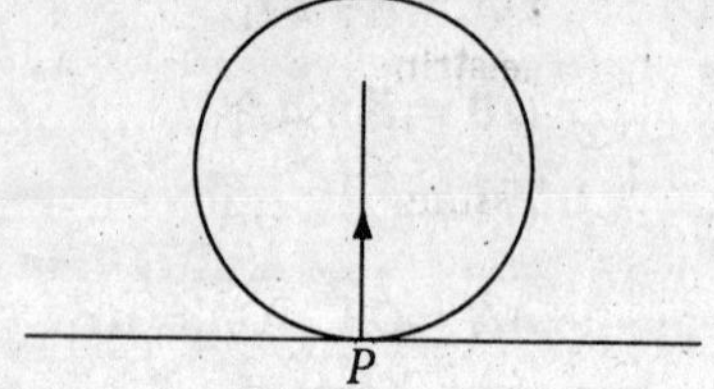

Fig. 3.8

4. **The reaction between two bodies of the material system considered.**

Since by Newton's third law of motion the action and the reaction are equal and opposite and therefore, the work done by one is balanced by that of the other. Hence, the total work done by them is zero.

5. **The reaction at a fixed point or a fixed axis of rotation.**

   Suppose that a body is compelled to move about a point (or an axis of rotation). Now the point where the reaction acts, is a point of the body, so that the point, where the reaction acts, is a point of the body, so that the point of application dos not undergo any displacement. Hence, the virtual work of the reaction at the point (or, on the axis) is zero.

Theorem 1. *To calculate the work done by tension or thrust of a light rod (or a string).*

Proof. Let AB be a string such that the length $AB = l, T$ is the tension of the string. On account of small displacement, let AB take the position $A'B'$, where $A'B' = l + dl$ and $\angle B'A'N = \theta$ (small).

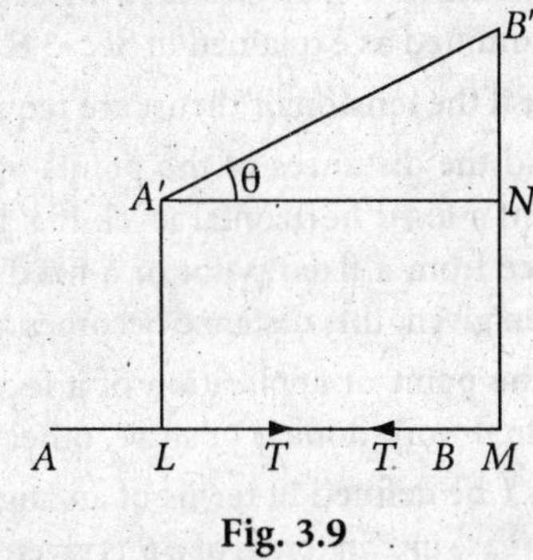

Fig. 3.9

Form A' and B' we draw $A'L$ and $B'M$ perpendicular to AB.

Here end A has moved a distance AL along the direction of the force and the end B has moved a distance BM in a direction opposite to its own direction.

Hence, the total work done by the tension T

$$= T \cdot AL - T \cdot BM = T(AL - BM)$$
$$= T[AB - LB - (LM - LB)] = T(AB - LM)$$
$$= T(AB - A'N) = T(AB - A'B'\cos\theta)$$
$$\left[\text{In } \Delta A'B'N,\ \cos\theta = \frac{A'N}{A'B'};\ A'N = A'B'\cos\theta\right]$$
$$= T(AB - A'B')\ [\because\ \theta \text{ is small, } \therefore\ \cos\theta = 1]$$
$$= T[l - (l + dl)] = -T\,dl.$$

Similarly, the work done by the thrust T of a light rod of length $l = T \cdot dl$.

Note. If the string is elastic, it is consistent with the geometrical conditions of the system to assume a virtual variation in its length.

But when the string is inelastic or a light rod is given, and it is required to find the tension in the string or the thrust in the rod, it is not consistent with the geometrical conditions of the system to assume that a virtual variation in their lengths occur. In such cases:

1. Replace the string by two forces T and T acting inwards. If AB be of length l, then the virtual work done by the tension T of the string is $-T \cdot \delta l$.

A T T B

Fig. 3.10

2. Replace the rod by two forces T and T acting outwards. If AB be the length l, then the virtual work done by the thrust T of the rod $T \cdot \delta l$.

A T T B

Fig. 3.11

3.6 Rules for Solving Problems in Virtual Work

1. Draw the figure and mark the forces.
2. If all the forces act in one direction, give a simple linear displacement in that direction.
3. If there is some symmetry in the given system, then if possible, give a displacement in which symmetry is preserved.
4. If the tension of an elastic string or thrust of a light rod are not required, these forces may be omitted as explained in Sec. 3.6.

   But if the tension or thrust are required, proceed as in note of Sec. 3.7.
5. Find the distances of the points of application of various forces from a fixed point or from a fixed horizontal level. If x be the distance of the point of application of certain force from a fixed point or a fixed horizontal level, then after virtual displacement has been given, this distance becomes $x+\delta x$ and thus the virtual displacement is δx.
6. If the point of application of a force is below the fixed point or fixed horizontal level, virtual work done is positive, otherwise (i.e., if it is above) virtual work done is negative.

   Let x be defined in terms of an angle θ by the equation $x = a\cos\theta$, then δx, the small increase in x in terms of $\delta\theta$ is given by $\delta x = \delta(a\cos\theta) = -a\sin\delta\theta\,\delta\theta$, i.e., δx is treated as dx.

WORKED-OUT EXAMPLES

1. A regular hexagon $ABCDEF$ consists of six equal uniform rods, each of weight W, freely jointed together. The hexagon rests in a vertical plane and AB is in contact with a horizontal table, if C and F be connected by a light string, prove that its tension is $W\sqrt{3}$.

Solution: Let each side of the hexagon be a and G its CG where the total weight $6W$ acts.

Let T be the tension of the string CF. Here AB is a fixed horizontal level.

From A and B we draw AL and BM perpendicular to CF.

In ΔALF,

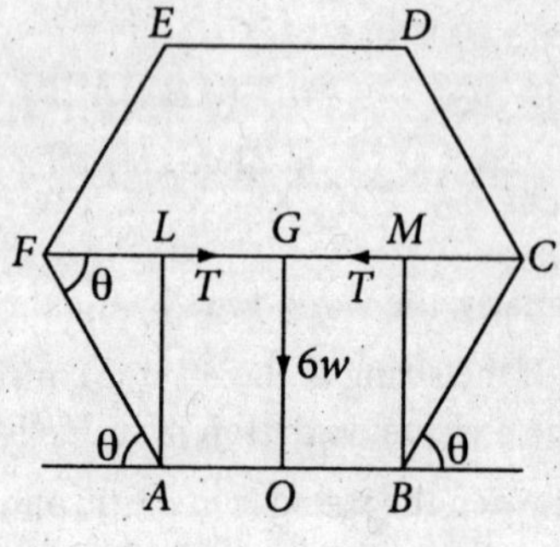

Fig. 3.12

$$\sin\theta = \frac{AL}{AF} = \frac{AL}{a}.$$

$AL = a\sin\theta$. $\therefore$ $OG = AL = a\sin\theta$.

Also in ΔALF,

$$\cos\theta = \frac{LF}{AF} = \frac{LF}{a}.$$

$$LF = a\cos\theta = CM;\ LM = AB = a.$$

$$\therefore\ CF = CM + ML + LF = 2CM + LM = 2a\cos\theta + a.$$

Let the system be given a symmetrical small displacement about the vertical GO, such that θ becomes $\theta+\delta\theta$, and the length of the string alters.

Then the equation of virtual work is

$$-6W\delta(OG)-T\delta(CF)=0$$

$$-6W\delta(a\sin\theta)-T\delta(a+2a\cos\theta)=0,$$

$$\text{i.e., } (-6Wa\cos\theta+2Ta\sin\theta)\delta\theta=0.$$

But $\delta\theta\neq 0$. $\therefore$ $T=3W\cos\theta$.

In the position of equilibrium, $\theta=60°$.

$$T=3W\cot 60°=3W\times\frac{1}{\sqrt{3}}=\sqrt{3}W \text{ (Proved).}$$

2. Four equal jointed rods, each of length a, are hung from an angular point, which is connected by an elastic string with the opposite point. If the rods hang in the form of a square and if the modulus of elasticity of the string be equal to the weight of a rod, show that unstretched length of the string is $a\sqrt{2}/3$. [JPU(H) 2010, MU(H) 2006]

Solution: Let $ABCD$ be the square formed by the four equal jointed rods. It hangs from the angular point A. Here A is the fixed point and distances will be measured from A downwards. Let W be the weight of each rod.

Let $\angle BAC=\theta$. Let T be the tension of the elastic string AC and a be the length of each side of the square. Let G be the CG of the square, where the total weight $4W$ of the four rods act. Let l be the natural length of the string.

Let the system be given a small symmetrical displacement so that θ becomes $\theta+\delta\theta$ and the length of the string alters. Then the equation of virtual work is

$$4W\delta(AG)-T\delta(AC)=0. \qquad (3.1)$$

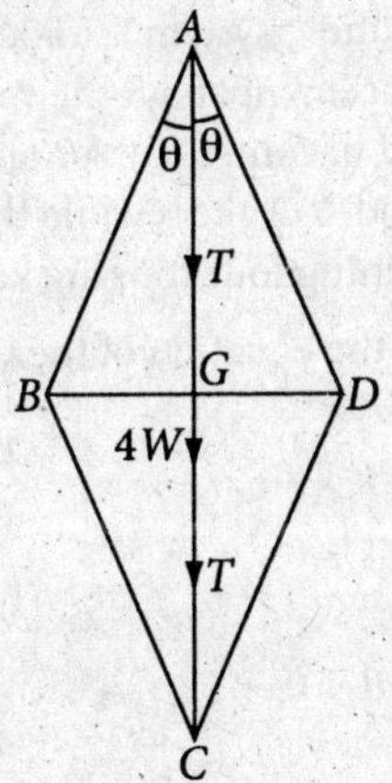

Fig. 3.13

In $\triangle AGD$,

$$\cos\theta=\frac{AG}{AD}=\frac{AG}{a}.$$

$$AG=a\cos\theta$$

$$AC=2AG=2a\cos\theta.$$

So from (1), we get

$$4W\delta(a\cos\theta)-T\delta(2a\cos\theta)=0$$

$$[-4Wa\sin\theta+T\delta 2a\sin\theta]\delta\theta=0.$$

But $\delta\theta\neq 0$.

$$2T=4W \text{ or, } T=2W.$$

Also by Hooke's law

$$T=\frac{\lambda(AC-l)}{l} \text{ or, } 2W=\frac{W(2a\cos\theta-l)}{l} \quad [\text{Given } \lambda=W,\ T=2W]$$

$$2l=2a\cos\theta-l\Rightarrow 3l=2a\cos\theta.$$

In the position of equilibrium, $\theta = 45°$.

$$\therefore\ 3l = 2a\cos 45° = \frac{2a}{\sqrt{2}} = \sqrt{2}a,\ \therefore\ l = \frac{a\sqrt{2}}{3}\ \textbf{(Proved)}.$$

3. Five weightless rods of equal lengths are jointed together W as to form a rhombus $ABCD$ with one diagonals BD. If a weight W is attached to C and the system be suspended from A, show that there is a thrust in BD equal to $W/\sqrt{3}$.

[BVABU(H) 2006, IAS 2002]

Solution: Let the length of each rod be a and the thrust in BD be T.

Since the force at the point of suspension, A, is balanced by the weight W at C, therefore, AC must be vertical.

Consequently, BD is horizontal.

If $\angle BAC = \theta$, $OA = a\cos\theta$.

And $OB = a\sin\theta$.

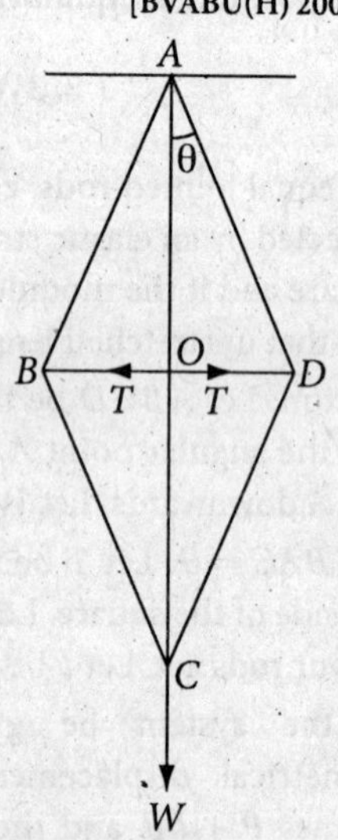

Fig. 3.14

$$AC = 2OA = 2a\cos\theta$$

$$\text{and } BD = 2OB = 2a\sin\theta.$$

Here A is the fixed point.

Let the system undergo a small displacement about the vertical AC, such that θ becomes $\theta + \delta\theta$ and the length of the rod BD alters while the length of the remaining four rods are kept unaltered.

Then the equation of the virtual work is

$$W\delta(AC) + T\delta(BD) = 0$$

$$W\delta(2a\cos\theta) + T\delta(2a\sin\theta) = 0$$

$$(-2aW\sin\theta + T2a\cos\theta)\delta\theta = 0.$$

But $\delta\theta \neq 0$.

$$\therefore\ T = W\tan\theta.$$

In the position of equilibrium, $\theta = 30°$.

$$T = W\tan 30° = W/\sqrt{3}.$$

4. A string of length a forms the shorter diagonal of a rhombus of four uniform rods, each of length b and weight W, which are hinged together. If one of the rods be supported on a horizontal position, prove that tension of the string is $\frac{2W(2b^2-a^2)}{b\sqrt{4b^2-a^2}}$.

[IAS 1999, LNMU(H) 1996, MU(H) 2007, PU(H) 2011, TMBU(H) 2009, BVABU(H) 1998, VU(H) 2004]

Solution: Let $ABCD$ be the rhombus formed of four uniform rods, each of weight w. Let the rod CD be fixed in a horizontal position, O is the CG of the four rods where the total weight $4W$ acts vertically downwards.

Let $\angle BDC = \theta$.

Let the system be given a slight displacement so that the angle θ changes into $\theta + \delta\theta$ and the length of the string DB is altered.

Then the equation of virtual work is

$$4W\delta(OE) - T\delta(BD) = 0. \quad (1)$$

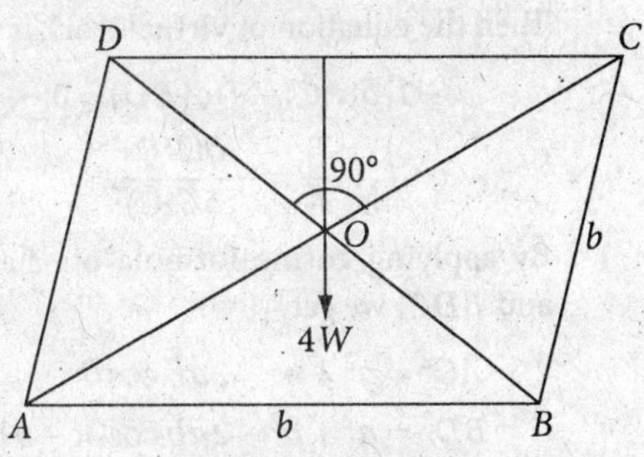

Fig. 3.15

In ΔDOC, $DO = DC\cos\theta = b\cos\theta$.

$$\therefore\ DB = 2DO = 2b\cos\theta.$$

In ΔOED,

$$OE = OD\sin\theta = b\cos\theta\sin\theta = \frac{b}{2}\sin 2\theta.$$

From (1),

$$4W\delta\left(\frac{b}{2}\sin 2\theta\right) - T\delta(2b\cos\theta) = 0.$$

$$\left(4W\frac{b}{2}2\cos 2\theta + 2bT\sin\theta\right)\delta\theta = 0.$$

But $\delta\theta \neq 0$,

$$\therefore\ 4Wb\cos 2\theta = -2bT\sin\theta \text{ or, } T = \frac{-2W\cos 2\theta}{\sin\theta}. \quad (2)$$

In the position of equilibrium $DB = a$; $\therefore\ 2b\cos\theta = a$.

$$\cos\theta = \frac{a}{2b};\ \therefore\ \sin\theta = \sqrt{1-\cos^2\theta} = \sqrt{1-\frac{a^2}{4b^2}} = \frac{\sqrt{4b^2-a^2}}{2b}.$$

So from (2),

$$T = \frac{-2W(2\cos^2\theta - 1)}{\frac{\sqrt{4b^2-a^2}}{2b}} = -4bW\left[2\cdot\frac{a^2}{4b^2} - 1\right]\sqrt{4b^2-a^2}$$

$$= \frac{4bW(2b^2-a^2)}{2b^2\sqrt{4b^2-a^2}} = \frac{2W(2b^2-a^2)}{b\sqrt{4b^2-a^2}} \text{ (Proved).}$$

5. Four rods are jointed to form a parallelogram. The opposite joints are jointed by strings forming the diagonal and the whole system is placed on a smooth horizontal table. Show that their tensions are in the same ratio as their lengths.

Solution: Let $ABCD$ be a parallelogram formed by four rods.

Let $AB = a$, $AD = b$ and $\angle ADC = \theta$.

Let T_1 and T_2 be the tensions of the two strings. Let us give a displacement to the system such that θ changes into $\theta + \delta\theta$ and lengths of the strings AC and BD alter.

Then the equation of virtual work is

$$-T_1\delta(AC)-T_2\delta(BD)=0;$$

$$\therefore\ \frac{T_1}{T_2}=-\frac{\delta(BD)}{\delta(AC)}. \qquad (1)$$

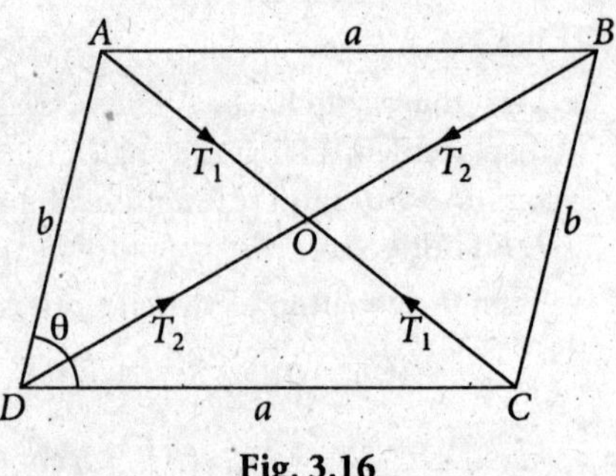

Fig. 3.16

By applying cosine formula on $\triangle ADC$ and BDC, we get

$$AC^2=a^2+b^2-2ab\cos\theta$$

$$BD^2=a^2+b^2-2ab\cos(\pi-\theta)$$

$$BD^2=a^2+b^2+2ab\cos\theta.$$

$$\therefore\ AC^2+BD^2=2\left(a^2+b^2\right)=\text{constant}.$$

$$\therefore\ 2AC\delta(AC)+2BD\delta(BD)=0 \text{ or, } -\frac{\delta(BD)}{\delta(AC)}=\frac{AC}{BD}=\frac{T_1}{T_2} \text{ [from (1)]}.$$

6. The middle points of the opposite sides of a quadrilateral are connected by light rods of lengths l and l'. If T and T' be the tensions in these rods, prove that $\frac{T}{l}+\frac{T'}{l'}=0$.

[MU(H) 2011, PU(H) 2010, RU(H) 1993, BVABU(H) 2000, IAS 2006, LNMU(H) 2011, TMBU(H) 2010, CU(H) 04]

Solution: Let P, Q, R and S be the middle points of the sides of the quadrilateral $ABCD$. Then $PQRS$ is a parallelogram whose diagonal $PR=l$ and $QS=l'$.

Let the system be given a small virtual displacement such that the angles of the quadrilateral change but the sides do not.

Fig. 3.17

The equation of virtual works

$$T\delta l+T'\delta l'=0 \text{ or, } \frac{T}{T'}=-\frac{\delta l'}{\delta l}. \qquad (1)$$

From the properties of quadrilateral, we have

$$PR=l=\tfrac{1}{2}(BC+AD) \text{ and } QS=l'=\tfrac{1}{2}(AB+CD).$$

$$\therefore\ l^2-l'^2=\tfrac{1}{4}\left[(BC+AD)^2-(AB+CD)^2\right]=\text{constant}.$$

Differentiating, we get

$$2l\delta l-2l'\delta l'=0,\ \frac{l}{l'}=\frac{\delta l'}{\delta l}=-\frac{T}{T'} \text{ [from (1)]}$$

$$\therefore\ \frac{T}{l}+\frac{T'}{l'}=0 \text{ (Proved)}.$$

7. Six equal heavy rods, hinged at their ends, form a regular hexagon $ABCDEF$, which when hung up by the point A is kept from altering its shape by two light rods BF and CE. Prove that the thrusts of these rods are $5\sqrt{\frac{3}{2}}W$ and $\frac{\sqrt{3}}{2}W$, where W is the weight of each rod.

[PU(H) 1995, VU(H) 2006, TMBU(H) 2006]

Solution: The point A is fixed. So we measure the distances from A downwards. Let each slant rod make an angle θ with the vertical in the position of equilibrium and we suppose that T_1 and T_2 are the thrusts in the light rods BF and CE. Let $2a$ be the length of each rod. $6W$ is the weight of the regular hexagon which acts at CG of the hexagon.

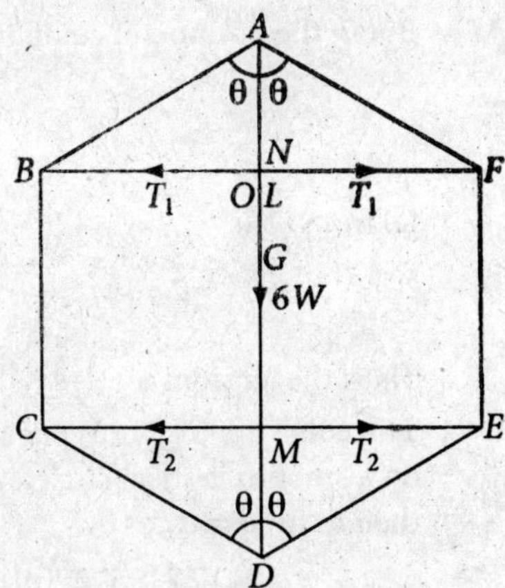

Fig. 3.18

We first choose a small virtual displacement of the system about the vertical AD, so that θ changes into $\theta + \delta\theta$ and the lengths of the light rods BF and CE also change.

The equation of the virtual work is

$$6W\delta(AG) + T_1\delta(BF) + T_2\delta(CE) = 0. \qquad (1)$$

We have

$$AG = AO + OG = AO + a = 2a\cos\theta + a$$
$$BF = 2OF = 2\cdot 2a\sin\theta = 4a\sin\theta = CE.$$

So from (1),

$$6W\delta(2a\cos\theta + a) + T_1\delta(4a\sin\theta) + T_2\delta(4a\sin\theta) = 0.$$
$$[-6W\cdot 2a\sin\theta + 4aT_1\cos\theta + 4aT_2\cos\theta]\,\delta\theta = 0.$$

But $\delta\theta \neq 0$.

$$-12W\sin\theta + 4T_1\cos\theta + 4T_2\cos\theta = 0$$
$$\text{or, } 4\cos\theta(T_1 + T_2) = 12W\sin\theta$$
$$\text{or, } T_1 + T_2 = 3W\tan\theta.$$

In the position of equilibrium of the hexagon, $\theta = 60°$.

$$\therefore\ T_1 + T_2 = 3W\sqrt{3}. \qquad (2)$$

Next we choose a symmetrical small virtual displacement about AD such that θ changes into $\theta + \delta\theta$ at A only and the length of the light rod BF alters, but the lower slant rods and the length CE are not disturbed. So equation of virtual work is

$$2W\delta(AN) + T_1\delta(BF) = 0$$
$$\Rightarrow 2W\delta(a\cos\theta) + T_1\delta(4a\sin\theta) = 0$$
$$\therefore (-2wa\sin\theta + T_1 4a\cos\theta)\delta\theta = 0.$$

But $\delta\theta \neq 0$,

$$\therefore\ T_1 = \frac{2Wa\sin\theta}{4a\cos\theta} = \frac{W}{2}\tan\theta.$$

But in the position of equilibrium $\theta = 60°$,

$$T_1 = \frac{W}{2}\tan 60° = \frac{W}{2}\sqrt{3}.$$

So from (2), $$\frac{W}{2}\sqrt{3} + T_2 = 3W\sqrt{3} \Rightarrow T_2 = W\sqrt{3}\left(3 - \frac{1}{2}\right) = \frac{5}{2}W\sqrt{3}.$$

Thus, the tensions are $\frac{5\sqrt{3}}{2}W$ and $\frac{\sqrt{3}}{2}W$. **(Proved).**

8. Two equal uniform rods AB and AC, each of length $2b$, are freely jointed at A and rest on a smooth vertical circle of radius a. Show that if 2θ be the angle between them, then $b\sin^3\theta = a\cos\theta$.

[IAS 1991, PU(H) 2009, RU(H) 2000, BVABU(H) 2007, TMBU(H) 2011, MU(H) 2012]

Solution: Let D and E be the middle points of AB and AC respectively. Then

$$AD = \tfrac{1}{2}AB = b \text{ and } AE = \tfrac{1}{2}AC = b.$$

G is the middle point of DE, where the total weight $2w$ may be supposed to act vertically downwards.

From $\triangle AGD$,

$$AG = AD\cos\theta = b\cos\theta.$$

In $\triangle OLA$,

$$\sin\theta = \frac{OL}{OA} = \frac{a}{OA}.$$

$\therefore\ OA = a\,\text{cosec}\,\theta.$

Also $OG = OA - AG = a\,\text{cosec}\,\theta - b\cos\theta$.

Here the centre O of the given circle may be taken as the fixed point.

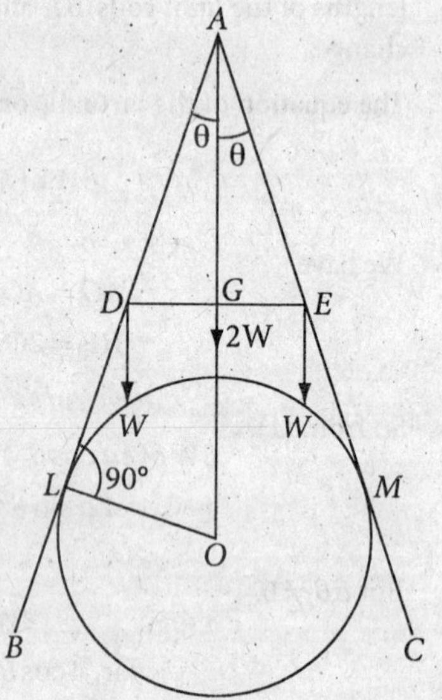

Fig. 3.19

Let us suppose that the rods undergo a small symmetrical displacement about the vertical through the centre O such that the angle θ changes into $\theta + \delta\theta$.

Then the equation of virtual work is

$$-2W\delta(OG) = 0 \Rightarrow \delta(a\,\text{cosec}\,\theta - b\cos\theta) = 0, \text{ as } W \neq 0$$
$$\Rightarrow (-a\,\text{cosec}\,\theta\cot\theta + b\sin\theta)\delta\theta = 0.$$

But $\delta\theta \neq 0$.

$$\therefore -a\,\text{cosec}\,\theta\cot\theta + b\sin\theta = 0 \Rightarrow \frac{a\cos\theta}{\sin^2\theta} = b\sin\theta;$$

$\therefore\ a\cos\theta = b\sin^3\theta$ **(Proved).**

9. A smooth rod passes through a smooth ring at the focus of an ellipse whose major axis is horizontal and rests with its lower end on the quadrant of the curve which is furthest removed from the focus. Find the position of equilibrium and show that its length must be at least $\frac{3a}{4}+\frac{1}{4}\sqrt{1+8e^2}$, where $2a$ is the major axis and e the eccentricity.

[IAS 1994]

Solution: Let W be the weight of the smooth rod LM, of length $2c$, which passes through a smooth ring at the focus S of the ellipse, G be the mid-point of LM, then

$$LG = GM = \tfrac{1}{2}LM = c.$$

Let GN be perpendicular to AA' and $GN = y$.

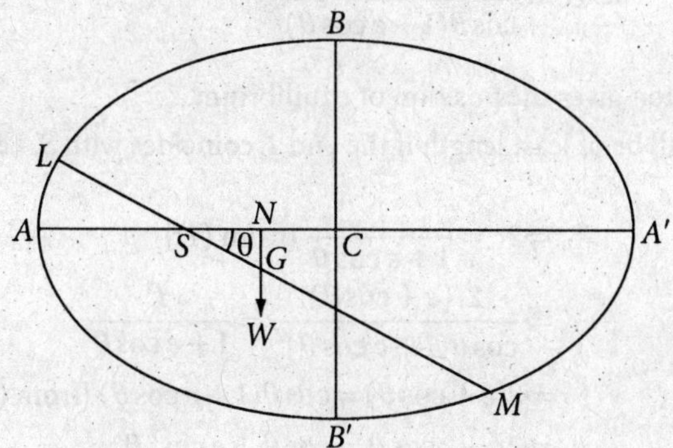

Fig. 3.20

Then the equation of virtual work is

$$W\delta y = 0 \text{ or, } \delta y = 0 \text{ as } W \neq 0.$$

The equation of the ellipse referred to S as pole is

$$l/r = 1 + e\cos\theta. \tag{1}$$

Let M be (r,θ), then $SM = r$.
Now

$$\therefore\ SG = SM - GM = r - c.$$

In ΔGNS,

$$\sin\theta = \frac{GN}{GS} = \frac{y}{r-c}.$$

$$\therefore\ y = (r-c)\sin\theta.$$

Now $\delta y = 0$

$$\Rightarrow \delta[(r-c)\sin\theta] = 0$$

$$\Rightarrow \delta\left[\left(\frac{l}{1+e\cos\theta} - c\right)\sin\theta\right] = 0 \text{ [using (1)]}$$

$$\Rightarrow \delta\left[\frac{l\sin\theta}{1+e\cos\theta} - c\sin\theta\right] = 0$$

$$\Rightarrow \left[\frac{l\left\{\cos\theta(e+\cos\theta)+\sin^2\theta\, e\right\}}{(1+e\cos\theta)^2} - c\cos\theta\right]\delta\theta = 0.$$

But $\delta\theta \neq 0$.

$$\Rightarrow \frac{l\{\cos\theta(1+e\cos\theta)+\sin^2\theta\, e\}}{(1+e\cos\theta)^2} - c\cos\theta = 0$$

$$\Rightarrow l\cos\theta(1+e\cos\theta)+l\sin^2\theta\, e = c\cos\theta(1+e\cos\theta)^2$$

$$\Rightarrow el(\cos^2\theta+\sin^2\theta)+l\cos\theta = c\cos\theta(1+e\cos\theta)^2$$

$$\Rightarrow el + l\cos\theta = c\cos\theta(1+e\cos\theta)^2$$

$$\Rightarrow 2c = \frac{2l(e+\cos\theta)}{\cos\theta(1+e\cos\theta)^2}. \qquad (2)$$

This equation gives the position of equilibrium.

The rod will be of least length if the end L coincides with S, i.e., when $r = 2c$

$$\Rightarrow 2c = \frac{l}{1+e\cos\theta} \text{ [from (1)]}$$

$$\Rightarrow \frac{2l(e+\cos\theta)}{\cos\theta(1+e\cos\theta)^2} = \frac{l}{1+e\cos\theta}$$

$$\Rightarrow 2(e+\cos\theta) = \cos\theta(1+e\cos\theta) \text{ [from (2)]}$$

$$\Rightarrow 2e+2\cos\theta = \cos\theta + e\cos^2\theta$$

$$\Rightarrow e\cos^2\theta - \cos\theta - 2e = 0$$

$$\Rightarrow \cos\theta = \frac{1\pm\sqrt{1^2+4.2e^2}}{2e} = \frac{1\pm\sqrt{1+8e^2}}{2e}.$$

The value $\frac{1}{2e}\left[1+\sqrt{1+8e^2}\right]$ is inadmissible.

$$\therefore \cos\theta = \frac{1}{2e}\left[1-\sqrt{1+8e^2}\right].$$

If l be the semi-latus rectum of the ellipse, then

$$l = \frac{b^2}{a} = \frac{a^2(1-e^2)}{a} = a\left(1-e^2\right).$$

Hence, the least length of the rod, r

$$= \frac{l}{1+e\cos\theta} = \frac{a(1-e^2)}{1+\frac{e}{2e}\left(1-\sqrt{1+8e^2}\right)}$$

$$= \frac{2a(1-e^2)}{3-\sqrt{1+8e^2}} \times \frac{3+\sqrt{1+8e^2}}{3+\sqrt{1+8e^2}} = \frac{2a(1-e^2)\left(3+\sqrt{1+8e^2}\right)}{9-(1+8e^2)}$$

$$= \frac{2a(1-e^2)\left(3+\sqrt{1+8e^2}\right)}{8(1-e^2)} = \frac{a}{4}\left[3+\sqrt{1+8e^2}\right]$$

$$= \frac{3a}{4} + \frac{a}{4}\sqrt{1+8e^2} \text{ (Proved).}$$

10. A heavy uniform rod of length $2a$ rests with its end in contact with two smooth inclined planes of inclinations α and β to the horizon. If θ be the inclination of the rod to the horizon, prove by principle of virtual work that $\tan\theta = \frac{1}{2}(\cot\alpha - \cot\beta)$.

[VU(H) 2000, IAS 2007]

Solution: Let W be the weight of the rod AB acting through its CG. Then

$$AG = BG = \tfrac{1}{2}AB = a.$$

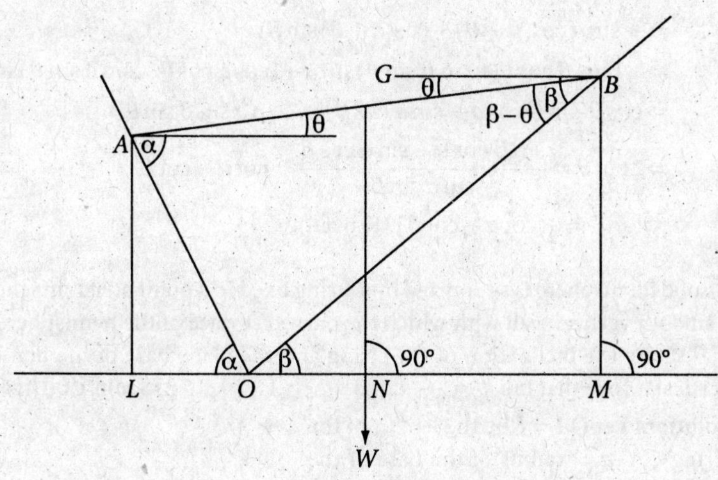

Fig. 3.21

From Fig. 3.21, we have

$$\angle OAB = \alpha + \theta,\ \angle OBA = \beta - \theta.$$
$$\therefore\ \angle AOB = 180^\circ - (\alpha + \theta + \beta - \theta) = 180^\circ - (\alpha + \beta).$$

In ΔOAB,

$$\frac{OA}{\sin(\beta - \theta)} = \frac{OB}{\sin(\alpha + \theta)} = \frac{AB}{\sin\{180 - (\alpha + \beta)\}}$$

$$\therefore\ OA = \frac{2a\sin(\beta - \theta)}{\sin(\alpha + \beta)},\ \text{as } AB = 2a;\ OB = \frac{2a\sin(\alpha + \theta)}{\sin(\alpha + \theta)}$$

$$GN = \frac{1}{2}(AL + BM) = \frac{1}{2}[OA\sin\alpha + OB\sin\beta]$$
$$= \frac{1}{2}\left[\frac{2a\sin(\beta - \theta)\sin\alpha}{\sin(\alpha + \beta)} + \frac{2a\sin(\alpha + \theta)\sin\beta}{\sin(\alpha + \beta)}\right]$$
$$= \frac{2a}{2\sin(\alpha + \beta)}[\sin(\beta - \theta)\sin\alpha + \sin(\alpha + \theta)\sin\beta].$$

We give a displacement to the system such that θ becomes $\theta + \delta\theta$. So equation of virtual work is

$$W\delta(GN)=0,$$
$$\Rightarrow \delta(GN)=0\ [\because\ W\neq 0]$$
$$\Rightarrow \delta\left[\frac{a}{\sin(\alpha+\beta)}\{\sin(\beta-\theta)\sin\alpha+\sin(\alpha+\theta)\sin\beta\}\right]=0$$
$$\Rightarrow \delta\theta[-\cos(\beta-\theta)\sin\alpha+\cos(\alpha+\theta)\sin\beta]=0.$$

But $\delta\theta\neq 0$,

$$\Rightarrow -\sin\alpha\cos(\beta-\theta)+\cos(\alpha+\theta)\sin\beta=0$$
$$\Rightarrow -(\cos\beta\cos\theta+\sin\beta\sin\theta)\sin\alpha+(\cos\alpha\cos\theta-\sin\alpha\sin\theta)\sin\beta=0$$
$$\Rightarrow \cos\theta(\sin\beta\cos\alpha-\sin\alpha\cos\beta)=2\sin\alpha\sin\beta\sin\theta$$
$$\Rightarrow 2\tan\theta=\frac{\sin\beta\cos\alpha-\sin\alpha\cos\beta}{\sin\alpha\sin\beta}=\cot\alpha-\cot\beta.$$
$$\therefore\ \tan\theta=\tfrac{1}{2}(\cot\alpha-\cot\beta)\ \textbf{(Proved)}.$$

11. A solid hemisphere is supported by a string fixed to a point on its rim and to a point on a smooth vertical wall with which the curved surface of the hemisphere is in contact. If θ,ϕ are the inclination of the string and the plane base of the hemisphere to the vertical, prove that $\tan\phi=\frac{3}{8}+\tan\theta$. [IAS 2010, CU(H) 2006, VU(H) 2001]

Solution: Let $OA=l$ be the length of the string CA. a = radius of the base of the hemisphere of weight w.

Let O be the point of suspension of the hemisphere by the string, C is the centre of the hemisphere and G the CG.

$$\therefore\ \text{CG}=\tfrac{3}{8}a.$$

Given $\angle AOE=\theta, \angle AFD=\phi$, we draw GL perpendicular to ED.

Then

$$z=OL=OE+ED+DL. \qquad (1)$$

In $\triangle OAE$,

$$\cos\theta=\frac{OE}{OA}=\frac{OE}{l}\Rightarrow OE=l\cos\theta.$$

In $\triangle ACM$,

$$\cos\phi=\frac{AM}{AC}=\frac{AM}{a}\Rightarrow AM=a\cos\phi=ED.$$

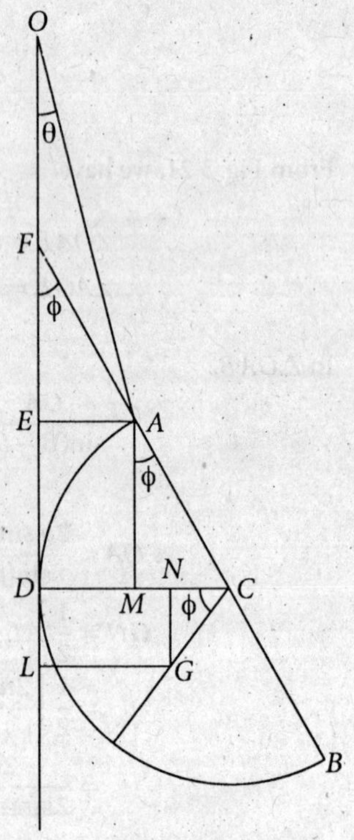

Fig. 3.22

In ΔGCN,

$$\sin\phi = \frac{GN}{GC} \Rightarrow GN = GC\sin\phi \Rightarrow \frac{3a}{8}\sin\phi \text{ [from (1)]}.$$

$$z = l\cos\theta + a\cos\phi + \frac{3a}{8}\sin\phi. \qquad (2)$$

Let the system be given a small virtual displacement such that θ changes into $\theta + \delta\theta$ and ϕ into $\phi + \delta\phi$, while the length of the string remains unchanged. Then the equation of virtual work is

$$W\delta z = 0 \Rightarrow \delta z = 0\ (\because\ W \neq 0)$$

$$\Rightarrow \delta\left(l\cos\theta + a\cos\phi + \frac{3a}{8}\sin\phi\right) = 0 \text{ [from (1)]}$$

$$\Rightarrow -l\sin\theta\,\delta\theta - a\sin\phi\,\delta\phi + \frac{3a}{8}\cos\phi\,\delta\phi = 0$$

$$\Rightarrow a\left(\frac{3}{8}\cos\phi - \sin\phi\right)\delta\phi = l\sin\theta\,\delta\theta. \qquad (3)$$

In ΔACM,

$$\sin\phi = \frac{CM}{AC} = \frac{CM}{a} \Rightarrow CM = a\sin\phi.$$

In ΔAEO,

$$\sin\theta = \frac{AE}{AO} = \frac{AE}{l} \Rightarrow AE = l\sin\theta.$$

$$\therefore\ a = CD = CM + MD = CM + AE = a\sin\phi + l\sin\theta \Rightarrow a\cos\phi\,\delta\phi + l\cos\theta\,\delta\theta = 0.$$

$$\therefore\ \delta\phi = \frac{-l\cos\theta}{a\cos\phi}\delta\theta. \qquad (4)$$

Using (4) in (3), we get

$$a\left(\frac{3}{8}\cos\phi - \sin\phi\right)\left(\frac{-l\cos\theta}{a\cos\phi}\right)\delta\theta = l\sin\theta\,\delta\theta$$

$$\Rightarrow \delta\theta\left[l\sin\theta + a\left(\frac{3}{8}\cos\phi - \sin\phi\right)\frac{l\cos\theta}{a\cos\phi}\right] = 0.$$

But $\delta\theta \neq 0$.

$$\therefore\ \sin\theta + a\left(\frac{3}{8}\cos\phi - \sin\phi\right)\frac{\cos\theta}{a\cos\phi} = 0,$$

$$\Rightarrow \sin\theta + \frac{3}{8}\cos\theta - \tan\phi\cos\theta = 0,$$

$$\Rightarrow \tan\phi\cos\theta = \frac{3}{8}\cos\theta + \sin\theta,$$

$$\Rightarrow \tan\phi = \frac{3}{8} + \tan\theta \text{ (Proved)}.$$

12. A smooth parabolic wire is fixed with its axis vertical and vertex downwards, and in it is placed a uniform rod of length $2l$ with its ends resting on the wire. Show that for the equilibrium, the rod is either horizontal or makes with the horizontal an angle θ given by $\cos^2\theta = 2a/l$, where $4a$ is the latus rectum of the parabola.

Solution: Let O be the vertex of the parabolic wire and $AB = 2l$ is the rod whose weight W acts through the centre G.

Let the equation of the parabola be $x^2 = 4ay$.

Let coordinates of A and B be $(2at_1, at_1^2)$ and $(2at_2, at_2^2)$ respectively.

Then

Fig. 3.23

$$\tan\theta = \frac{a\left(t_2^2 - t_1^2\right)}{2a(t_2 - t_1)} = \frac{t_2 + t_1}{2}. \quad (1)$$

Now $AB = 2l \Rightarrow AB^2 = 4l^2$.

$$\therefore (2at_2 - 2at_1)^2 + \left(at_2^2 - at_1^2\right)^2 = 4l^2$$

$$\Rightarrow 4a^2(t_2 - t_1)^2 + a^2(t_1 + t_2)^2(t_2 - t_1)^2 = 4l^2$$

$$\Rightarrow (t_2 - t_1)^2 a^2 \left[4 + (t_1 + t_2)^2\right] = 4l^2$$

$$\Rightarrow (t_2 - t_1)^2 a^2 [4 + 4\tan^2\theta] = 4l^2 \text{ [from (1)]}$$

$$\Rightarrow (t_2 - t_1)^2 a^2 \sec^2\theta = l^2.$$

$$\therefore\ t_2 - t_1 = \frac{l}{a}\cos\theta. \quad (2)$$

Now

$$GL = \frac{1}{2}(AM + BN) = \frac{1}{2}\left[at_1^2 + at_2^2\right] = \frac{a}{2}\left[(t_1 + t_2)^2 + (t_1 - t_2)^2\right]$$

$$= \frac{a}{4}\left[4\tan^2\theta + \frac{l^2\cos^2\theta}{a^2}\right] \text{ [from (1) and (2)].}$$

Let the system be given a slight symmetrical displacement so that θ changes into $\theta + \delta\theta$ and the ends of the rod AB always remain in contact with the wire.

Then the equation of virtual work is $-W\delta(GL) = 0$.

$$\Rightarrow \delta\left[\frac{a}{4}\left(4\tan^2\theta + \frac{l^2\cos^2\theta}{a^2}\right)\right] = 0\ (\because W \neq 0)$$

$$\Rightarrow \left[\frac{a}{4}\left(8\tan\theta\sec^2\theta - \frac{2l^2}{a^2}\cos\theta\sin\theta\right)\right]\delta\theta = 0.$$

But $\delta\theta \neq 0$.

$$\therefore \frac{4\sin\theta}{\cos\theta}\frac{1}{\cos^2\theta} - \frac{l^2}{a^2}\cos\theta\sin\theta = 0;$$

$$\Rightarrow \sin\theta\left[\frac{4}{\cos^3\theta} - \frac{l^2}{a^2}\cos\theta\right] = 0.$$

Either

$$\sin\theta = 0 \text{ or, } \frac{4}{\cos^3\theta} - \frac{l^2}{a^2}\cos\theta = 0$$

If $\sin\theta = 0 \Rightarrow \theta = 0$.

So the radius horizontal.

If
$$\frac{4}{\cos^2\theta} - \frac{l^2}{a^2}\cos\theta = 0 \Rightarrow \frac{4}{\cos^3\theta} = \frac{l^2}{a^2}\cos\theta$$
$$\Rightarrow \cos^4\theta = \frac{4a^2}{l^2} \Rightarrow \cos^2\theta = \frac{2a}{l} \text{ (Proved).}$$

13. Four uniform rods are joined to form a rectangle $ABCD$. AB is fixed in a vertical position with A uppermost; and the rectangle is kept in shape by a string joining AC. Find the tension of the string.

Solution: Let $\angle ABC = \theta$, $AB = a$, $BC = b$ and the weight per unit length of the rods $= w$. Let T be the tension of the string.

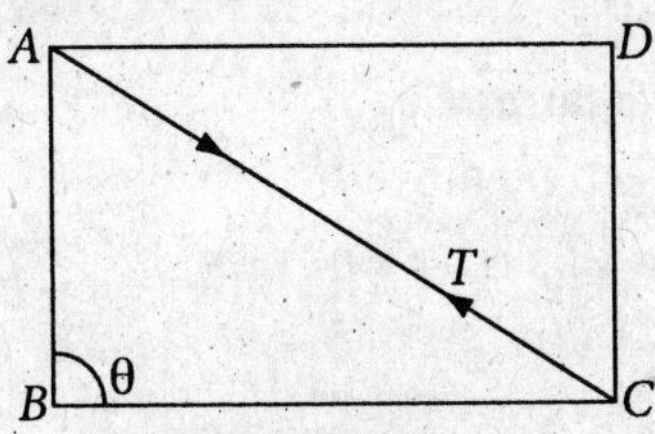

Fig. 3.24

Then the equation of virtual work is

$$bw\delta\left(\frac{1}{2}b\cos\theta\right) + aw\delta(b\cos\theta) + bw\delta\left(\frac{1}{2}b\cos\theta\right) + T\delta(AC) = 0$$
$$\Rightarrow \left[-\frac{b^2 w\sin\theta}{2} - abw\sin\theta - \frac{b^2}{2}w\sin\theta\right]\delta\theta = -T\delta(AC)$$
$$\Rightarrow \frac{bw\sin\theta}{2}[b+2a+b]\delta\theta = T\delta\left(\sqrt{a^2+b^2-2ab\cos\theta}\right)$$
$$\Rightarrow \frac{bw\sin\theta}{2}(2b+2a)\delta\theta = \frac{T2ab\sin\theta\,\delta\theta}{2\sqrt{a^2+b^2-2ab\cos\theta}}$$
$$\Rightarrow b\sin\theta\,\delta\theta\left[(b+a)w - \frac{Ta}{\sqrt{a^2+b^2-2ab\cos\theta}}\right] = 0.$$

But $b\sin\theta\,\delta\theta \neq 0$

$$\text{i.e., } (b+a)w - \frac{Ta}{\sqrt{a^2+b^2-2ab\cos\theta}} = 0$$
$$\Rightarrow T = \frac{(a+b)w}{a}\sqrt{a^2+b^2-2ab\cos\theta}.$$

But in the position of equilibrium, $\theta = 90°$.

$$T = \frac{(a+b)}{a}w\sqrt{a^2+b^2}.$$

14. A heavy elastic string, whose natural length is $2\pi a$ is placed round a smooth cone whose axis is vertical and whose semi-vertical angle is α. If w be the weight and λ

the modulus of elasticity of the string, prove that it will be in equilibrium when in the form of the circle whose radius is $a\left(1+\frac{w}{2\pi\lambda}\cos\alpha\right)$. [IAS 1998, TMBU(H) 2000]

Solution: Here the fixed point is the vertex O. Let x be the distance of the centre C of the string from O, then radius of the string $= 2\pi x\tan\alpha$.

The two forces which contribute to the sum of virtual works are:

1. Weight W of the string placed at C whose depth from the fixed point O is x.
2. Tension T in the string whose length is $2\pi x\tan\alpha$.

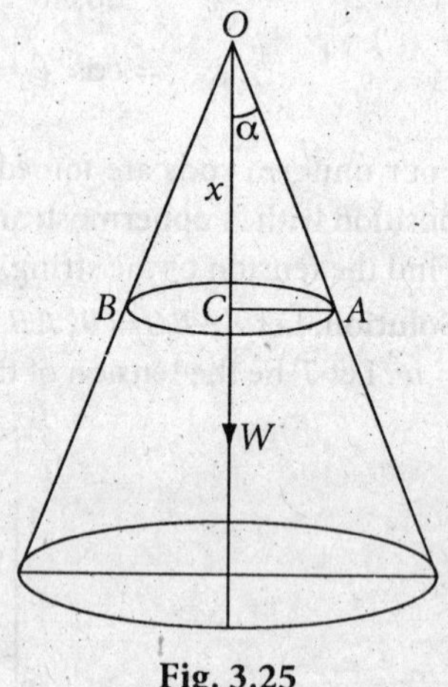

Fig. 3.25

Equation of the virtual work is

$$W\delta x - T\delta(2\pi x\tan\alpha)=0$$
$$\Rightarrow [W - T2\pi\tan\alpha]\delta x = 0$$
$$\Rightarrow W - 2T\pi\tan\alpha = 0, \text{ as } \delta x \neq 0$$
$$\Rightarrow T = \frac{W}{2\pi} w\tan\alpha. \quad (1)$$

Also by Hooke's law

$$T=\lambda\left(\frac{2\pi x\tan\alpha-2\pi a}{2\pi a}\right)=\lambda\left(\frac{x\tan\alpha-a}{a}\right). \quad (2)$$

From (1) and (2), we get

$$\frac{W}{2\pi}w\tan\alpha=\frac{\lambda(x\tan\alpha-a)}{a}\Rightarrow x\tan\alpha-a=\frac{aW\cot\alpha}{2\lambda\pi}$$
$$\Rightarrow x\tan\alpha=a\left(1+\frac{W}{2\lambda\pi}\cot\alpha\right).$$

This gives radius of the string in equilibrium form.

15. $ABCDEF$ is a regular hexagon formed of light rods smoothly jointed at their ends with a diagonal AD horizontally. Four equal forces P act inwards at the middle points of the rods AB, CD, DE, FA and at right angles to the respective sides. Prove that the thrust on the diagonal AD is $P/\sqrt{3}$.

Solution: Let $2a$ be the length of each side of the hexagon. From B and C, we draw BM and CN perpendicular to AD. Let $\angle BAM=\theta$.

In ΔABM,

$$\cos\theta=\frac{AM}{AB}=\frac{AM}{2a}\Rightarrow AM=2a\cos\theta=DN.$$

$$\therefore AD=AM+MN+DN=2AM+2a=4a\cos\theta+2a=2a(2\cos\theta+1)$$

and the virtual work done by T is

$$T\delta\{2a(1+2\cos\theta)\}=-4aT\sin\theta\,\delta\theta.$$

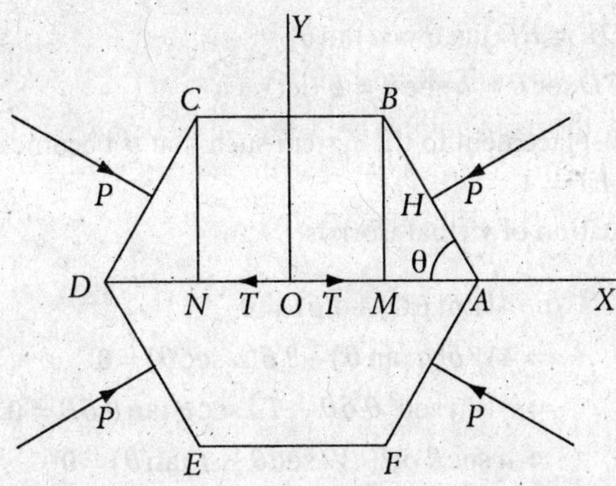

Fig. 3.26

Taking the centre O of the hexagon as origin and OX, OY as x-axis and y-axis, the coordinates of the middle point H of AB are $(a+a\cos\theta, a\sin\theta)$ and the components of P are $-P\sin\theta$ and $-P\cos\theta$.

Hence, the virtual work done by P at H

$$= X\delta x + Y\delta y = Pa\sin^2\theta\,\delta\theta - P\cos^2\theta\, a\delta\theta = -Pa\cos 2\theta\,\delta\theta.$$

Similarly, for the other P's.

Hence, the equation of virtual work is

$$-4aT\sin\theta\,\delta\theta - 4Pa\cos 2\theta\,\delta\theta = 0 \Rightarrow T\sin\theta = -P\cos 2\theta.$$

In case of equilibrium, $\theta = \pi/3$.

$$\Rightarrow T\frac{\sqrt{3}}{2} = -P\left(-\frac{1}{2}\right) \Rightarrow T = \frac{P}{\sqrt{3}}.$$

Since T is positive, so the stress is a thrust **(Proved).**

16. Four uniform rods each of weight W are freely jointed so as to form a square $ABCD$, the side AB is fixed in a vertical position with A uppermost and the figure is kept in shape by a string, joining the middle points of AD, DC. Find the tension of the string.

Solution: The total weight $4W$ of the square acts at its centre O. Let E and F be the mid-points of AD and CD. Then EF is connected by a string T. Let T the tension of the string. Let $\angle FED = \theta$.

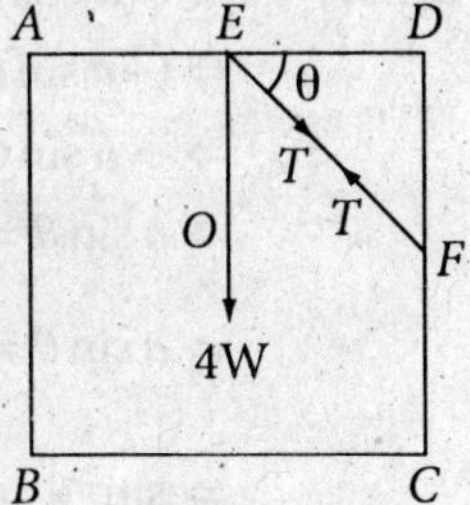

Fig. 3.27

The side AB being fixed, A is fixed. Since AB is a fixed vertical line, so AD is a fixed horizontal line. Let the distances be measured from this fixed horizontal line. Let $2a$ be the length of each side of the square.

$\therefore\ EO = DF = ED\tan\theta = a\tan\theta$

and $EF = ED\sec\theta = a\sec\theta = b$ (let).

Let us give a displacement to the system such that θ becomes $\theta+\delta\theta$ and l becomes $l+\delta l$, where $EF = l$.

Hence, the equation of virtual work is

$$4W\delta(EO) - T\delta l = 0$$
$$\Rightarrow 4W\delta(a\tan\theta) - T\delta(a\sec\theta) = 0$$
$$\Rightarrow 4Wa\sec^2\theta\,\delta\theta - Ta\sec\theta\tan\theta\,\delta\theta = 0$$
$$\Rightarrow a\sec\theta\,\delta\theta(4W\sec\theta - T\tan\theta) = 0$$
$$\Rightarrow 4W\sec\theta - T\tan\theta = 0 \text{ as } a\sec\theta\,\delta\theta \neq 0$$
$$\Rightarrow T = \frac{4W\sec\theta}{\tan\theta} = \frac{4W}{\cos\theta}\frac{\cos\theta}{\sin\theta} = 4W\operatorname{cosec}\theta.$$

In the position of equilibrium $\theta = 45°$.

$\therefore\ T = 4W\operatorname{cosec}45° = 4W\sqrt{2} = 4\sqrt{2}W.$

17. A uniform beam of length $2a$ rests in equilibrium against a smooth vertical wall and upon a smooth peg at a distance b from the wall. Show by principle of virtual work that the rod is inclined to the wall at an angle $\sin^{-1}(b/a)^{1/3}$.

Solution: Let AB be the beam and G be its CG where its weight W acts. Let the peg be at P. Let the distances be measured from the fixed point P.

Let θ be the inclination of the beam with the wall, which is vertical.

Height of G above P

$$= LM = AM - AL = AG\cos\theta - LP\cot\theta$$
$$= a\cos\theta - b\cot\theta.$$

Let us give a displacement in the beam such that θ becomes $\theta+\delta\theta$. The reactions at A and P do not work.

So equation of virtual works

$$-W\delta(LM) = 0 \Rightarrow \delta(a\cos\theta - b\cot\theta) = 0 \text{ as } W \neq 0$$
$$\Rightarrow (-a\sin\theta + b\operatorname{cosec}^2\theta)\delta\theta = 0$$
$$\Rightarrow -a\sin\theta + b\operatorname{cosec}^2\theta = 0 \text{ as } \delta\theta \neq 0$$
$$\Rightarrow a\sin\theta = b\operatorname{cosec}^2\theta$$
$$\Rightarrow a\sin\theta = \frac{b}{\sin^2\theta}$$
$$\Rightarrow \sin^3\theta = \frac{b}{a} \Rightarrow \sin\theta = \left(\frac{b}{a}\right)^{1/3}$$
$$\Rightarrow \theta = \sin^{-1}\left(\frac{b}{a}\right)^{1/3} \text{ (Proved).}$$

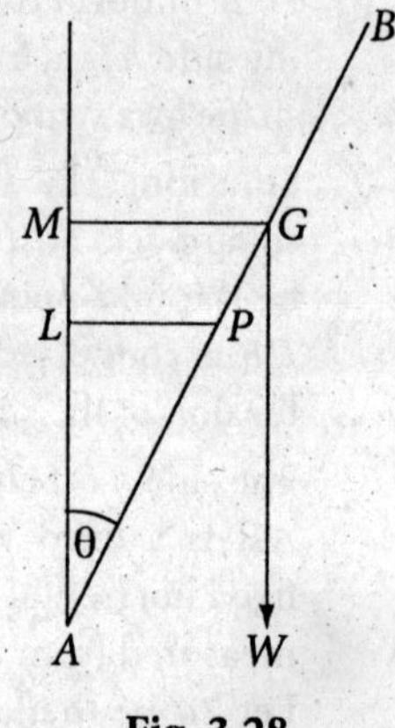

Fig. 3.28

18. Four equal rods each of weight W form a rhombus $ABCD$ with smooth hinges at the joints. The frame is suspended by the end A and a weight W' is attached at C. A stiffening rod of negligible weight joins the middle points of AB, AD. Keeping these inclined at an angle α to AC. Show that the thrust in the stiffening rod is $(4W+2W')\tan\alpha$. [VU(H) 2007, CU(H) 2002]

Solution: Let O be the point of intersection of the diagonals, then the weight $4W$ of the four rods act vertically downwards at O.

Let E and F be the mid-points of AB and AD.

Then EF is the stiffening rod. A weight W' is attached at C.

It is required to find the thrust in the rod EF and T be the thrust in the rod EF and a be the length of each side of the rhombus.

Let us give a displacement to the system such that α becomes $\alpha+\delta l$ and length of the string changes. The length of the rods remains unchanged.

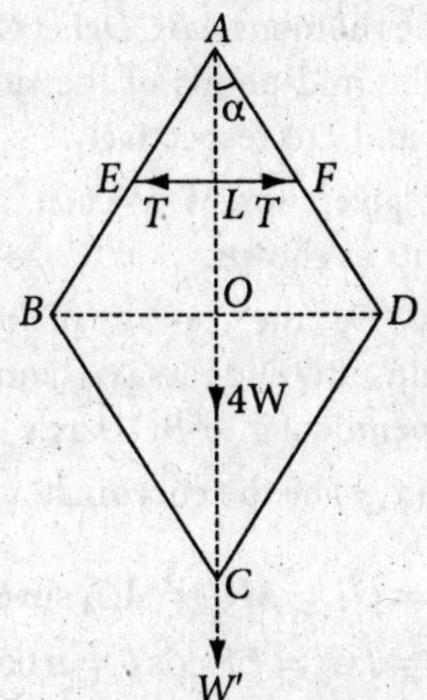

Fig. 3.29

So, equation of the virtual work is

$$4W\delta(AO)+W'\delta(AC)+T\delta(EF)=0. \quad (1)$$

In $\triangle AOD$,

$$\cos\alpha=\frac{AO}{AD}=\frac{AO}{a}\Rightarrow AO=a\cos\alpha.$$

$$\therefore\ AC=2AO=2a\cos\alpha.$$

Also in $\triangle ALF$,

$$\sin\alpha=\frac{LF}{AF}=\frac{LF}{a/2}\Rightarrow LF=\frac{a}{2}\sin\alpha.$$

$$\therefore\ EF=2LF=2\frac{a}{2}\sin\alpha=a\sin\alpha.$$

Putting the values of AO, AC EF in (1), we get

$$4W\delta(a\cos\alpha)+W'(2a\cos\alpha)+T\delta(a\sin\alpha)=0$$
$$\Rightarrow(-4Wa\sin\alpha-W'2a\sin\alpha+Ta\cos\alpha)\delta\alpha=0$$
$$\Rightarrow Ta\cos\alpha-2a\sin\alpha(2W+W')=0 \text{ as } \delta\alpha\neq0$$
$$\Rightarrow T\cos\alpha=2\sin\alpha(2W+W')$$
$$\Rightarrow T=2\tan\alpha(2W+W')=(4W+2W')\tan\alpha \text{ (Proved).}$$

19. A frame consists of five bars forming the sides of a rhombus $ABCD$ with diagonal AC. If four forces (equal) P act inwards at the middle points of the sides, and at right angles to the respective sides, prove that the tension in AC is $\frac{P\cos 2\theta}{\sin\theta}$, where θ is the angle BAC. The bars are weightless. [VU(H) 2006]

Solution: Let $2a$ be the length of each side of the rhombus $ABCD$. Let G_1, G_2, G_3, G_4 be the mid-points of the sides AB, BC, CD and DA respectively.

The given forces P each, act at these points as shown.

Take O, the centre of rhombus, as origin, BOD as x-axis and OY, a line perpendicular to BOD as y-axis.

Let (x, y) be the coordinates of G_1. Then

$$x = OL = MG_1 = AG_1 \sin\theta = a\sin\theta$$
$$y = LG_1 = LB\cos\theta = a\cos\theta.$$

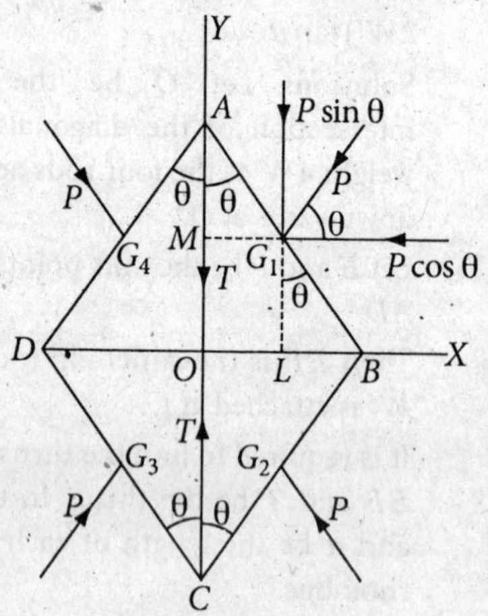

Fig. 3.30

$\therefore$ coordinates of G_1 are $(a\sin\theta, a\cos\theta)$.

Resolved parts of force P at G_1 are $-P\cos\theta$ along OX and $-P\sin\theta$ along OY.

Let T be the tension in the string AC.

Let us give a displacement to the system such that θ becomes $\theta + \delta\theta$, x becomes $x + \delta x$, y becomes $y + \delta y$ and the length of the string changes. The length of the bars remains unchanged.

Virtual work done by the force P at G_1

$$= \text{sum of the virtual works done by } P\sin\theta \text{ and } P\cos\theta$$
$$= -P\sin\theta \cdot \delta y - P\cos\theta \cdot \delta x$$
$$= -P\sin\theta \cdot \delta(a\cos\theta) - P\cos\theta \cdot \delta(a\sin\theta)$$
$$= P(\sin\theta\, a\sin\theta - P\cos\theta\cos\theta)\delta\theta$$
$$= -aP(\cos^2\theta - \sin^2\theta)\delta\theta$$
$$= -aP\cos 2\theta\,\delta\theta.$$

On account of the symmetry of the forces about the axes, same amount of work is done by each force.

$\therefore$ total virtual work done by all the four forces $= -4aP\cos 2\theta\,\delta\theta$.

Also in $\triangle OAB$,

$$\cos\theta = \frac{OA}{AB} = \frac{OA}{2a} \Rightarrow OA = 2a\cos\theta.$$
$$\therefore AC = 2AO = 4a\cos\theta.$$

The equation of virtual work is

$$-T\delta(AC)-4P\cdot a\cos 2\theta\,\delta\theta=0 \Rightarrow -T\delta(4a\cos\theta)-4Pa\cos 2\theta\,\delta\theta=0$$
$$\Rightarrow (T4a\sin\theta-4Pa\cos 2\theta)\delta\theta=0$$
$$\Rightarrow T4a\sin\theta-4Pa\cos 2\theta=0 \text{ as } \delta\theta\neq 0$$
$$\Rightarrow T4a\sin\theta=4Pa\cos 2\theta$$
$$\Rightarrow T=\frac{P\cos 2\theta}{\sin\theta} \textbf{ (Proved).}$$

20. A regular pentagon $ABCDE$ is formed of five uniform heavy rods each of weight W freely jointed at their extremities. It is freely suspended from A and is maintained in its regular pentagonal form by a light rod joining B and E. Prove that the stress in this rod is $W\cot 18°$. [VU(H) 2008]

Solution: Let T be the stress in the rod BE.

The weights W, W at G_1 and G_5, the mid-points of AB and AE respectively are equivalent to a weight $2W$ at F, the mid-point of G_1G_5.

The weights W, W at G_2 and G_4 the mid-points of BC and DE respectively are equivalent to a weight $2W$ at K, the mid-point of G_2G_4.

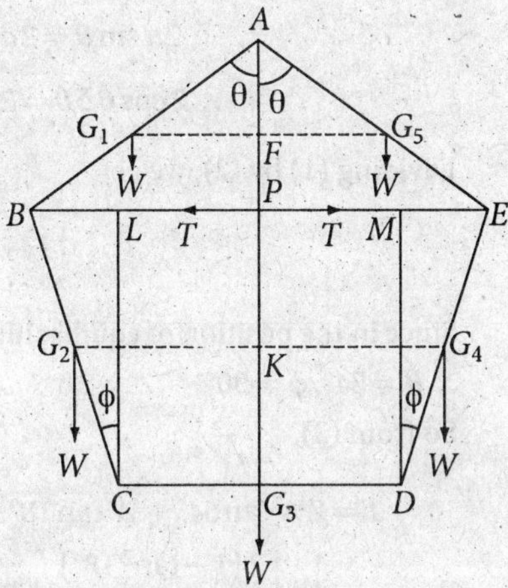

Fig. 3.31

Let θ be the angle which each of the upper slant rods makes with the vertical and ϕ, the angle which each of the lower slant rods makes with the vertical.

Since the point of suspension A is fixed, let the distances be measured from A.

Let $2a$ be the length of each rod.

$\therefore AF=AG_1\cos\theta=a\cos\theta.$

$$AK=AP+PK=AB\cos\theta+\tfrac{1}{2}CL=2a\cos\theta+\tfrac{1}{2}2a\cos\phi=2a\cos\theta+a\cos\phi$$
$$AG_3=AP+PG_3=AB\cos\theta+CL=2a\cos\theta+2a\cos\phi$$
$$BE=2BP=2AB\sin\theta=4a\sin\theta.$$

Let the system be given a small virtual displacement such that θ becomes $\theta+\delta\theta$ and ϕ becomes $\phi+\delta\phi$.

So, the equation of virtual work is

$$2W\cdot\delta(AF)+2W\cdot\delta(AK)+W\cdot\delta(AG_3)+T\cdot\delta(BE)=0$$
$$\Rightarrow 2W\cdot\delta(a\cos\theta)+2W\cdot\delta(2a\cos\theta+a\cos\phi)+W\cdot\delta(2a\cos\theta+2a\cos\phi)$$
$$+T\cdot\delta(4a\sin\theta)=0$$

$$\Rightarrow -2Wa\sin\theta\,\delta\theta + 2W(-2a\sin\theta\,\delta\theta - a\sin\phi\,\delta\phi)$$
$$+W(-2a\sin\theta\,\delta\theta - 2a\sin\phi\,\delta\phi) + T(4a\cos\theta\,\delta\theta) = 0$$
$$\Rightarrow (-2Wa\sin\theta - 4Wa\sin\theta - W2a\sin\theta + T4a\cos\theta)\delta\theta$$
$$+(-2Wa\sin\phi - 2Wa\sin\phi)\delta\phi = 0$$
$$\Rightarrow (-8Wa\sin\theta + 4aT\cos\theta)\delta\theta - 4Wa\sin\phi\,\delta\phi = 0$$
$$\Rightarrow 4a(T\cos\theta - 2W\sin\theta)\delta\theta = 4aW\sin\phi\,\delta\phi$$
$$\Rightarrow (T\cos\theta - 2W\sin\theta)\delta\theta = W\sin\phi\,\delta\phi. \quad (1)$$

Now,

$$BP = AB\sin\theta = 2a\sin\theta.$$

Also,

$$BP = BL + LP = BC\sin\phi + CG_3 = 2a\sin\phi + a.$$

$$\therefore\ 2a\sin\theta = 2a\sin\phi + a \Rightarrow 2\sin\theta = 2\sin\phi + 1.$$

$$\therefore\ 2\cos\theta\,\delta\theta = 2\cos\phi\,\delta\phi \Rightarrow \cos\theta\,\delta\theta = \cos\phi\,\delta\phi. \quad (2)$$

Dividing (1) by (2), we get

$$(T - 2W\tan\theta) = W\tan\phi. \quad (3)$$

Since in the position of equilibrium, each interior angle of a regular pentagon $= 108°$.
$\therefore\ \theta = 54°, \phi = 90° - 72° = 18°$.
So from (3),

$$T = 2W\tan 54° + W\tan 18° = W[2\cot 36° + \tan 18°]$$
$$= W\left[\frac{2(1-\tan^2 18°)}{2\tan 18°} + \tan 18°\right] = W\left[\frac{1}{\tan 18°}\right] = W\cot 18° \textbf{ (Proved)}.$$

21. Five equal uniform rods, each of weight W are freely jointed together to form a pentagon $ABCDE$, which is suspended from the point A and maintained in the shape of a regular pentagon by two strings joining A to C and D. Show that the tension of either string is $2W\cos 18°$. [NBU(H) 2004]

Solution: Let T be the tension of each of the strings AC and AD.

The weights W, W at G_1 and G_2, the mid-points of AB and AE are equivalent to a weight $2W$ at F, the mid-point of G_1G_2.

The weights W, W at G_2 and G_4, the mid-points of BC and DE respectively are equivalent to a weight $2W$ at K, the mid-point of G_2G_4.

Let θ be the angle which each of the upper slant rods makes with the vertical and ϕ, the angle which each of the lower slant rods makes with the vertical.

Since the pint of suspension A is fixed, let the distances be measured from A.

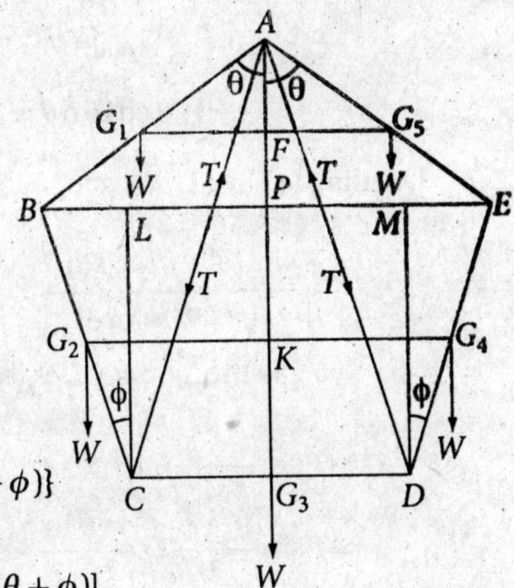

Fig. 3.32

Let $2a$ be the length of each rod.

$\therefore AF = AG_1 = a\cos\theta.$

$$AK = AP + PK = AB\cos\theta + \tfrac{1}{2}CL$$
$$= 2a\cos\theta + \tfrac{1}{2}2a\cos\phi$$
$$= 2a\cos\theta + a\cos\phi$$
$$AG_3 = AP + PG_3 = AB\cos\theta + CL$$
$$= 2a\cos\theta + 2a\cos\phi.$$

In $\triangle ABC$,

$$AC^2 = AB^2 + BC^2 - 2AB\cdot BC\cos\{\pi - (\theta+\phi)\}$$
$$\Rightarrow AC^2 = (2a)^2 + (2a)^2 + 2\cdot 2a\cdot 2a\cos(\theta+\phi)$$
$$\Rightarrow AC^2 = 8a^2 + 8a^2\cos(\theta+\phi) = 8a^2[1+\cos(\theta+\phi)]$$
$$\Rightarrow AC^2 = 8a^2 2\cos^2\frac{\theta+\phi}{2} = 16a^2\cos^2\frac{\theta+\phi}{2}$$
$$\Rightarrow AC = 4a\cos\frac{\theta+\phi}{2}.$$

Let system be given a small virtual displacement such that θ becomes $\theta+\delta\theta$ and ϕ becomes $\phi+\delta\phi$.

So, the equation of virtual work is

$$2W\cdot\delta(AP) + 2W\cdot\delta(AK) + W\cdot\delta(AG_3) - T\cdot\delta(AC) - T\cdot\delta(AC) = 0$$
$$\Rightarrow 2W\cdot\delta(a\cos\theta) + 2W\cdot\delta(2a\cos\theta + a\cos\phi) + W\cdot\delta(2a\cos\theta + 2a\cos\phi)$$
$$-2T\cdot\delta\left(4a\cos\frac{\theta+\phi}{2}\right) = 0$$
$$\Rightarrow -2Wa\sin\theta\delta\theta + 2W(-2a\sin\theta\delta\theta - a\sin\phi\delta\phi) + W(-2a\sin\theta\delta\theta - 2a\sin\phi\delta\phi)$$
$$+8aT\sin\frac{\theta+\delta}{2}\left(\frac{\delta\theta+\delta\phi}{2}\right) = 0$$
$$\Rightarrow \delta\theta\left[-2Wa\sin\theta - 4Wa\sin\theta - 2Wa\sin\theta + 4aT\sin\frac{\theta+\delta}{2}\right]$$
$$+\delta\phi\left(-2Wa\sin\phi - 2W\sin\phi + 4aT\sin\frac{\theta+\phi}{2}\right) = 0$$
$$\Rightarrow \delta\theta\left[-8Wa\sin\theta + 4aT\sin\frac{\theta+\phi}{2}\right]$$
$$+\delta\phi\left(-4aW\sin\phi + 4aT\sin\frac{\theta+\phi}{2}\right) = 0$$
$$\Rightarrow \delta\theta\left(T\sin\frac{\theta+\phi}{2} - 2W\sin\theta\right) = \delta\phi\left(W\sin\phi - T\sin\frac{\theta+\phi}{2}\right). \tag{1}$$

Now,

$$BP = AB\sin\theta = 2a\sin\theta.$$

Also,

$$BP = BL + LP = BC\sin\phi + CG_3 = 2a\sin\phi + a.$$

$$\therefore\ 2a\sin\theta = 2a\sin\phi + a \Rightarrow 2\sin\theta = 2\sin\phi + 1.$$

$$\therefore\ 2\cos\theta\,\delta\theta = 2\cos\phi\,\delta\phi \Rightarrow \cos\theta\,\theta\,\delta\theta = \cos\phi\,\delta\phi. \qquad (2)$$

Dividing (1) by (2), we get

$$\frac{T\sin\frac{\theta+\phi}{2} - 2W\sin\theta}{\cos\theta} = \frac{W\sin\phi - T\sin\frac{\theta+\phi}{2}}{\cos\phi}$$

$$\Rightarrow T\sin\frac{\theta+\phi}{2}\cos\phi - 2W\sin\theta\cos\phi = W\sin\phi\cos\theta - T\sin\frac{\theta+\phi}{2}\cos\theta$$

$$\Rightarrow T\sin\frac{\theta+\phi}{2}(\cos\phi + \cos\theta) = W(\sin\phi\cos\theta + 2\sin\theta\cos\phi)$$

$$\Rightarrow T\sin\frac{\theta+\phi}{2}2\cos\frac{\theta+\phi}{2}\cos\frac{\theta-\phi}{2} = W[\sin(\theta+\phi) + \sin\theta\cos\phi]$$

$$\Rightarrow T = \frac{W[\sin(\theta+\phi) + \sin\theta\cos\phi]}{\sin(\theta+\phi)\cos\frac{\theta-\phi}{2}}.$$

In equilibrium, $\theta = \frac{2\pi}{5}, \phi = \frac{\pi}{5}$.

$$\therefore\ T = \frac{W\left(\sin\frac{2\pi}{5}\cos\frac{\pi}{5} + \sin\frac{3\pi}{5}\right)}{\sin\frac{3\pi}{5}\cos\frac{\pi}{10}} = \frac{W\left[\sin\left(\pi - \frac{3\pi}{5}\right)\cos\frac{\pi}{5} + \sin\frac{3\pi}{5}\right]}{\sin\frac{3\pi}{5}\cos\frac{\pi}{10}}$$

$$= \frac{W\sin\frac{3\pi}{5}\left(\cos\frac{\pi}{5} + 1\right)}{\sin\frac{3\pi}{5}\cos\frac{\pi}{10}} = \frac{W2\cos^2\frac{\pi}{10}}{\cos\frac{\pi}{10}} = 2W\cos\frac{\pi}{10} = 2W\cos 18^\circ \text{ (Proved).}$$

22. A regular hexagon $ABCDEF$ is composed of six heavy uniform rods jointed together and two opposite angles C and F are connected by a string, which is horizontal, AB being in contact with a horizontal plane. A weight W_1 is placed at the middle point of DE. If W be the weight of each rod, show that the tension of the string is $\frac{3W+W_1}{\sqrt{3}}$.

[CU(H) 1997, KU(H) 2008]

Solution: Let T be the tension in the string.

The weight $6W$ acts at O, the centre of the hexagon.

Since AB is fixed in a horizontal position, let distances be measured from AB.

Let a be the length of each side of the hexagon.

Let θ be the inclination of lower slant rods with the horizontal.

$\therefore\ ON = MB = BC\sin\theta = a\sin\theta.$

$\therefore\ HN = 2ON = 2a\sin\theta.$

$FC = FL + LM + MC = 2FL + a = 2a\cos\theta + a.$

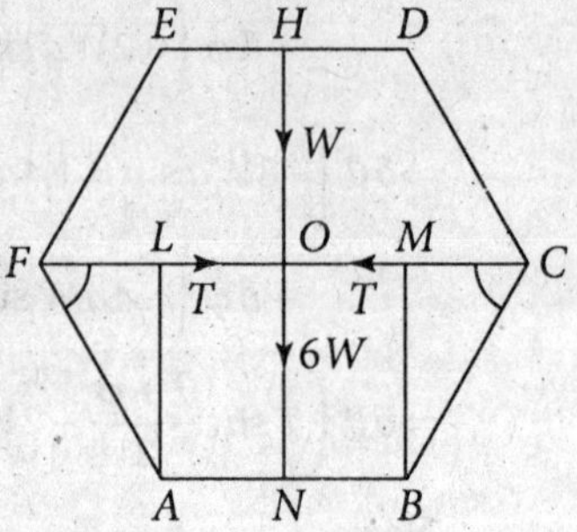

Fig. 3.33

Let the system be given a small virtual displacement such that θ becomes $\theta + \delta\theta$ and the length of the string changes. The length of the rod remains unchanged.

So equation of virtual work is

$$
\begin{aligned}
&-6W\cdot\delta(ON)-W_1\cdot\delta(HN)-T\cdot\delta(FC)=0\\
&\Rightarrow 6W\cdot\delta(a\sin\theta)+W_1\delta(2a\sin\theta)+T\cdot\delta(a+2a\cos\theta)=0\\
&\Rightarrow (6Wa\cos\theta+W_1 2a\cos\theta-T2a\sin\theta)\delta\theta=0\\
&\Rightarrow 6Wa\cos\theta+2aW_1\cos\theta-T2a\sin\theta=0 \text{ as } \delta\theta\neq 0\\
&\Rightarrow T\sin\theta=(2W+W_1)\cos\theta\\
&\Rightarrow T=(3W+W_1)\cot\theta.
\end{aligned}
$$

In the position of equilibrium, $\theta=60°$.

$$\therefore\ T=(3W+W_1)\cot 60°=\frac{(3W+W_1)}{\sqrt{3}}.$$

23. A uniform rectangular plate $ABCD$ of weight W hangs from A. If a weight W' be suspended from B, prove that the inclination θ of the diagonal AC to the vertical is given by

$$\tan\theta=\frac{2W'ab}{W(a^2+b^2)+2W'a^2}, \text{ where } AB=a \text{ and } AD=b. \qquad \text{[CU(H) 1987]}$$

Solution: Let the weight W of the plate be acting at G, the centre of the plate.

Let $\angle CAB=\alpha=\angle ACD$, so that

$$\tan\alpha=\frac{AC}{AD}=\frac{b}{a}. \qquad (1)$$

Since AC makes an angle θ with the vertical, so if AK be the vertical line through A, then $\angle GAK=\theta$.

$\therefore\ \angle BAK=\alpha-\theta$.

Now, let the system be given a small virtual displacement such that θ changes into $\theta+\delta\theta$, then the equation of virtual work is

$$W'\delta(BN)+W\delta(GM)=0, \qquad (2)$$

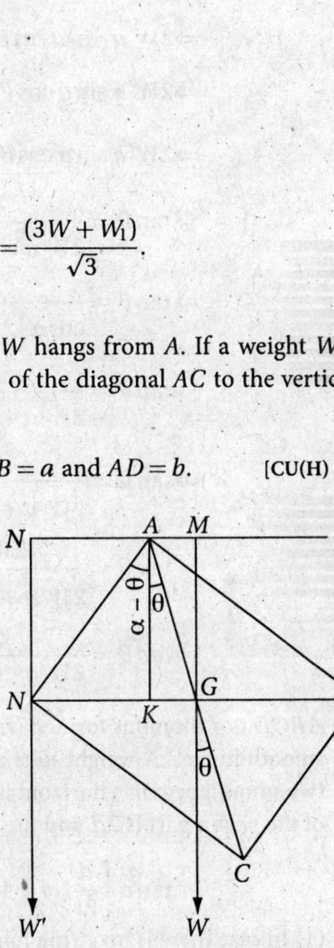

Fig. 3.34

where BN and GM are perpendiculars from B and G respectively on the horizontal line through A.

$$\therefore\ BN=AB\cos(\alpha-\theta)=a\cos(\alpha-\theta)$$
$$\text{and } GM=AG\cos\theta=\tfrac{1}{2}\sqrt{a^2+b^2}\cos\theta.$$

From (2), we get

$$W'\delta[a\cos(\alpha-\theta)]+W\left[\delta\tfrac{1}{2}\left(\sqrt{a^2+b^2}\right)\cos\theta\right]=0$$

$$\Rightarrow\left[W'a\sin(\alpha-\theta)-\frac{W}{2}\sqrt{a^2+b^2}\sin\theta\right]\delta\theta=0$$

$$\Rightarrow W'a\sin(\alpha-\theta)-\frac{W}{2}\sqrt{a^2+b^2}\sin\theta=0 \text{ as } \delta\theta\neq0$$

$$\Rightarrow 2W'a[\sin\alpha\cos\theta-\cos\alpha\sin\theta]-W\sqrt{a^2+b^2}\sin\theta=0$$

$$\Rightarrow 2W'a\sin\alpha\cos\theta-\sin\theta\left[2W'a\cos\alpha+W\sqrt{a^2+b^2}\right]=0$$

$$\Rightarrow 2W'a\sin\alpha\cos\theta=\sin\theta\left[2W'a\cos\alpha+W\sqrt{a^2+b^2}\right]$$

$$\Rightarrow\tan\theta=\frac{2W'a\sin\alpha}{2W'a\cos\alpha+W\sqrt{a^2+b^2}}$$

$$\Rightarrow\tan\theta=\frac{2W'a\sin\alpha}{\cos\alpha\left[2W'a+W\sqrt{a^2+b^2}\sec\alpha\right]}$$

$$\Rightarrow\tan\theta=\frac{2W'a\tan\alpha}{2W'a+W\sqrt{a^2+b^2}\sqrt{1+\tan^2\alpha}}$$

$$\Rightarrow\tan\theta=\frac{2W'a\frac{b}{a}}{2W'a+W\sqrt{a^2+b^2}\sqrt{1+\frac{b^2}{a^2}}}\quad\text{[from (1)]}$$

$$\Rightarrow\tan\theta=\frac{2W'b}{2W'a+\frac{W(a^2+b^2)}{a}}$$

$$\Rightarrow\tan\theta=\frac{2W'ab}{2W'a^2+W(a^2+b^2)}.$$

24. $ABCD$ is a rhombus formed with four rods each of length l and weight W joined by smooth hinges. A weight w is attached to the lowest hinge C and the frame rests on two smooth pegs in a horizontal line and B and D are joined by a string. If the distance of the pegs a part is $2d$ and the angle at A is 2α, show that the tension in the string is

$$\tan\alpha\left[\frac{d}{2l}(w+4W)\operatorname{cosec}^2\alpha-(w+2W)\right].$$ **[CU(H) 1984]**

(1) In case there is no string joining B and D, prove that

$$\sin^2\alpha=\frac{d(4W+w)}{l(4W+2w)}$$

(2) In case the rods were weightless prove that the tension in the string BD will be $W\tan\alpha\left[\frac{d}{2l}\operatorname{cosec}^3\alpha-1\right]$. **[KU(H) 2002]**

Solution: Let $ABCD$ be the rhombus, which rests on two smooth pegs P and Q at a distance $2d$ apart. Since PQ is fixed, so all the distances will be measured from PQ. The weight $4W$ of the rhombus act at its $C\cdot G'G'$. Also a weight w is attached to the lowest hinge C. Let $< DAC=\alpha$, $AB=BC=CD=DA=l$, $PQ=2d$.

Let us give a small virtual displacement to the system such that α changes into $\alpha+\delta\alpha$ and the length of the string changes. The length of the rods remain unchanged. So equation of virtual work is

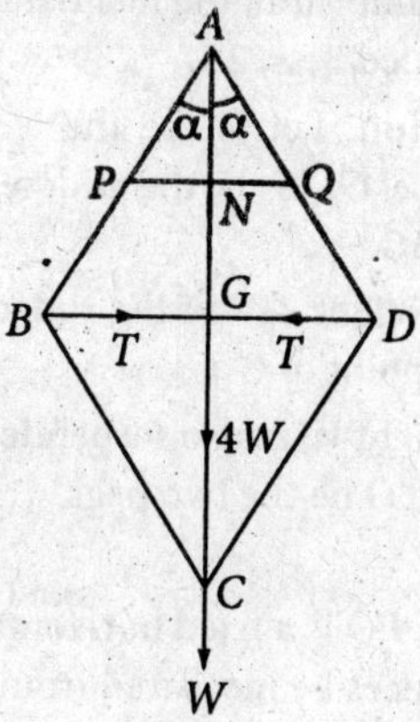

$$4W\delta(NG)+W\delta(NC)-T\delta(BD)=0. \tag{1}$$

Now,

$$NG=AG-AN=l\cos\alpha-NQ\cot\alpha$$
$$=l\cos\alpha-d\cot\alpha$$

and

$$NC=AC-AN=2AG-AN$$
$$=2l\cos\alpha-d\cot\alpha.$$

Also,

Fig. 3.35

$$BD=2GD=2AD\sin\alpha=2l\sin\alpha.$$

Putting the values of NG, NC and BD in (1), we get

$$4W\delta(l\cos\alpha-d\cot\alpha)+w\delta(2l\cos\alpha-d\cot\alpha)-T\delta(2l\sin\alpha)=0$$
$$\Rightarrow 4W\left(-l\sin\alpha+d\operatorname{cosec}^2\alpha\right)\alpha+W\left[-2l\sin\alpha+d\operatorname{cosec}^2\alpha\right]\delta\alpha-T2l\cos\alpha\delta\alpha=0$$
$$\Rightarrow\left[4W\left(d\operatorname{cosec}^2\alpha-l\sin\alpha\right)+W\left(d\operatorname{cosec}^2\alpha-2l\sin\alpha\right)-T2l\cos\alpha\delta\alpha\right]=0$$
$$\Rightarrow d\operatorname{cosec}^2\alpha(4W+w)-2l\sin\alpha(2W+W)-T2l\cos\alpha=0 \text{ as } \delta\alpha\neq 0$$
$$\Rightarrow d\operatorname{cosec}^3\alpha(4W+w)-2l(2W+w)-T2l\cot\alpha=0$$
$$\Rightarrow\frac{d}{2l}(4W+w)\operatorname{cosec}^3\alpha-(2W+w)-T\cot\alpha=0$$
$$\Rightarrow T=\tan\alpha\left[\frac{d}{2l}(4W+w)\operatorname{cosec}^3\alpha-(2W+w)\right]. \tag{1}$$

(1) In this case there will be no tension, so putting $T=0$ in (1), we get

$$0=\frac{d}{2l}(4W+w)\operatorname{cosec}^3\alpha-(2W+w)\Rightarrow\frac{d}{2l}(4W+w)\operatorname{cosec}^3\alpha=(2W+w)$$
$$\Rightarrow\operatorname{cosec}^3\alpha=\frac{2l(2W+w)}{d(4W+w)}$$
$$\Rightarrow\sin^3\alpha=\frac{d(4W+w)}{2l(2W+w)}=\frac{d(4W+w)}{l(4W+2w)}.$$

(2) Since the rods are weightless, so putting $W=0$ in (1), we get

$$T=\tan\alpha\left[\frac{d}{2l}(w+0)\operatorname{cosec}^3\alpha-(w+0)\right]\Rightarrow T=w\tan\alpha\left[\frac{d}{2l}\operatorname{cosec}^3\alpha-1\right].$$

25. Two equal uniform rods AB and AC of length a are freely jointed at A and placed symmetrically over two smooth pegs on the same horizontal level, distance c apart. A

weight equal to that of one rod is suspended from the joint A. Show that in the position of equilibrium, the inclination of either rod with the horizontal is $\theta = \cos^{-1}\left(\frac{3c}{2a}\right)^{1/2}$, provided $3c < 2a$.

[IAS 2005]

Solution: Let W be the weight of each rod. Let G_1, G_2 be the middle points of AB and AC.

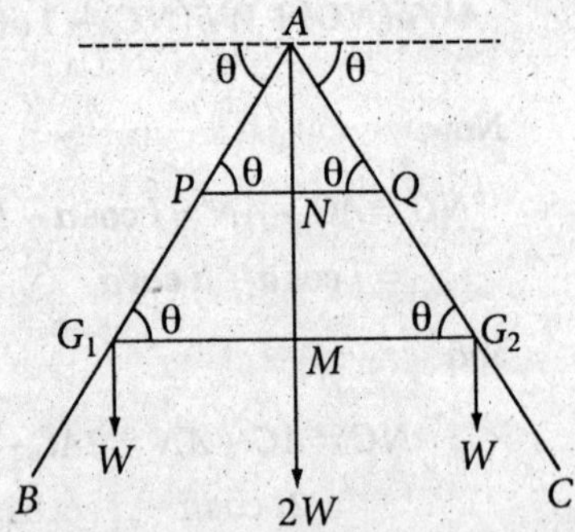

Fig. 3.36

The weight $2W$ of the rod acts at M, the mid-point of G_1G_2.

A weight W is also suspended from A. Let P and Q be the two pegs.

$\therefore PQ = c$.

Since PQ is a fixed horizontal line, let the distances be measured from PQ.

$$AN = PN\tan\theta = \frac{c}{2}\tan\theta.$$

$$AM = AG_1\sin\theta = \frac{a}{2}\sin\theta.$$

$$\therefore NM = AM - AN = \frac{a}{2}\sin\theta - \frac{c}{2}\tan\theta.$$

Let the system be given a small virtual displacement such that θ becomes $\theta + \delta\theta$ and the length of the rods remain unchanged.

So equation of virtual work is

$$2W\delta(NM) - W\delta(AN) = 0$$

$$\Rightarrow 2W\delta\left[\frac{1}{2}(a\sin\theta - c\tan\theta)\right] - W\delta\left(\frac{c}{2}\tan\theta\right) = 0$$

$$\Rightarrow \frac{2}{2}\left[a\cos\theta - c\sec^2\theta\right]\delta\theta - \frac{c}{2}\sec^2\theta\,\delta\theta = 0$$

$$\Rightarrow \left[a\cos\theta - \frac{c}{\cos^2\theta} - \frac{c}{2\cos^2\theta}\right]\delta\theta = 0$$

$$\Rightarrow 2a\cos^3\theta - 2c - c = 0 \text{ as } \delta\theta \neq 0$$

$$\Rightarrow \cos^3\theta = \frac{3c}{2a} \Rightarrow \cos\theta = \left(\frac{3c}{2a}\right)^{1/3}$$

$$\Rightarrow \theta = \cos^{-1}\left(\frac{3c}{2a}\right)^{1/3} \quad \text{(Proved).}$$

26. Weights W_1, W_2 are fastened to a light inextensible string ABC at the points B, C, the end A being fixed. Prove that, if a horizontal force P is applied at C and in equilibrium, AB, BC are inclined at angles θ, ϕ to the vertical, then $P = (W_1 + W_2)\tan\theta = W_2\tan\phi$.

[CU(H) 2005]

Solution: From B and C, we draw BM and CK perpendiculars to the vertical lines through A. From B, we draw BD and BM perpendicular to CK and AK respectively. Let $AB = l_1$ and $BC = l_2$.

Then $AM = l_1 \cos\theta, BM = l_1 \sin\theta = DK$.
$MK = BD = l_2 \cos\phi, CD = l_2 \sin\phi$.

$\therefore AK = AM + MK = l_1 \cos\theta + l_2 \cos\phi$

and $CK = CD + DK = l_2 \sin\phi + l_1 \sin\theta$.

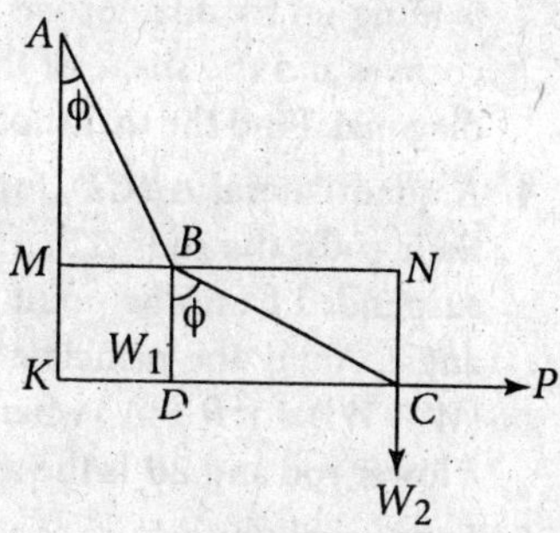

We measure the distances from the fixed point A downwards. Let us give a displacement to the system such that θ becomes $\theta + \delta\theta$ and ϕ becomes $\phi + \delta\phi$. So equation of virtual works

Fig. 3.37

$$W_1\delta(AM) + W_2\delta(AK) + P\delta(KC) = 0$$
$$\Rightarrow W_1\delta(l_1\cos\theta) + W_2\delta(l_1\cos\theta + l_2\cos\phi) + P\delta(l_2\sin\phi + l_1\sin\theta) = 0$$
$$\Rightarrow - W_1 l_1 \sin\theta\delta\theta - W_2(l_1\sin\theta\delta\theta + l_2\sin\phi\delta\phi) + P(l_2\cos\phi\delta\phi + l_1\cos\theta\delta\theta) = 0$$
$$\Rightarrow l_1\delta\theta[-(W_1 + W_2)\sin\theta + P\cos\theta] - l_2\delta\phi[W_2\sin\phi - P\cos\phi] = 0$$
$$\Rightarrow l_1\delta\theta[(W_1 + W_2)\sin\theta - P\cos\theta] = [P\cos\phi - W_2\sin\phi]\, l_2\delta\phi. \tag{1}$$

Now, we give the system another small virtual displacement such that B remains fixed and ϕ changes into $\phi + \delta\phi$, the distances being measured from the fixed point B downwards, then the equation of virtual work is

$$W_2\delta(BD) + P\delta(CD) = 0$$
$$\Rightarrow W_2\delta(l_2\cos\phi) + P\delta(l_2\sin\phi) = 0$$
$$\Rightarrow (-W_2 l_2\sin\phi + P l_2\cos\phi)\delta\phi = 0$$
$$\Rightarrow - W_2\sin\phi + P\cos\phi = 0 \text{ as } \delta\phi \neq 0$$
$$\Rightarrow P\cos\phi = W_2\sin\phi \Rightarrow P = W_2\tan\phi. \tag{2}$$

Also from (1) and (2), we get

$$[(W_1 + W_2)\sin\theta - P\cos\theta]\, l_1\delta\theta = 0$$
$$\Rightarrow P = (W_1 + W_2)\tan\theta \text{ as } \delta\theta \neq 0$$
$$\Rightarrow P = (W_1 + W_2)\tan\theta = W_2\tan\phi.$$

EXERCISE 3

1. Six equal rods, each of weight W are freely jointed at their extremities so as to form a regular hexagon $ABCDEF$, the rod AB is fixed in a horizontal position and the middle points of AB and DE are joined by a string, prove that the tension of the string is $3W$.

   [MU(H) 2008, PU(H) 2013]

2. Six equal bars are jointed at their ends forming a regular hexagon $ABCDEF$ which is kept in shape by vertical strings joining the middle points of BC, CD and AF, FE respectively, the side AB being held horizontal and uppermost. Prove that the tension of each string is three times the weight of each bar.

3. A square framework formed of uniform heavy rods of equal weight W jointed together, is hung up by one corner. A weight W is suspended from each of the three lower corners and the shape of the square is preserved by a light rod along the horizontal diagonal. Find the thrust of the light rod. [PU(H) 1996, LNMU(H) 1998]
4. A quadrilateral $ABCD$, formed of four uniform rods freely jointed to each other at their ends; the rods AB, AD being equal and the rods BC, CD being equal is freely suspended from the point A. A string joins A to C and is such that ABC is a right angle. Apply the principle of virtual work to show that the tension of the string is $(W+W')\sin^2\theta + W'$ where W is the weight of an upper rod and W' is the weight of a lower rod and 2θ is the angle BAD. [IAS 1989, CU(H) 1995, VU(H) 2008]
5. Four equal uniform rods are jointed to form a rhombus $ABCD$, which is placed in a vertical plane with AC vertical and A resting on a horizontal plane. The rhombus is kept in position in which $\angle BAC = \theta$ by a light string joining B and D. Show that its tension is $2W\tan\theta$, where W is the weight of a rod.
6. A rhombus $ABCD$ is formed of four equal uniform rods freely jointed together and suspended from A. It is kept in position by a light rod joining the middle point of BC and CD. Prove that if T be the thrust in this rod and W the weight of the rhombus, then $T = W\tan A/2$. [TMBU(H) 1996, BVBAU(H) 2011, CU(H) 2006, NBU(H) 2003]
7. Five equal uniform rods are freely jointed together at their ends to form a closed chain $ABCDEA$ and the system hangs freely from A. Prove that if AB and BC are inclined to the vertical at an angle θ and ϕ respectively, then $\tan\phi = 2\tan\theta$.
8. Two equal beams AB, AC each of weight W connected by a hinge at A are placed in a vertical plane with their extremities B, C resting on a horizontal plane. They are kept from falling by strings connecting B and C with the middle points of the opposite sides. Show that the tension of either string is $\frac{W}{8}\sqrt{1+9\cot^2\theta}$, where θ is the inclination of the beam to the horizon.
9. A heavy rod AB of length $2l$ rests upon a fixed smooth peg at C and with its end B upon a smooth curve. If it rests in all positions, show that the curve is one whose polar equation with C as origin is $r = l + \frac{a}{\sin\theta}$. [RU(H) 1992, BVABU(H) 2010]
10. A uniform smooth rod passes through a ring at the focus of a fixed parabola whose axis is vertical and vertex below the focus, and rests with one end on the parabola. Prove that the rod will be in equilibrium if it makes the vertical an angle θ given by the condition $\cos^4\frac{\theta}{2} = \frac{a}{2c}$ where $4a$ is the latus rectum of the parabola and $2c$ the length of the rod. [TMBU(H) 1998, BVABU(H) 2008]
11. Two small smooth rings of equal weight slide on a fixed elliptic wire, whose major axis is vertical and they are connected by a string which passes over a small smooth peg at the upper focus. Show that the weights will be in equilibrium wherever they are placed. What are the positions of equilibrium if the weights are unequal. [VU(H) 2003]
12. Two uniform rods AB and AC jointed at A are in equilibrium in a vertical plane. B and C rest on a smooth horizontal plane and the middle points of AB and AC are connected by a string. Show that the tension of the string is $\frac{W}{\tan B+\tan C}$, where W is the total weight of the rods AB and AC. [IAS 1995]
13. A rod AB is movable about a pivot at A and to B is attached a string whose other end is tied to a ring. The ring slides along a smooth horizontal wire passing through A.

Prove by the principle of virtual work that the horizontal force necessary to keep the ring at rest is $\frac{W\cos\alpha\cos\beta}{2\sin(\alpha+\beta)}$, where W is the weight of the rod and α, β the inclinations of the rod and string to the horizontal. [TMBU(H) 1991, CU(H) 1994]

14. Two particles P, Q, connected by a string, rest on the convex side of a smooth cycloid. Prove that in the position $\sin\frac{\theta}{2} = \frac{Q}{P+Q}\frac{l}{4a}$ where l = length of the string, a = radius of the generating circle and θ = angle, the radius to the point on generating circle corresponding to P makes with vertical.

15. Four uniform rods are freely jointed at their extremities and form a parallelogram $ABCD$ which is suspended by the joint A and is kept in shape by a string AC, prove that the tension of the string is equal to half the weight of the four rods.

16. Two equal rods each of weight $l\,w$ and length l are hinged together and placed outside a smooth horizontal cylindrical peg of radius r. Then the lower ends are tied together by a string and the rods are left at the same inclination ϕ to the horizontal. Find the tension in the string and if the string is slack, show that ϕ satisfies the equation $\tan^3\phi + \tan\phi = \frac{l}{2r}$. [IAS 1992]

17. Six equal heavy beams are freely jointed at their ends to form a heavy hexagon and are placed in a vertical plane with one beam resting on a horizontal plane, the middle points of the two upper slant beams which are inclined at an angle θ to the horizon are connected by a light cord. Show that its tension is $6W\cot\theta$, where W is the weight of each beam. [IAS 1985]

18. Four equal rods, each of weight W, are freely jointed together so as to form a rhombus, the whole hangs from a point A which is fixed and is kept in position by a horizontal thrust BD. A weight W' hangs from C. If 2α be the angle of the rhombus at A, prove that the stress in BD is $(2W+W')\tan\alpha$. [IAS 1980]

19. A prism whose cross section is an equilateral triangle rests with two edges on smooth planes inclined at α and β to the horizon. If θ be the angle which the plane containing these edges makes with the vertical, show that

$$\tan\theta = \frac{2\sqrt{3}\sin\alpha\sin\beta + \sin(\alpha+\beta)}{\sqrt{3}\sin(\beta-\alpha)}.$$

20. A regular octahedron formed by twelve equal rods, each of weight W, freely jointed together, is suspended from one corner. Find the thrust in each horizontal rod.

[TMBU(H) 1993]

21. Two uniform rods AB, BC of weights W and W' are smoothly jointed at B and their middle points are joined across by a cord. The rods are tightly held in a vertical plane with their ends A, C resting on a smooth horizontal plane. Show that the principle of virtual work that the tension in the cord is $(W+W')\cos A\cos C\sin B$. [CU(H) 2008]

22. Two particles of masses m and m' are connected by a string of length l resting on a smooth cycloid with its vertex upwards and base horizontal. Prove that in the equilibrium position the distance of the particle m from the vertex measured along the arc is $\frac{m'l}{m+m'}$. [CU(H) 1999, VU(H) 2005]

23. Four uniform rods AB, BC, CD, DE each of length a and weight W are smoothly jointed together at their ends B, C and D, and the ends A, E are smoothly jointed to

fixed points at a distance $2a$ apart in the same horizontal line. If AB, BC make angles θ and ϕ respectively with the horizontal when the system hangs in equilibrium, show by the principle of virtual work that $3\cot\theta = \cot\phi$. If B and D are connected by an inextensible string of length a, then show that the tension of the string is $\frac{W}{\sqrt{3}}$.

[CU(H) 2001]

24. A freely jointed framework is formed of five equal uniform rods each of weight W. The framework is suspended from one corner which is also jointed to the middle point of the opposite sides by an inextensible string; if the two upper and two lower rods make angles θ and ϕ respectively with the vertical, prove that the tension of the string to the weight of rod is $(4\tan\theta + 2\tan\phi) : (\tan\theta + \tan\phi)$.

Answers

3. $4W$. 11. If the weights are unequal the equilibrium is possible only when at least one ring must be either at the end points of the major axis. 20. $\frac{3\sqrt{2}}{2}W$.

Chapter 4

Equilibrium of Strings and Chains

4.1 Introduction

In the present chapter we shall consider strings which are perfectly flexible, i.e., which offer no resistance to bending at any point. In such case, the resultant action, across any section of the string, consists of a single force whose line of action is along the tangent to the curve formed by the string. The normal section of the string is taken to be so small that it may be regarded as a curved line. A chain, whose links are short and perfectly smooth, behaves like a flexible string.

4.2 The Catenary [TMBU(H) 2010, RU(H) 1996, BVABU(H) 2006, JPU(H) 2010, MU(H) 2013]

When a uniform string or chain hangs freely between two points, not in the same vertical lines, the curve in which it hangs under the action of gravity, is called a *catenary*. If the weight per unit length of the chain or string is constant, the catenary is called the *uniform* or *common catenary*. The catenary in general is taken to mean the common catenary.

4.3 Intrinsic Equation of Common Catenary

[TMBU(H) 2010, RU(H) 1996, BVABU(H) 2006, MU(H) 2012, JPU(H) 2012]

Let a string hang between two points A and B not lying in the same vertical line. Let C be the lowest point of the catenary and P be any point on it, where $CP = s$. Let T and T_0 be the tensions at P and C respectively. If w be the weight per unit length of the string, then for equilibrium of CP, T, T_0 and ws must meet in a point. Let the tangent at P make angle ψ with horizontal.

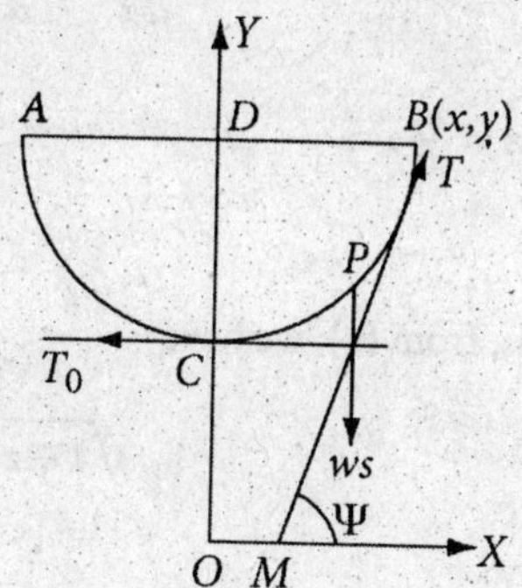

Fig. 4.1

Resolving the forces horizontally and vertically, we get

$$T\cos\psi = T_0 \quad (1)$$

$$T\sin\psi = ws. \quad (2)$$

Dividing (2) by (1), we get

$$\frac{T\sin\psi}{T\cos\psi}=\frac{ws}{T_0};\ \tan\psi=\frac{ws}{T_0}. \quad (3)$$

Let us assume that tension T_0 at C is equal to the weight of the string of length c. $T_0=wc$. So from (3),

$$\tan\psi=\frac{ws}{wc}=\frac{s}{c};$$

$$\therefore\ s=c\tan\psi.$$

This is the required equation of catenary.

4.4 Cartesian Equation of the Common Catenary

[TMBU(H) 2011, RU(H) 2000, BVABU(H) 2009, MU(H) 2013, KU(H) 2004]

The intrinsic equation of common catenary is

$$s=c\tan\psi. \quad (1)$$

Also

$$\tan\psi=\frac{dy}{dx}.$$

From (1), we have

$$s=c\frac{dy}{dx}.$$

Differentiating both sides w.r.t. x, we get

$$\frac{ds}{dx}=c\frac{d^2y}{dx^2}\ \text{or,}\ \sqrt{1+\left(\frac{dy}{dx}\right)^2}=c\frac{d^2y}{dx^2}. \quad (2)$$

Let $z=\frac{dy}{dx}$.

$$\therefore\ \frac{dz}{dx}=\frac{d^2y}{dx^2}.$$

Thus, from (2), we have

$$\sqrt{1+z^2}=\frac{c\,dz}{dx}\ \text{or,}\ \frac{dx}{c}=\frac{dz}{\sqrt{1+z^2}}.$$

Integrating, we get

$$\frac{x}{c}=\sinh^{-1}z+C_1\ \text{or,}\ \frac{x}{c}=\sinh^{-1}\left(\frac{dy}{dx}\right)+C_1. \quad (3)$$

To determine the constant C_1 let us choose the axis of y to be a vertical line through the lowest point C so that for the point C, $x=0$, $\frac{dy}{dx}=0$, hence $C_1=0$.

So from (3), we have

$$\frac{x}{c}=\sinh^{-1}\left(\frac{dy}{dx}\right) \text{ or, } \frac{dy}{dx}=\sinh\frac{x}{c} \text{ or, } dy=\sinh\frac{x}{c}\,dx.$$

Integrating, we get $$y=c\cosh\frac{x}{c}+c_2. \tag{4}$$

At C, $y=c$, $x=0$, hence $c_2=0$.

So from (4), we get $$y=c\cosh\frac{x}{c},$$

which is the Cartesian equation of the common catenary.

[MU(H) 2002, KU(H) 2006, LNMU(H) 2011]

4.5 Definitions

Since $\cosh\frac{x}{c}$ is an even function of x, the curve is symmetrical w.r.t. the y-axis, or the vertical through the lowest point of the catenary. This vertical line of symmetry is called the *axis of the catenary.*

The linear magnitude is called the *parameter of the catenary* and determines its size. Thus, two arcs of catenaries which have their parameters equal are said to be arcs of equal catenaries.

The lowest point C is called the *vertex of the catenary.* The horizontal line at a depth c below the vertex, i.e., the axis of x, is called the *directrix of the catenary.*

CD is called the *sag of the catenary*, AB is called the *span of the catenary.*

Clearly, span $=AB=2BD=2x$, sag $=CD=OD-OC=y-c$.

4.6 Some Important Relations for the Common Catenary

1. *Relation between x and s*

   For a catenary, we have $$s=c\tan\psi=c\frac{dy}{dx};\ \therefore\ \frac{dy}{dx}=\frac{s}{c}. \tag{1}$$

   Also $$y=c\cosh\frac{x}{c}.$$

   Differentiating, we have $$\frac{dy}{dx}=c\sinh\frac{x}{c}\cdot\frac{1}{c}=\sinh\frac{x}{c}. \tag{2}$$

   From (1) and (2), we have $$\frac{s}{c}=\sinh\frac{x}{c};\ s=c\sinh\frac{x}{c},$$

   which is a relation between x and s.

2. *Relation between y and s*

   For a catenary, we have $$y=c\cosh\frac{x}{c}.$$

   Also $$s=c\sinh\frac{x}{c}.$$

Squaring and subtracting, we get

$$y^2 - s^2 = c^2\left[\cosh^2\frac{x}{c} - \sinh^2\frac{x}{c}\right] = c^2$$

$$\Rightarrow y^2 = c^2 + s^2,$$ [TMBU(H) 2007, BVABU(H) 2011, MU(H) 2002]

which is the required relation between y and s.

3. *Relation between y and ψ*

We have $y^2 = c^2 + s^2$ and $s = c\tan\psi$

$$\therefore\ y^2 = c^2 + c^2\tan^2\psi = c^2(1+\tan^2\psi)$$

$$\Rightarrow y^2 = c^2\sec^2\psi$$

$$\Rightarrow y = c\sec\psi.$$ [KU(H) 2001, BVABU(H) 1999]

4. *Relation between x and ψ or, x, y and s*

For any curve we have

$$\frac{dx}{ds} = \cos\psi$$

$$\therefore\ \frac{dx}{d\psi} = \frac{dx}{ds}\frac{ds}{d\psi} = \cos\psi\frac{d}{d\psi}(c\tan\psi) = c\cos\psi\sec^2\psi = c\sec\psi,$$

$$\text{or,}\ dx = c\sec\psi\, d\psi.$$

Integrating, we get

$$x = c\log(\sec\psi + \tan\psi) + c_1.$$

At the lowest point C, $x = 0$, $\psi = 0$, $c_1 = 0$. Hence, c_1

$$\therefore\ x = c\log(\sec\psi + \tan\psi),$$ [RU(H) 2000, BVABU(H) 2011]

which is the required relation between x and ψ.

5. *Relation between tension and ordinate*

From (1) of Theorem 4.3, we have

$$T\cos\psi = T_0 \text{ and } T_0 = wc$$

$$T\cos\psi = wc$$

$$T = wc\sec\psi = wy, \tag{1}$$

which is the relation between T and y. The relation (1) shows that the tension at any point of a catenary varies as the height of the point above the directrix. [TMBU(H) 2006]

Hence, the tension at any point of a catenary is equal to the weight of the portion of string whose length is the vertical distance between the point and the directrix. [MU(H) 2008]

6. *Radius of curvature at any point of a catenary*

For a catenary, we have $s = c\tan\psi$

$$\rho = \frac{ds}{d\psi} = c\sec^2\psi.$$

7. *Relation between x, y and s.*

We have

$$y + s = c\cosh\frac{x}{c} + c\sinh\frac{x}{c} = c\left(\cosh\frac{x}{c} + \sinh\frac{x}{c}\right) \Rightarrow y + s = ce^{x/c}.$$

4.7 Approximations of the Common Catenary

1. The Cartesian equation of common catenary is

$$\begin{aligned} y = c\cosh xc &= \frac{c}{2}\left(e^{\frac{x}{c}} + e^{-\frac{x}{c}}\right) \\ &= \frac{c}{2}\left[\left\{1+\frac{x}{c}+\frac{1}{2!}\left(\frac{x}{c}\right)^2+\ldots\right\}+\left\{1-\frac{x}{c}+\frac{1}{2!}\left(\frac{x}{c}\right)^2\ldots\right\}\right] \\ &= \frac{c}{2}\cdot 2\left[1+\frac{1}{2!}\left(\frac{x}{c}\right)^2+\ldots\right]. \end{aligned} \tag{1}$$

If x/c is small, then neglecting the powers of x/c higher than two, the (1) reduces to

$$y = c\left[1+\frac{1}{2!}\left(\frac{x}{c}\right)^2\right] \text{ or, } y = c+\frac{x^2}{2c} \Rightarrow y-c = \frac{x^2}{2c} \Rightarrow x^2 = 2c(y-c),$$

which is the equation of a parabola of latus rectum $2c$ or $\frac{2T_0}{w}$.

This shows that so long as $\frac{x}{c}$ is small, i.e., x is small and c large, the curve coincides nearly with a parabola of latus rectum $2c$.

Examples of such a case are the electric transmission wires and the telegraphic wires tightly stretched between the poles.

2. We also obtain an approximation to the shape of the catenary when x is large, i.e., at points far removed from the lowest point. When x is large, $e^{-x/c}$ becomes small.

   Hence, $y = \frac{1}{2}ce^{x/c}$.

   Thus, for large values of x, the catenary behaves as an exponential curve.

4.8 Catenary of Uniform Strength

Theorem 1. *The form of a curve in which a chain hangs when its weight (or mass) at each point is proportional to the tension at the point is called the catenary of uniform strength.*

[KU(H) 2007]

Proof. Let (x, y) be the coordinates of a point P whose arcual distance from the lowest point O (origin) is S and w the weight of the element at that point P. Then since the tension at any point is proportional to the weight of the element at that point, we have

$$\text{tension } T = wc,$$

where c is a constant.

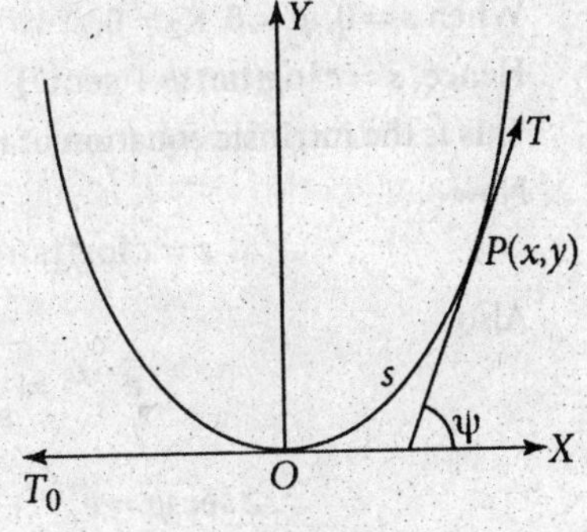

Fig. 4.2

We consider the equilibrium of the portion OP. Resolving horizontally and vertically, we have

$$T\cos\psi = T_0 \text{ and } T\sin\psi = \int_0^S w\,ds = \int_0^S \frac{T}{c}ds \; (\because T = wc).$$

$\therefore$ dividing, we get

$$\tan\psi = \frac{\frac{1}{c}\int_0^S T\,ds}{T_0} \Rightarrow T_0\tan\psi = \frac{1}{c}\int_0^s T_0\sec\psi\,ds\ (\because\ T\cos\psi = T_0)$$

$$\Rightarrow c\tan\psi = \int_0^s \sec\psi\,ds.$$

Differentiating w.r.t. s, we get

$$c\sec^2\psi\frac{d\psi}{ds} = \sec\psi \Rightarrow \frac{ds}{d\psi} = c\sec\psi. \qquad (1)$$

Also
$$\frac{dx}{d\psi} = \frac{dx}{ds}\cdot\frac{ds}{d\psi} = \cos\psi\, c\sec\psi = c.$$

Integrating, we get $x = c\psi + K$.
But when $x = 0, \psi = 0, K = 0$.
Hence,

$$x = c\psi \text{ or, } \psi = \frac{x}{c}. \qquad (2)$$

Again,
$$\frac{dy}{d\psi} = \frac{dy}{ds}\cdot\frac{ds}{d\psi} = c\sin\psi$$

Integrating, we get $y = c\log\sec\psi + K_1$ when $y = 0, \psi = 0$. $\therefore K_1 = 0$.

$$\therefore\ y = c\log\sec\psi \Rightarrow y = c\log\sec\frac{x}{c}\ \text{[from (2)]}$$

This is the Cartesian equation of the catenary of uniform strength.
Again, from (1),

$$ds = c\sec\psi\,d\psi. \qquad \text{[KU(H) 2005]}$$

Integrating, we get

$$s = c\log(\sec\psi + \tan\psi) + K_2.$$

When $s = 0, \psi = 0, K_2 = 0$.
Hence, $s = c\log(\tan\psi + \sec\psi)$.
This is the intrinsic equation of the catenary of uniform strength.
Now

$$s = c\log(\tan\psi + \sec\psi) \Rightarrow \sec\psi + \tan\psi = e^{s/c}.$$

Also

$$e^{-s/c} = \frac{1}{\sec\psi + \tan\psi} = \sec\psi - \tan\psi.$$

$$\therefore\ 2\sec\psi = e^{s/c} + e^{-s/c} \Rightarrow \sec\psi = \frac{e^{s/c} + e^{-s/c}}{2} = \cosh\frac{s}{c}.$$

So from (1),

$$\rho = \frac{ds}{d\psi} = c\sec\psi = c\cosh\frac{s}{c}$$

= radius of curvature of the catenary of uniform
= strength at a point P, where arc $OP = s$.

Also

$$T = wc \Rightarrow w = \frac{T}{c} = \frac{T_0 \sec\psi}{c} = \frac{T_0}{c}\cosh\frac{s}{c} \qquad \text{[KU(H) 2003]}$$

$\therefore$ $w =$ weight per unit length $\infty \cosh\frac{s}{a}$.

Theorem 2. *A string hangs under gravity and it is loaded so that the weight of each element of it is proportional to the horizontal projection of that element. Show that it will hang in the form of a parabola.* [CU(H) 1993, KU(H) 2008]

Proof. Let the lowest point O of the chain be taken as origin.

Let (x,y) be the coordinates of a point P whose arcual distance from the lowest point O is s and let T be the tension.

Let T_0 be the tension at the lowest point O. The weight of the portion of the chain from origin to the point (x,y) is wx.

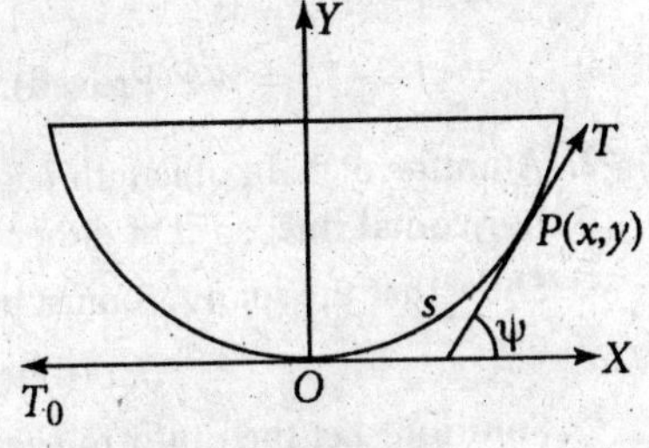

Fig. 4.3

Resolving horizontally and vertically, for the equilibrium of the portion OP of the chain $T\cos\psi = T_0 = wc$ (say) and $T\sin\psi = wx$.

Dividing, we get

$$\tan\psi = \frac{x}{c} \Rightarrow \frac{dy}{dx} = \frac{x}{c} \Rightarrow dy = \frac{x}{c}dx.$$

Integrating, we get

$$y = \frac{x^2}{2c} + K.$$

Since at the origin $O, x=0, y=0, K=0$, therefore,

$$y = \frac{1}{2}\cdot\frac{x^2}{c} \Rightarrow x^2 = 2cy,$$

which is a parabola of latus rectum $2c$.

WORKED-OUT EXAMPLES

1. If T is the tension at any point P of a catenary and T_0 that at the lowest point C, prove that $T^2 - T_o^2 = W^2$, W being the weight of the arc CP of the catenary.

[IAS 2000; TMBU(H) 2005, 10]

Solution: Let the tangent at P make an angle ψ with the x-axis and T be the tension of the catenary at P. Let s be the length of the arc CP whose weight is W. The arc CP is in equilibrium under the action of the three forces.

(i) Tension T at P.

(ii) Tension T_0 at C.

(iii) Weight W of the arc CP acting vertically downwards.

Resolving the forces horizontally and vertically, we have

$$T\cos\psi = T_0 \qquad (1)$$

$$\text{and } T\sin\psi = W. \qquad (2)$$

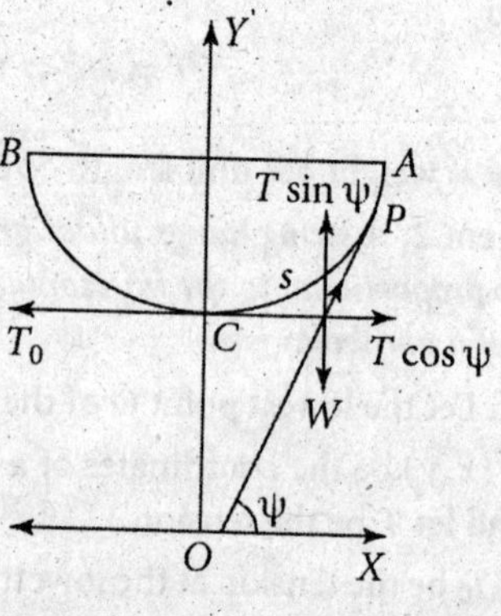

Fig. 4.4

Squaring and adding (1) and (2), we get

$$T^2\left(\cos^2\psi + \sin^2\psi\right) = T_0^2 + W^2$$

$$\Rightarrow T^2 = T_0^2 + W^2$$

$$\Rightarrow T^2 - T_0^2 = W^2 \text{ (Proved).}$$

2. A uniform chain of length l is to be suspended from two point A and B in the same horizontal line, so that either terminal tension is n times than at the lowest point. Show that the span AB must be $\frac{l}{\sqrt{n^2-1}}\log\left(n+\sqrt{n^2-1}\right)$.

[MU(H) 2013, TMBU(H) 2009, BVABU(H) 2010, KU(H) 2007, JPU(H) 2010]

Solution: Let the uniform chain ACB of length l be suspended from two points A, and B in the same horizontal line.

Let (x, y) be the coordinates of the point A and ψ be the angle which the tangent at A makes with the x-axis. Let T and T_0 be the tensions of the string at A and C, then given

$$T = nT_0$$

$$wy = nwc \quad [T_0 = wc]$$

$$y = nc$$

$$c\sec\psi = nc \quad [\because y = c\sec\psi]$$

$$n = \sec\psi.$$

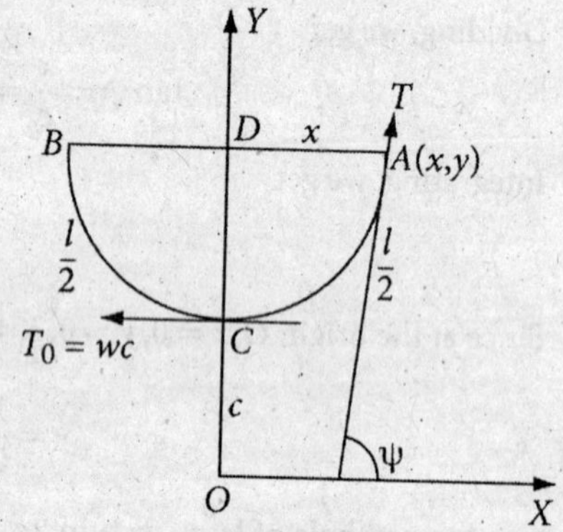

Fig. 4.5

Now at the point A, $s = \text{arc } CA = \frac{l}{2}$.

$\therefore$ using formula, $s = c\tan\psi$, for A, we get

$$\frac{l}{2} = c\sqrt{\sec^2\psi - 1} \text{ or, } \frac{l}{2} = c\sqrt{n^2-1}.$$

$$\therefore c = \frac{l}{2\sqrt{n^2-1}}. \qquad (1)$$

Solution: Let w be the weight per unit length of the string. Let the tangent at A be inclined at an angle ψ to the x-axis.

Since the force acting at B is horizontal and the string is hanging freely from A,

$\therefore$ B is the lowest point of the catenary of which BA is the part.

$$\therefore OB = c \text{ and } T_0 = wc.$$

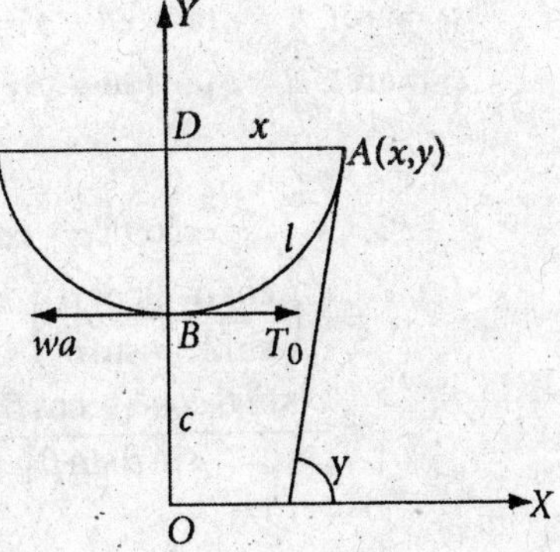

Fig. 4.8

At B we have $T_0 = wa$.

$$\therefore wa = wc.$$

$$\Rightarrow c = a. \qquad (1)$$

Also using the formula, $s = c\tan\psi$ for A, we have $l = a\tan\psi$.

From (1) and $s = l$ for A.

$$\therefore \frac{l}{a} = \tan\psi.$$

So

$$\sec\psi = \sqrt{1+\tan^2\psi} = \sqrt{1+\frac{l^2}{a^2}} = \frac{\sqrt{l^2+a^2}}{a}.$$

Let coordinates of A be (x, y). Then using the formula

$$s = c\sinh\frac{x}{c} \text{ for } A, \text{ we get}$$

$$l = a\sinh\frac{x}{a}, \text{ i.e., } \frac{l}{a} = \sinh\frac{x}{a}, \text{ i.e., } \frac{x}{a} = \sinh^{-1}\frac{l}{a}, \; x = a\sinh^{-1}\frac{l}{a}.$$

This is the required horizontal distance between A and B.

Again, using the formula, $y^2 = c^2 + s^2$ for $A(x, y)$, we obtain

$$y^2 = a^2 + l^2 \text{ or, } y = \sqrt{a^2+l^2}.$$

Hence, the required vertical distance between A and B

$$= BD = y - c = \sqrt{a^2+l^2} - a.$$

6. A string of length l hangs between two points (not in the same vertical) and makes angles α, β with the vertical at the points of support. Prove that the height one point above the other is $\frac{\cos\frac{\alpha+\beta}{2}}{\cos\frac{\alpha-\beta}{2}}$, the two extremities being on one side of the vertex.

Solution: Let AB be the given portion of length l of the catenary and C the lower point of the catenary. Let arc $CA = s_1$, arc $CB = s_2$.

Let coordinates of A and B be (x_1, y_1) and (x_2, y_2).

From A and B we draw AL and BM perpendicular to OX.

Applying the formula, $s = c\tan\psi$ for A and B, we get

$$s_1 = \text{arc}\,CA = c\tan(90^\circ - \alpha) = c\cot\alpha$$
$$s_2 = \text{arc}\,CB = c\tan(90^\circ - \beta) = c\cot\beta.$$

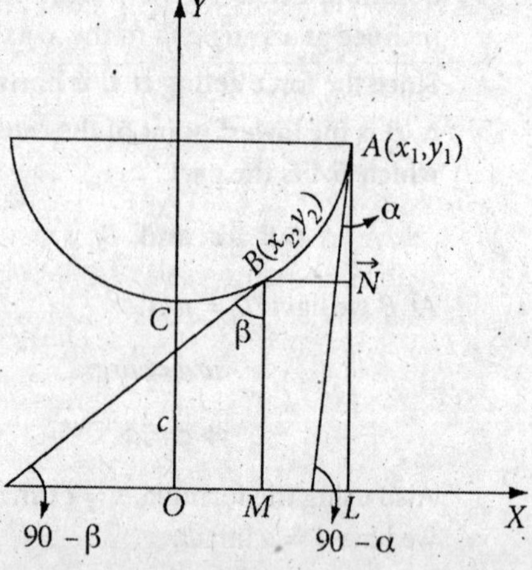

Fig. 4.9

$$\therefore\ l = s_1 - s_2 = c(\cot\alpha - \cot\beta)$$
$$= c\left(\frac{\cos\alpha}{\sin\alpha} - \frac{\cos\beta}{\sin\beta}\right)$$
$$= \frac{c(\sin\beta\cos\alpha - \cos\beta\sin\alpha)}{\sin\alpha\sin\beta}.$$

$$\Rightarrow l = \frac{c\sin(\beta-\alpha)}{\sin\alpha\sin\beta}. \quad (1)$$

Applying the formula $y = c\sec\psi$ for A and B, we get

$$y_1 = c\sec(90^\circ - \alpha) = c\,\text{cosec}\,\alpha$$
$$y_2 = c\sec(90^\circ - \beta) = c\,\text{cosec}\,\beta.$$

Now, height of one extremity above the other

$$BM = LN = AL - AN = y_1 - y_2 = c(\text{cosec}\,\alpha - \text{cosec}\,\beta)$$
$$= c\left(\frac{1}{\sin\alpha} - \frac{1}{\sin\beta}\right) = c\,\frac{(\sin\beta - \sin\alpha)}{\frac{c\sin(\beta-\alpha)}{l}} \quad \text{[from (1)]}$$
$$= \frac{l\,2\cos\frac{\alpha+\beta}{2}\sin\frac{\beta-\alpha}{2}}{2\sin\frac{\beta-\alpha}{2}\cos\frac{\beta-\alpha}{2}} = \frac{l\cos\frac{\alpha-\beta}{2}}{\cos\frac{\alpha+\beta}{2}}$$

Dividing (1) by (2), we get

$$\frac{l}{K} = \frac{c\sin(\beta-\alpha)}{c(\sin\beta - \sin\alpha)} = \frac{2\sin\frac{\beta-\alpha}{2}\cos\frac{\beta-\alpha}{2}}{2\cos\frac{\beta+\alpha}{2}\sin\frac{\beta-\alpha}{2}} \text{ or, } \frac{l}{K} = \frac{\cos\frac{\alpha-\beta}{2}}{\cos\frac{\alpha+\beta}{2}}.$$

where $k = AN$.

Hence,

$$K\cos\frac{\alpha-\beta}{2} = l\cos\frac{\alpha+\beta}{2}.$$

7. Prove that the length of a heavy endless chain which will hang over a circular pulley of radius a so as to be in contact with two-third of circumference is

$$a\left\{\frac{4\pi}{3} + \frac{3}{\log\left(2+\sqrt{3}\right)}\right\}.$$

[IAS 2009, TMBU(H) 2008, BVABU(H) 2000, MU(H) 2010, VU(H) 2006, LNMU(H) 2011]

Solution: Let $ALBEA$ be the circular pulley of a radius a and $ALBCA$ be the endless chain hanging over the pulley. The portion ALB of the chain is in contact with the pulley and its remaining portion ACB is hanging in the form of a catenary.

Given, the portion ALB of the chain

$$= \frac{2}{3} \times \text{circumference of the pulley}$$

$$= \frac{2}{3} \times 2\pi a = \frac{4}{3}\pi a.$$

Now the portion ABC is one-third and hence, the angle subtended by it at the centre is $= \frac{1}{3}(2\pi) = 120°$.

$$\therefore \angle AO_1D = \tfrac{1}{2}\angle AO_1B = 60°$$
$$= \angle DAF = \angle AFX.$$

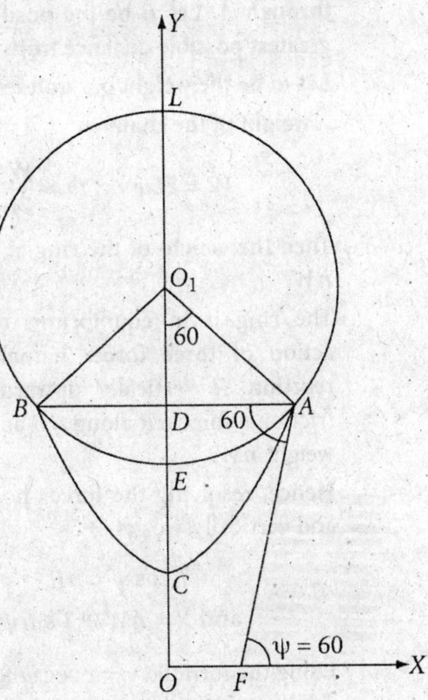

Fig. 4.10

In ΔO_1AD,

$$\sin 60° = \frac{AD}{O_1A} = \frac{AD}{a}.;\ AD = \frac{\sqrt{3}}{2}a.$$

Applying the formula $x = c\log(\sec\psi + \tan\psi)$ for A, we get

$$\frac{\sqrt{3}}{2}a = c\log(\sec 60° + \tan 60°)$$

$$\frac{\sqrt{3}}{2}a = c\log\left(2+\sqrt{3}\right)$$

$$c = \frac{\sqrt{3}a}{2\log\left(2+\sqrt{3}\right)}.$$

Applying the formula, $s = c\tan\psi$ for A, we get

$$\text{arc}\,CA = c\tan 60° = c\sqrt{3}.$$

$$\text{The arc}\,ACB = 2\times \text{arc}\,CA = 2c\sqrt{3} = \frac{2\sqrt{3}\sqrt{3}a}{2\log\left(2+\sqrt{3}\right)} = \frac{3a}{\log\left(2+\sqrt{3}\right)}.$$

Hence, the length of the chain

$$= \text{the portion } ALB + \text{the arc} ABC$$

$$= \frac{4}{3}\pi a + \frac{3a}{\log\left(2+\sqrt{3}\right)} = a\left\{\frac{4\pi}{3} + \frac{3}{\log\left(2+\sqrt{3}\right)}\right\} \text{ (Proved).}$$

8. A heavy chain of length $2l$, has one end tied at A and the other is attached to a small heavy ring which can slide on a rough horizontal rod which passes through A. If the weight of the ring be n times the weight of the chain, show that its greatest possible distance from A is $\frac{2l}{\lambda}\log\left[\lambda + \sqrt{1+\lambda^2}\right]$, where $1/\lambda = \mu(2n+1)$ and μ is the coefficient of friction. **[IAS 1989; TMBU(H) 1996; RU(H) 1992, 2000]**

Solution: Let one end of a heavy chain of length $2l$ be fixed at A and the other end be attached to a small heavy ring which can slide on a rough horizontal rod ADB

through A. Let B be the position of limiting equilibrium of the ring when it is at greatest possible distance from A.

Let w be the weight per unit length of the chain ACB of length $2l$.

$\therefore$ weight of the chain

$$W = 2lw \text{ or, } w = \frac{W}{2l}. \quad (1)$$

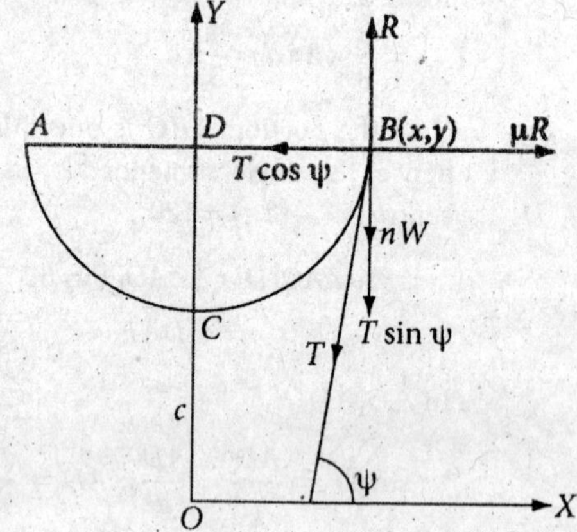

Fig. 4.11

Then the weight of the ring at $B(x, y)$ is nW.

The ring is in equilibrium under the action of three forces namely (i) the reaction R vertically upward (ii) the frictional force μR along AB and (iii) the weight nW.

Hence, resolving the forces horizontally and vertically, we get

$$T\cos\psi = \mu R$$
$$\text{and } R = nW + T\sin\psi.$$

Using the formula $y = c\sec\psi$ for B, we get $T = wy = wc\sec\psi$.
Also using the formula $s = c\tan\psi$ for B, we get

$$l = c\tan\psi. \quad (2)$$

$$\therefore T\cos\psi = \mu R = \mu(nW + T\sin\psi) \Rightarrow T\cos\psi = n\mu W + T\mu\sin\psi$$
$$\Rightarrow \frac{W}{2l}c\sec\psi\cos\psi = n\mu W + \frac{W}{2l}c\sec\psi\mu\sin\psi \Rightarrow \frac{W}{2l}c = n\mu W + \frac{W}{2l}c\mu\tan\psi$$
$$\Rightarrow c = 2l\mu n + c\mu\tan\psi \Rightarrow c = 2\mu nc\tan\psi + c\mu\tan\psi \text{ [from (2)]}$$
$$\Rightarrow 1 = \mu\tan\psi(2n+1).$$

But by question,

$$\frac{1}{\lambda} = \mu(2n+1) \text{ or, } 1 = \frac{\tan\psi}{\lambda} \text{ or, } \tan\psi = \lambda.$$

$$\therefore \sec\psi = \sqrt{1+\tan^2\psi} = \sqrt{1+\lambda^2}.$$

Hence, the required greatest possible distance of the ring from $A = AB$

$$= 2x = 2c\log(\sec\psi + \tan\psi) = 2c\log\left(\lambda + \sqrt{\tan^2\psi + 1}\right)$$
$$= 2c\log\left(\lambda + \sqrt{1+\lambda^2}\right) = \frac{2l}{\tan\psi}\log\left(\lambda + \sqrt{1+\lambda^2}\right) \text{ [from (2)]}$$
$$= \frac{2l}{\lambda}\log\left(\lambda + \sqrt{1+\lambda^2}\right) \text{ (Proved).}$$

9. A telegraph wire, stretched between two poles at a distance a metre apart, sags n metre in the middle. Prove that the tension at the end is approximately $w\left(\frac{a^2}{8n}+\frac{7}{6}n\right)$, where w is the weight per unit length. [RU(H) 1998]

Solution: Let ACB be the telegraph wire stretched between two poles at A and B. Let coordinates of A be (x,y). Then $x = a/2$.

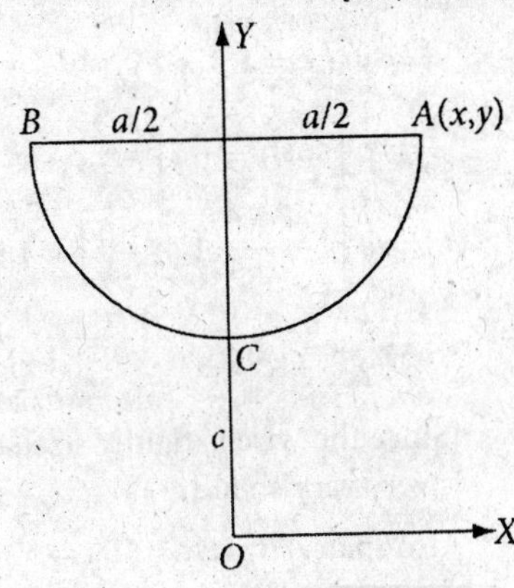

Fig. 4.12

Using $y = c\cosh\frac{x}{c}$ for A,

$$y = c\cosh\frac{a}{2c}$$
$$= c\left[1+\frac{1}{2!}\left(\frac{a}{2c}\right)^2+\frac{1}{4!}\left(\frac{a}{2c}\right)^4+\ldots\right]$$
$$= c\left[1+\frac{1}{8c^2}+\frac{a^4}{24\cdot 16c^4}+\ldots\right]$$
$$y = c+\frac{a^2}{8c}+\frac{a^4}{24\cdot 16c^3}+\ldots.$$

Also for A, $y = c+n$,

$$c+n = c+\frac{a^2}{8c}+\frac{a^4}{24\cdot 16c^3}+\ldots$$
$$\text{or, } n = \frac{a^2}{8c}+\frac{a^4}{24\cdot 16c^3}+\ldots. \tag{1}$$

Since the wire is tightly stretched, therefore, c is very large.

$$n = \frac{a^2}{8c} \text{ to the first approximation.}$$
$$\therefore\ c = \frac{a^2}{8n}.$$

From (1),

$$n = \frac{a^2}{8c}+a^4 24\cdot 16\frac{8\cdot 8\cdot 8n^3}{a^6} = \frac{a^2}{8c}+\frac{4n^3}{3a^2} \Rightarrow \frac{a^2}{8c} = n-\frac{4n^3}{3a^2} = n\left(1-\frac{4n^2}{3a^2}\right)$$
$$\Rightarrow c = \frac{a^2}{8n}\left(1-\frac{4n^2}{3a^2}\right)^{-1} = \frac{a^2}{8n}\left(1+\frac{4n^2}{3a^2}\right) \text{ [nearly]} = \frac{a^2}{8n}+\frac{n}{6}.$$

Hence, at A,

$$T = wy = w(n+c) = w\left(\frac{a^2}{8n}+\frac{n}{6}+n\right) = w\left(\frac{a^2}{8n}+\frac{7n}{6}\right).$$

10. A telephonic wire weighing 0.04 lb per foot has a horizontal span of 150 feet and sag of 1.5 feet. Find the length of the wire and also find maximum tension. [IAS 2000]

Solution: Here $W = 0.04$ lb per foot to and span $AB = 150$ feet.
The sag $SD = 1.5$ feet.

$$\Rightarrow y - c = 1.5 \Rightarrow c\cosh\frac{x}{c} - c = 1.5$$

$$\Rightarrow c\left[\cosh\frac{x}{c} - 1\right] = 1.5$$

$$\Rightarrow c\left[1 + \frac{x^2}{2c^2} + \cdots - 1\right] = 1.5$$

$$\Rightarrow \frac{x^2}{2c} = 1.5. \qquad (1)$$

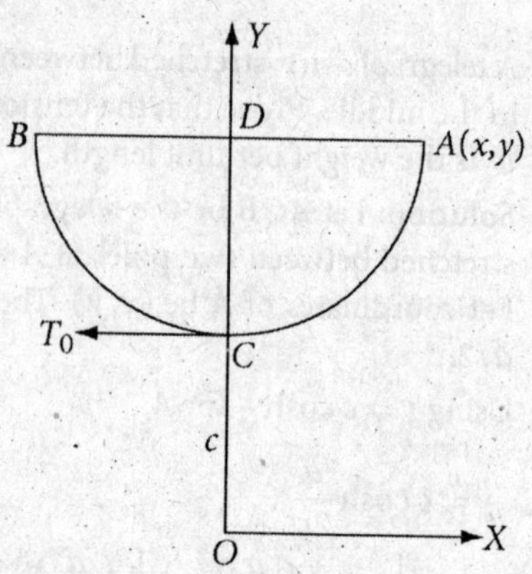

Fig. 4.13

[Since the wire is tightly stretched and $\frac{x}{c}$ is very very small].

Also span $AB = 2x = 150 \Rightarrow x = 75$.

Hence, from (1),

$$\frac{75 \times 75}{2c} = \frac{3}{2} \Rightarrow c = 75 \times 25 = 1875.$$

For arcAC, $s = \frac{l}{2}$,

$$\text{sag } CD = 1.5 \Rightarrow y - c = 1.5 \Rightarrow y - 1875 = 1.5 \Rightarrow y = 1975.5.$$

Now

$$y^2 = c^2 + s^2 \Rightarrow (1975.5)^2 - (1875)^2 = s^2 \Rightarrow 3850.5 \times 100.5 = s^2 \Rightarrow s = 386.98.$$

$\therefore$ arc$AB = l = 25 = 773.96$ feet.

The maximum tension in the wire be at the extremities A or B.

$\therefore\ T = wy = 0.04 \times 1975.5 = 79.02$ lb-wt.

11. A uniform chain of the length l when tightly stretched between two points at the same level has sag k in the middle. Prove that the length of the chain exceeds the distance between the points by $\frac{8k^2}{3l}$ approximately. [TMBU(H) 1991, BVABU(H) 1998]

Solution: Let l be the length of the chain, $2a$ the distance between two points A and B; and k the sag. Then we are to prove that

$$l - 2a = \frac{8k^2}{3l} \text{ (nearly)}.$$

For the point A, using the formula $s = c\sinh\frac{x}{c}$,

$$s = \text{arc } AC = \frac{l}{2} = c\sinh\frac{x}{c} = c\left[\frac{x}{c} + \frac{1}{3!}\left(\frac{x}{c}\right)^3 + \ldots\right] \Rightarrow \frac{l}{2} = x + \frac{x^3}{6c^2},$$

for c is large as the chain is tightly stretched

$$\Rightarrow l - 2x = \frac{x^3}{3c^2}.$$

To first approximation, $x = \frac{l}{2}$.

But the sag at the middle is

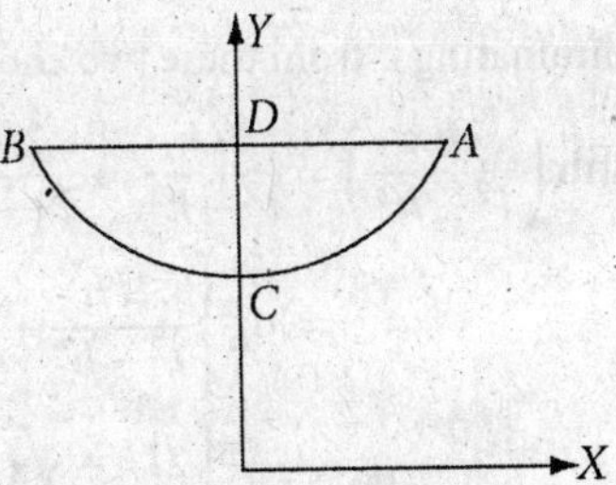

Fig. 4.14

$$k = DC = y - c = c\left(\cosh\frac{x}{c} - 1\right)$$
$$= c\left[1 + \frac{1}{2!}\left(\frac{x}{c}\right)^2 + \frac{1}{4!}\left(\frac{x}{c}\right)^4 + \cdots - 1\right]$$

$\therefore k = \frac{x^2}{2c}$ (nearly) or, $c = \frac{x^2}{2k}$ (nearly).

So from (1),

$$l - 2x = \frac{x^3}{3 \times \frac{x^4}{4k^2}} = \frac{4k^2}{3x} = \frac{4k^2}{3\left(\frac{l}{2}\right)} = \frac{8k^2}{3l} \text{ nearly (Proved).}$$

12. Prove that for a common catenary $x = c\log\left(\frac{y+s}{c}\right)$. [BVABU(H) 1999]

Solution: The parametric equations of a catenary are

$$x = c\log(\sec\psi + \tan\psi) \quad (1)$$
$$\text{and } y = c\sec\psi. \quad (2)$$

Also for a common catenary,

$$s = c\tan\psi. \quad (3)$$

From (2) and (3), we have

$$\sec\psi = \frac{y}{c} \text{ and } \tan\psi = \frac{s}{c}.$$

So (1) becomes $x = c\log\left(\frac{y+s}{c}\right)$ (Proved).

13. A uniform chain of length $2l$ has its extremities fastened to two fixed points at the same level and the sag in the middle. Prove that that span is $\frac{l^2-h^2}{h}\log\frac{l+h}{l-h}$.

[BVABU(H) 2011]

Solution: Let ACB be a uniform chain of length $2l$ whose extremities A and B are at the same level.

Here, span $AB = 2x$.

Now, from the formulae, $S = c\sinh\frac{x}{c}$ and $y^2 = c^2 + s^2$, we have

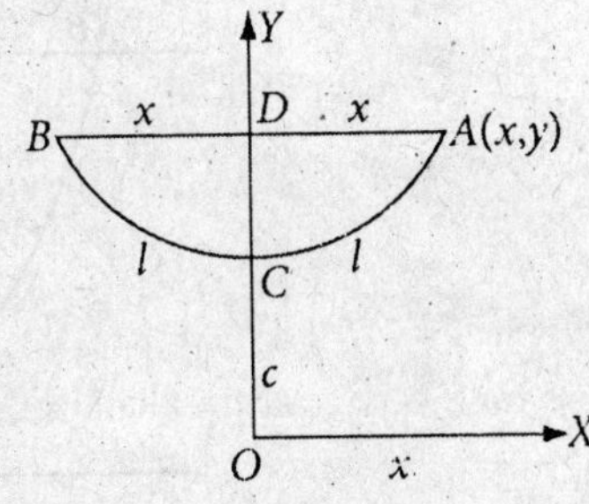

Fig. 4.15

$$l = c\sinh\frac{x}{c} \text{ and } (c+h)^2 = c^2 + l^2$$
$$\Rightarrow \sinh\frac{x}{c} = \frac{l}{c} \text{ and } h^2 + 2ch = l^2$$
$$\Rightarrow \sinh\frac{x}{c} = \frac{l}{c} \text{ and } c = \frac{l^2 - h^2}{2h}.$$

Eliminating c from these two equations, we have

$$\sinh\left(\frac{2hx}{l^2-h^2}\right)=\frac{2lh}{l^2-h^2}\Rightarrow\frac{2hx}{l^2-h^2}=\sinh^{-1}\left(\frac{2lh}{l^2-h^2}\right)$$

$$=\log\left[\frac{2lh}{l^2-h^2}+\sqrt{\frac{4l^2h^2}{(l^2-h^2)^2}+1}\right]\quad\left[\because\ \sinh^{-1}x=\log\left(x+\sqrt{x^2+1}\right)\right]$$

$$=\log\left[\frac{2lh+\sqrt{(l^2-h^2)^2+4l^2h^2}}{l^2-h^2}\right]=\log\left[\frac{2lh+l^2+h^2}{l^2-h^2}\right]$$

$$=\log\left[\frac{(l+h)^2}{(l+h)(l-h)}\right]\Rightarrow\frac{2hx}{l^2-h^2}=\log\frac{l+h}{l-h}$$

$$\Rightarrow 2x=\frac{l^2-h^2}{h}\log\frac{l+h}{l-h}=\text{span (Proved).}$$

14. The extremities of a heavy string of length $2l$ and weight $2lW$ are attached to two small rings which can slide on a fixed horizontal wire. Each of these rings is acted on by a horizontal force equal to lW. Show that the distance apart of the rings is $2l\log\left(1+\sqrt{2}\right)$. **[KU(H) 2004]**

Solution: A and B are the rings. Let R be the normal reaction at each of the rings. C is the lowest-point of the string.

$\therefore$ arc $CA-l$.

Resolving the forces acting on the whole system vertically, we get

$$R+R=2lW\Rightarrow R=lW. \tag{1}$$

The ring at A is in equilibrium under the action of three forces, viz., tension T_A of the string, normal reaction R and the horizontal force lW.

Hence, tension T_A must balance the resultant of R and lW.

Therefore, if ϕ be the angle which the tangent at A makes with the horizontal, then

$$\tan\phi=\frac{R}{lW}=\frac{lW}{lW}\ \text{[from (1)]}$$

$$\Rightarrow\tan\phi=1\Rightarrow\phi=\frac{\pi}{4}. \tag{2}$$

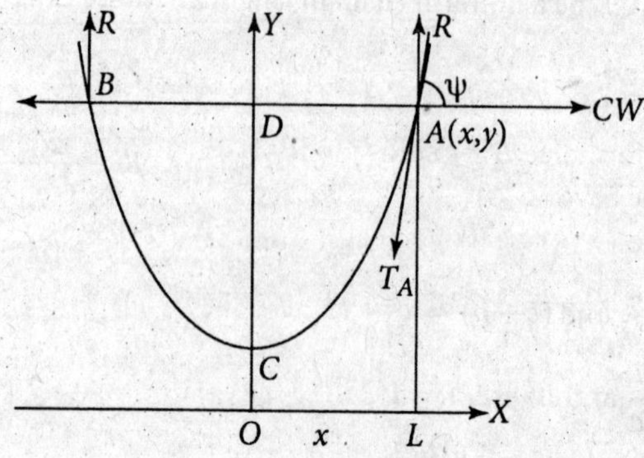

Fig. 4.16

Also for A, from the formula $s = c\tan\phi$,

$$l = c\tan\frac{\pi}{4} \Rightarrow l = c. \tag{3}$$

Let coordinates of A be (x, y), then the distance between the rings

$$= AB = 2x = 2c\log(\sec\phi + \tan\phi)$$
$$= 2l\log\left(\sec\frac{\pi}{4} + \tan\frac{\pi}{4}\right) = 2l\log\left(\sqrt{2}+1\right) \textbf{ (Proved).}$$

15. A heavy string of uniform density and thickness is suspended from two fixed points in the same horizontal line. A weight equal to $\frac{1}{n}$th of the weight of the string is attached to its lowest point. If θ and ϕ be the inclination to the vertical of the tangents at height and lower points of the string, prove that $\tan\phi = (1+n)\tan\theta$.

Solution: Let W be the weight and $2l$ be the length of the string suspended from two points A and B in the same horizontal line. Then the weight attached at the middle point C of the string is $\frac{W}{r}$.

The weight per unit length of the string is given by

$$w = \frac{W}{2l}. \tag{1}$$

If ϕ_C and ϕ_A are the angles of inclination of the tangents at C and A to the horizontal respectively, then

$$\psi_C = \frac{\pi}{2} - \psi \text{ and } \phi_A = \frac{\pi}{2} - \theta.$$

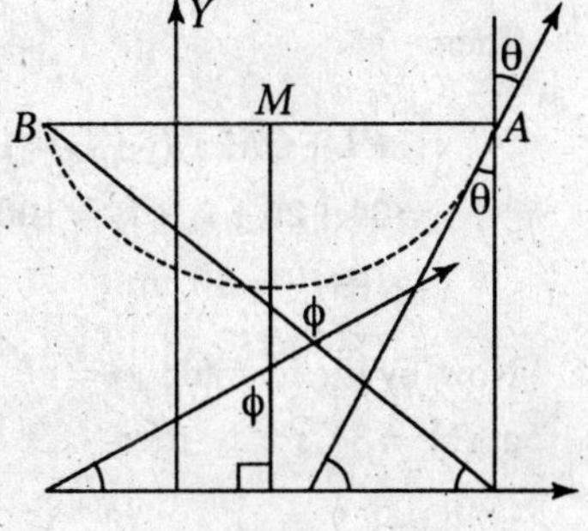

Fig. 4.17

For the equilibrium of the point C, resolving the forces acting on it vertically, we have

$$2T_C\sin\psi_C = \frac{W}{n}, \tag{2}$$

where T_C is the tension at C in each of the string AC and BC.

Also, we have $T_C\cos\psi_C = Wc$, where c is the parameter of the catenary of which AC is an arc.

$$\therefore\ T_C\cos\psi_C = \frac{W}{2lC} \text{ from (1).} \tag{3}$$

Dividing (2) by (3), we have

$$2\tan\psi = \frac{2l}{nc} \Rightarrow \tan(90° - \psi) = \frac{l}{nc} \Rightarrow \cot\psi = \frac{l}{nc} \Rightarrow c = \frac{l}{n}\tan\psi. \tag{4}$$

Now, from $s = c\tan\psi$, we have

$$\text{arc } DC = c\tan\psi_C \text{ and arc } DA = c\tan\psi_A$$

where D is the vertex of the catenary of which AC is an arc.

Subtracting, we have

$$\text{arc}\,DA - \text{arc}\,DC = c(\tan\psi_A - \tan\psi_C)$$

$$\Rightarrow \text{arc}\,CA = c\left[\tan\left(\frac{\pi}{2} - \theta\right) - \tan\left(\frac{\pi}{2} - \phi\right)\right]$$

$$\Rightarrow l = \frac{l}{n}\tan\phi[\cot\theta - \cot\phi] \text{ [from (4)]}$$

$$\Rightarrow n = \tan\phi\cot\phi - 1 \Rightarrow 1 + n = \frac{\tan\phi}{\tan\theta}$$

$$\Rightarrow \tan\phi = (1+n)\tan\theta. \tag{4.1}$$

16. A uniform string of length one metre hangs over two smooth pegs P and Q at different heights. The parts which hang vertically are of lengths 34 cm and 26 cm. Find the ratio in which the vertex of the catenary divides the whole string. [IAS 2007]

Solution: Let ACB be a whole string of length one metre 100 cm hangs over two smooth pegs at P and Q such that the height of A and B are respectively 34 cm and 26 cm.

Let the lengths of the string PC and CQ be l_1 and l_2 cm respectively.

Then

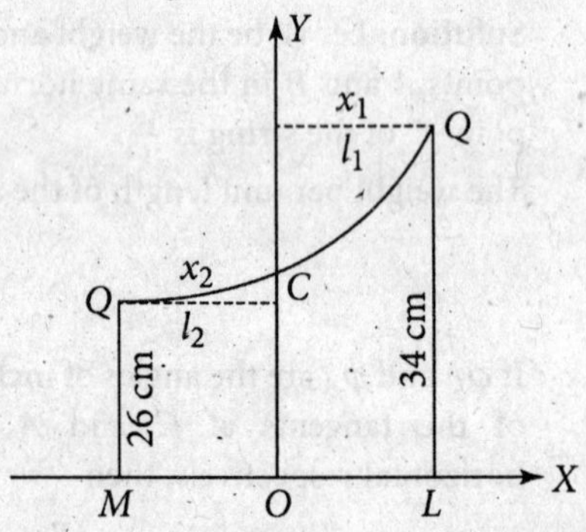

Fig. 4.18

$$PL + QM + l_1 + l_2 = 100 \text{ cm}$$

$$\Rightarrow 34 + 26 + l_1 + l_2 = 100 \text{ cm}$$

$$\Rightarrow l_1 + l_2 = 40 \text{ cm}. \tag{1}$$

Now, by the formula $y^2 = c^2 + s^2$, we have

at $P, y = 34, s = l_1$, then

$$(34)^2 = c^2 + l_1^2 \tag{2}$$

at $Q, y = 26, s = l_2$, then

$$(26)^2 = c^2 + l_2^2. \tag{3}$$

Subtracting (3) from (2), we get

$$(34)^2 - (26)^2 = l_1^2 - l_2^2 \Rightarrow (34+26)(34-26) = (l_1 - l_2)(l_1 + l_2)$$

$$\Rightarrow 60 \times 8(l_1 - l_2)40 \text{ [from (1)]}$$

$$\Rightarrow l_2 = l_1 - l_2 \Rightarrow l_1 - l_2 = l_2. \tag{4}$$

Solving (1) and (4), we get $l_1 = 26, l_2 = 14$.

$\therefore$ the vertex C divides the whole string in the ratio

$$= \frac{\text{arc}\,CP + PL}{\text{arc}\,CQ + QM} = \frac{l_1 + PL}{l_2 + QM} = \frac{26+34}{14+26} = \frac{60}{40} = \frac{2}{3}.$$

Hence, the required ratio 2 : 3.

EXERCISE 4A

1. A rope of length $2l$ feet is suspended between two points at the same level and the lowest point of the rope is b feet below the point of suspension. Show that the horizontal component of the tension is
$$\frac{w}{2b}\left(l^2-b^2\right),$$
w being the weight of the rope per foot of its length. [TMBU(H) 1976]

2. A heavy uniform string 90 inches long hangs over two smooth pegs at different heights. The parts which hang vertically are of lengths 30 inches and 33 inches. Prove that the vertex of the catenary divides the whole string in the ratio 4 : 5 and find the distance between the pegs.

3. The end links of a uniform chain slide along a fixed rough horizontal rod. Prove that the ratio of the maximum span to the length of the chain is
$$\mu\log\frac{1+\sqrt{1+\mu^2}}{\mu},$$
where μ is coefficient of friction.

4. If the ends of a uniform inextensible string of length l hanging freely under gravity slide on a fixed rough horizontal rod whose coefficient of friction is μ, show that at the most they can rest at a distance
$$\mu l\log\frac{1+\sqrt{1+\mu^2}}{\mu}.$$

5. A kite is flown with 600 ft of string from the hand to the kite, and a spring-balance held in hand shows a pull equal to the weight of 100 ft of the string inclined at 30° to the horizon. Find the vertical height of the kite above the hand. [IAS 1980]

6. A box-kite is flying at a height h with a length l of the wire paid out and with the vertex of the catenary on the ground. Show that at the kite the inclination of the wire to the ground is $2\tan^{-1}(\frac{h}{l})$ and that its tensions there and at the ground are
$$w\frac{(l^2+h^2)}{2h} \text{ and } w\frac{(l^2-h^2)}{2h},$$
where w is the weight of the wire per unit of length.

7. A weight W is suspended from a fixed point by a uniform string of length l and weight w per unit length. It is drawn aside by a horizontal force P. Show that in the position of equilibrium the distance of W from the vertical through the fixed point is
$$\frac{P}{w}\left\{\sinh^{-1}\left(\frac{W+lw}{P}\right)-\sinh^{-1}\left(\frac{W}{P}\right)\right\}.$$

8. A uniform string of weight W is suspended from two points at the same level and weight W' is attached to its lowest point. If α and β are now the inclinations to the horizontal of the tangents at the highest and the lowest points, prove that
$$\frac{\tan\alpha}{\tan\beta}=1+\frac{W}{W'}.$$

9. A uniform chain of length $2l$ and weight W is suspended from two points, A and B, in the same horizontal line. A load P is now suspended from the middle point D of the chain and the depth of the point below AB is found to be h. Show that each terminal tension

$$=\frac{1}{2}\left[\frac{Pl}{h}+W\frac{h^2+l^2}{2lh}\right].$$ [RU(H) 1994, BVABU(H) 2009]

10. A uniform chain of length l and weight W hangs between two fixed points at the same level and a weight W' is attached to the middle point. If K be the sag in the middle, prove that pull on either point of support is

$$\frac{K}{2l}W+\frac{l}{4K}W'+\frac{l}{8K}W.$$ [IAS 1999]

11. A and B are two points in the same horizontal line, distance $2a$ apart. AO, OB are two equal heavy strings tied together at O and carrying a weight at O. If l is the length of each string, d the depth of O below AB, show that the parameter c of the catenary in which either string hangs is given by

$$l^2-d^2=2c^2\left(\cosh\frac{a}{c}-1\right).$$

12. A telegraphic wire is made of a given material and such a length l is stretched between two posts, distance d apart and at the same height, as will produce the least possible tension at the posts. Show that

$$l=\frac{d}{\lambda}\sinh\lambda,$$

where λ is given by the equation $\lambda\tanh\lambda=1$. [NBU(H) 2006]

13. A ring of weight wl is attached to the middle point of a string of length l and weight wl which hangs over two smooth pegs symmetrically in the same horizontal distance $2a$ apart, the ends of the string hanging vertically. Show that the parameter c of the catenary is given by

$$b+l=e^{a/c}\left\{b+\sqrt{4c^2+b^2}\right\}$$

and show that the least value of b, for which equilibrium is possible, occurs when

$$\frac{1}{(c-a)^2}-\frac{1}{c^2}=\frac{4}{b^2}.$$

14. One end of a heavy rough uniform string is fastened to a point P at a height h above a table, and a length $z(2l-h)$ of the string rests on the table in a vertical plane through P. Show that the string is pulled as far away from P as is possible consistent with equilibrium when z is given by the equation

$$z^2-2(1+\mu h)z+l^2-h^2=0,$$

where μ is the coefficient of friction and l is the length of the string.

15. If the tangents at the points P and Q of a catenary are at right angles, prove that the tension at the middle point of the arc PQ is equal to the weight of a length of the string equal to half the arc PQ.

16. A and B are two points in the same horizontal line, distance $2a$ apart. **AD, DB are two** equal heavy strings tied together at D and carrying a weight at D. **If l is the length of** each string, d the depth of D below AB, $(l^2 > d^2 + a^2)$, show that **the parameter c of** the catenary in which either string hangs is given by

$$l^2 - d^2 = 2c^2\left(\cosh\tfrac{a}{c} - 1\right).$$ **[IAS 1988]**

17. If α, β be the inclinations to the horizon of the tangents at the extremities **of a portion** of a common catenary and l the length of the portion, show that the height of **one** extremity above the other is $l \sin\frac{\alpha+\beta}{2} \sec\frac{\alpha-\beta}{2}$, the two extremities being **on one side** of the lowest point.

Answers

2. $20\sqrt{2}\left[\cosh^{-1}\left(\frac{33}{40}\sqrt{2}\right) + \cosh^{-1}\left(\frac{3}{4}\sqrt{2}\right)\right]$. 5. 555.7 ft (nearly).

4.9 Equilibrium of a String under any given Field of Force

Let A be some fixed point on the curve, OX and OY are fixed perpendicular directions in the plane of the curve. Let $P(x, y)$ and $Q(x + \delta x, y + \delta y)$ be any two points on the string very near to each other such that arc $AP = s$ and arc $AQ = s + \delta s$.

Also, let T and $T + \delta T$ be the tensions at P and Q respectively. If m is the mass per unit length of the string and X and Y be sums of the resolved parts of forces per unit of mass in the direction of the axes, then the components of the forces on PQ parallel to the axes are $m\delta s X$ and $m\delta s Y$.

Resolved parts of tension T along the axes are $T\cos\phi$ and $T\sin\phi$. These components will be some functions of s. Let us suppose that

$$T\cos\psi = T\frac{dx}{ds} = f(s) \text{ as } \cos\psi = \frac{dx}{ds},$$

then the resolved part of $(T + \delta T)$ along OX

$$= f(s + \delta s) = f(s) + \delta s f'(s) \text{ (by Taylor's theorem)}$$

$$= T\frac{dx}{ds} + \delta s \frac{d}{ds}\left(T\frac{dx}{ds}\right),$$

neglecting terms containing square and highest powers of δs.

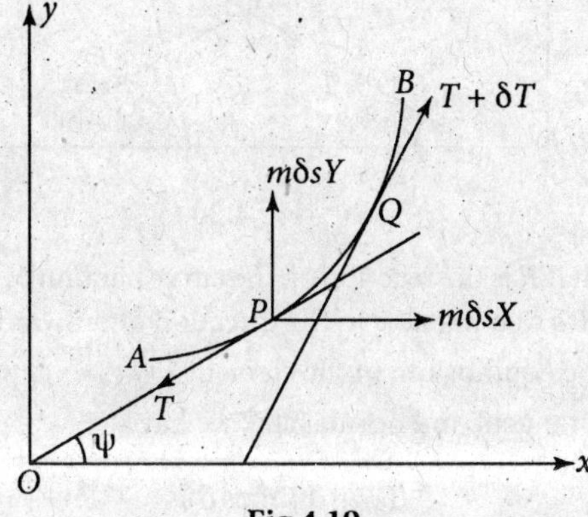

Fig 4.19

Considering equilibrium of the element PQ of the chain and resolving forces parallel to the axis of x, we have

$$T\frac{dx}{ds}=m\delta sX+T\frac{dx}{ds}+\delta s\frac{d}{ds}\left(T\frac{dx}{ds}\right).$$

Now as δs tends to zero, the above equation reduces to

$$\frac{d}{ds}\left(T\frac{dx}{ds}\right)+mX=0. \tag{1}$$

Similarly, resolving forces parallel to the axis of y, we have

$$\frac{d}{ds}\left(T\frac{dy}{ds}\right)+mY=0. \tag{2}$$

(1) and (2) are known as equations of equilibrium, from these we can determine the value of tension T at any point and the differential equation of the form of the string, provided we know, m, X and Y.

4.10 Equilibrium of a Light String in Contact with a Smooth Curve in a Vertical Plane

Let A be some fixed point on the chain and PQ be an element δs of the string. Let arc AP be s and tangents at P and Q make angles ψ and $\psi+\delta\psi$ with some fixed line. Let the tangents at P and Q be T and $T+\delta T$ as shown in the figure.

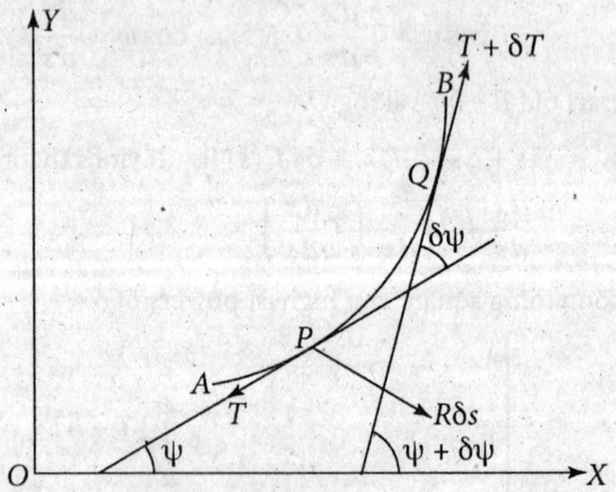

Fig 4.20

Again, we suppose that R is the reaction on the curve per unit of length, so that the reaction on the element PQ is $R\delta s$ acting at P in the direction of outward drawn normal at P.

Now, we consider the equilibrium of the element PQ.

Resolving along the tangent and normal at P, we have

$$(T+\delta T)\cos\delta\psi=T \tag{1}$$

$$\text{and } (T+\delta T)\sin\delta\psi=T\delta s. \tag{2}$$

Taking limit when $\delta\psi \to 0$, we get from (1)

$$T+\delta T = T \Rightarrow \delta T = 0$$
$$\text{or, } T = \text{constant} = c.$$

Thus, we conclude that the tension of a light inextensible string passing round a smooth curve is constant throughout.

Also, from (2), we get
$$(T+\delta T)\delta\psi = R\cdot\delta s \Rightarrow T = T\frac{d\psi}{ds} = \frac{c}{\rho}.$$

This shows that normal reaction varies inversely as radius of curvature.

4.11 Equilibrium of a Heavy String in Contact with a Smooth Curve in a Vertical Plane

Let A be some fixed point on the chain and PQ be an element δs of the string. Let arc AP be s and tangents at P and Q make angles ψ and $\psi+\delta\psi$ with some fixed line. Let the tensions at P and Q be T and $T+\delta T$ as shown in the figure.

Again, suppose that R is the reaction on the curve per unit of length so that the reaction on the element PQ is $R\delta s$ acting at P in the direction of outward drawn normal at P. If w be the weight per unit length, then the weight of element PQ is $w\delta s$, which is acting vertically downwards at P.

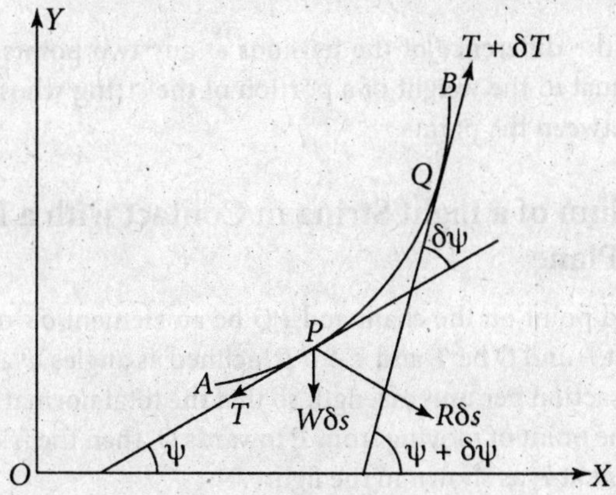

Fig 4.21

Considering the equilibrium of the element PQ and resolving along the tangent and normal at P,

$$(T+\delta T)\cos\delta\psi = T + W\delta s\sin\psi \tag{1}$$
$$\text{and } (T+\delta T)\sin\delta\psi = R\delta s + W\delta s\cos\psi \tag{2}$$

when $\delta\psi \to 0$, $\cos\delta\psi = 1$ and $\sin\delta\psi = \delta\psi$.

So from (1), we get

$$T+\delta T = T + W\delta s\sin\psi \Rightarrow \delta T = W\delta s\frac{dy}{ds} = w\delta y \Rightarrow \frac{dT}{dy} = W.$$

$\therefore\ T = Wy + C$, where c is a constant. (3)

Also, from (2), we get

$$(T+\delta T)\delta\psi = R\delta s + W\delta s\cos\psi$$
$$\Rightarrow T\delta\psi + \delta T\delta\psi = R\delta s + W\delta s\cos\psi$$
$$\Rightarrow T\delta\psi = (R + W\cos\psi)\delta s$$
$$\Rightarrow T\frac{d\psi}{ds} = R + W\cos\psi$$
$$\Rightarrow \frac{T}{\rho} = R + W\cos\psi$$
$$\Rightarrow R = \frac{T}{\rho} - W\cos\psi. \tag{4}$$

Results (3) and (4) give the tension and reaction at any point P.

If T_1 and T_2 be the tensions at points whose ordinates are y_1 and y_2, then by (3),

$$T_1 = Wy_1 + c$$
$$\text{and } T_2 = Wy_2 + c.$$
$$\therefore\ T_1 - T_2 = W(y_1 - y_2).$$

This shows that the difference of the tensions at any two points of a heavy string on a smooth curve is equal to the weight of a portion of the string whose length is equal to the vertical distance between the points.

4.12 Equilibrium of a Light String in Contact with a Rough Curve in a Vertical Plane

[KU(H) 2002]

Let A be some fixed point on the chain and PQ be an element δs of the string. Let arc AP be s and tangents at P and Q be T and $T+\delta T$ inclined as angles ψ and $\delta\psi$ with some fixed line. Let R be the reaction per unit of length so that the total normal reaction on PQ is $R\delta s$. If the string is on the point of moving from P towards Q, then the frictional force $\mu R\delta s$ will act along the tangent at P as shown in the figure.

Considering the equilibrium of the element PQ and resolving along the tangent and normal at P.

$$(T+\delta T)\cos\delta\psi = T + \mu R\delta s \tag{1}$$
$$\text{and } (T+\delta T)\sin\delta\psi = R\delta s. \tag{2}$$

When $\delta\psi \to 0, \cos\delta\psi = 1$ and $\sin\delta\psi = \delta\psi$, so from (1), we get

$$(T+\delta T)\cdot 1 = T + MR\delta s \Rightarrow \delta T = \mu R\delta s.$$

$$\therefore\ \frac{dT}{ds} = \mu R. \tag{3}$$

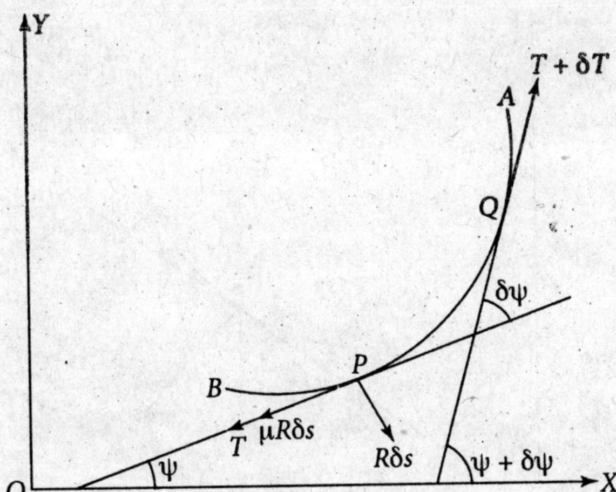

Fig 4.22

Also from (2),

$$(T+\delta T)\delta\psi = R\delta s \Rightarrow T\delta\psi = R\delta s.$$

$$\therefore\ T\frac{d\psi}{ds} = R. \tag{4}$$

Eliminating R from (3) and (4), we get

$$\frac{dT}{ds} = \mu T\frac{d\psi}{ds} \Rightarrow \frac{dT}{T} = \mu d\psi.$$

Integrating, we get

$$\log T = \mu\psi + c, \tag{5}$$

where c is a constant.

At the lowest point $\psi = 0$ and let $T = T_0$.

So from (5),

$$\log T_0 = Cc.$$

Hence, from (5), we get

$$\log T = \mu\psi + \log T_0 \Rightarrow \log T - \log T_0 = \mu\psi \Rightarrow \log\frac{T}{T_0} = \mu\psi \Rightarrow \frac{T}{T_0} = e^{\mu\psi} \Rightarrow T = T_0 e^{\mu\psi}.$$

This gives us the tension at any point in terms of tension at the lowest point. Having found T, we can find R from (4) as

$$R = \frac{T}{\rho} = \frac{T_0 e^{\mu\psi}}{\rho}.$$

4.13 Equilibrium of a Heavy String in Contact with a Rough Curve in a Vertical Plane

Let A be some fixed point on the chain and PQ be an element δs of the string. Let arc AP be s and tangents at P and Q be T and $T+\delta T$ inclined at angles ψ and $\psi+\delta\psi$ with some fixed line.

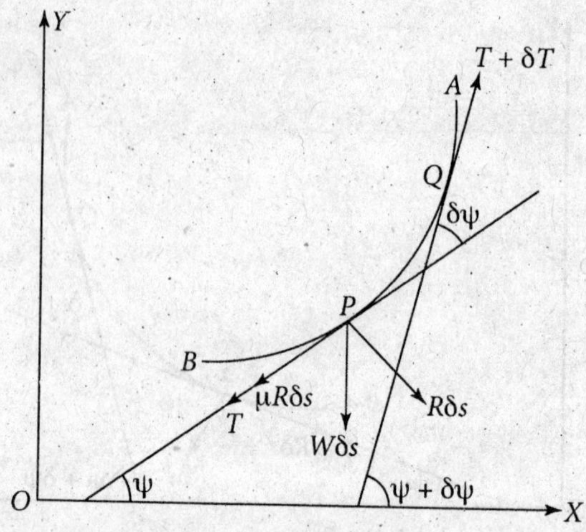

Fig 4.23

Let R be the reaction per unit of length so that the total normal reaction PQ is $R\delta s$. If the string is on the point of moving from P towards Q, then the frictional force $\mu R\delta s$ will act along the tangent at P as shown in the figure. Again, if w be the weight per unit length, then weight of element $PQ=(w\delta s)$ will be acting vertically downwards at P.

Considering the equilibrium of element PQ and resolving along the tangent and normal at P,

$$(T+\delta T)\cos\delta\psi = T+\mu R\delta s+W\delta s\sin\psi \quad (1)$$

$$\text{and } (T+\delta T)\sin\delta\psi = R\delta s+W\delta s\cos\psi. \quad (2)$$

When $\delta\psi\to 0, \cos\delta\psi=1$ and $\sin\delta\psi=\delta\psi$.

So from (1), we get

$$T+\delta T=T+\mu R\delta s+W\delta s\sin\psi \Rightarrow \frac{\delta T}{\delta s}=\mu R+W\sin\psi.$$

$$\therefore \frac{dT}{dS}=\mu R+W\sin\psi. \quad (3)$$

Also from (2), we get

$$(T+\delta T)\delta\psi=R\delta s+W\delta s\cos\psi \Rightarrow T\frac{\delta\psi}{\delta s}=R+W\cos\psi.$$

$$\therefore T\frac{d\psi}{ds}=R+W\cos\psi. \quad (4)$$

$$\Rightarrow T\frac{d\psi}{ds}-W\cos\psi=R.$$

Putting this value of R in (3), we get

$$\frac{dT}{dS}=\mu\left(T\frac{d\psi}{ds}-W\cos\psi\right)+W\sin\psi$$

$$\Rightarrow\frac{dT}{dS}-\mu T\frac{d\psi}{ds}=W(\sin\psi-\mu\cos\psi)$$

$$\Rightarrow\frac{dT}{d\psi}\cdot\frac{d\psi}{ds}-\mu T\frac{d\psi}{ds}=W(\sin\psi-\mu\cos\psi)$$

$$\Rightarrow\frac{dT}{dS}-\mu T=\rho W(\sin\psi-\mu\cos\psi)\left(\because\rho=\frac{ds}{d\psi}\right).$$

Above is a linear differential equation and its IF is $-e^{-\mu\int d\psi}=e^{-\mu\psi}$.

So solution is

$$T(\text{IF})=\int\rho W(\sin\psi-\mu\cos\psi)(\text{IF})\,d\psi$$

$$\Rightarrow Te^{-\mu\psi}=\int\rho W(\sin\psi-\mu\cos\psi)e^{-\mu\psi}\,d\psi+c. \tag{5}$$

If the equation of curve be given, then we can find the value of ρ and from above (5), we can find the value of T. Having found T and ρ, we can find out the value of R from (4).

WORKED-OUT EXAMPLES

1. A uniform heavy string rests on the upper surface of a rough vertical circle of radius a and partly hangs vertically. Prove that if one end be at the highest point of the circle, the greatest length that can hang freely is $\frac{2\mu a+(\mu^2-1)ae^{\frac{\mu+1}{2}}}{1+\mu^2}$, where μ is the coefficient of friction. [VU(H) 2001, KU(H) 2008]

   Solution: Suppose that A is the highest point of the circle, where there is one end of the string. Part of the string AB is on the circle and a portion x of its length is hanging vertically downwards.

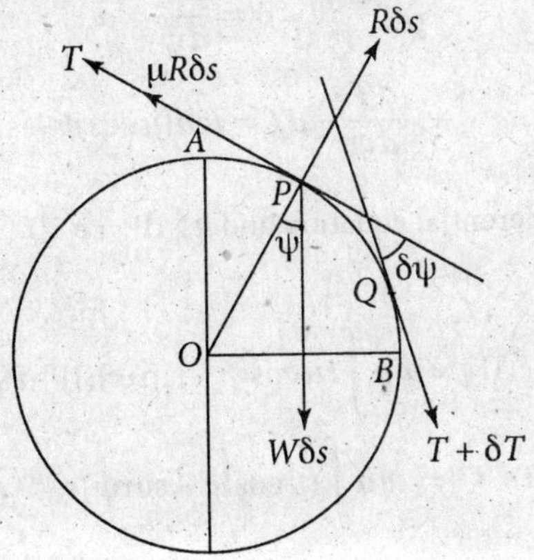

Fig. 4.24

Now consider an element $PQ = \delta s$. Let R be the normal reaction per unit length so that total reaction is $R\delta s$ acting along outwards drawn normal and the force of friction is $\mu R\delta s$ acting as shown in the figure.

Since the string is heavy, the weight of element δs is $w\delta s$ acting at P vertically downwards. Tangent at P makes an angle ψ with the horizontal and since angle between two lines is equal to angle between the perpendiculars on them, hence angle at P is also ψ.

Considering the equilibrium of the element PQ of the string we have on resolving along the tangent and normal

$$(T+\delta T)\cos\delta\psi + w\delta s\sin\psi - T - \mu R\delta s = 0 \tag{1}$$
$$\text{and } (T+\delta T)\sin\delta\psi + w\delta s\cos\psi - R\delta s = 0. \tag{2}$$

As $\delta\psi \to 0, \cos\delta\psi = 1$ and $\sin\delta\psi = \delta\psi$.

Hence, the above equations in limit reduce to

$$\frac{dT}{ds} = \mu R - W\sin\psi \text{ and } T\frac{d\psi}{ds} = r - w\cos\psi.$$

Eliminating R between the above, we get

$$\frac{dT}{ds} = \mu\left(T\frac{d\psi}{ds} + w\cos\psi\right) - w\sin\psi$$
$$\Rightarrow \frac{dT}{d\psi}\cdot\frac{d\psi}{ds} - \mu T\frac{d\psi}{ds} = \mu w\cos\psi - w\sin\psi$$
$$\Rightarrow \left(\frac{dT}{d\psi} - \mu T\right)\frac{d\psi}{ds} = \mu w\cos\psi - W\sin\psi$$
$$\Rightarrow \frac{dT}{d\psi} - \mu T = \rho(\mu w\cos\psi - w\sin\psi) \text{ as } \rho = \frac{ds}{d\psi}.$$

In a circle,

$$\rho = \frac{ds}{d\psi} = a.$$
$$\therefore \frac{dT}{d\psi} - \mu T = wa(\mu\cos\psi - \sin\psi).$$

This is a linear differential equation and its IF $= e^{-\int \mu d\psi} = e^{-\mu\psi}$.

So its solution is

$$T(\text{IF}) = wa\int(\mu\cos\psi - \sin\psi)(\text{IF})\,d\psi$$
$$\Rightarrow Te^{-\mu\psi} = wa\int(\mu\cos\psi - \sin\psi)e^{-\mu\psi}\,d\psi$$
$$\Rightarrow Te^{-\mu\psi} = \frac{wae^{-\mu\psi}}{\mu^2+1}\left[2\mu\sin\psi + \left(1-\mu^2\right)\cos\psi\right] + K. \tag{3}$$

At the highest point, $\psi=0$ and $T=0$.

So from (3),

$$0=\frac{Wa(1-\mu^2)}{1+\mu^2}+K. \tag{4}$$

Again, at the point $B, \psi=\frac{\pi}{2}$ and tension T at B supports the weight of the chain hanging downwards.

$\therefore\ T=wx.$

Putting for T and ψ in (3), we get

$$wxe^{-\frac{\mu\pi}{2}}=\frac{awe^{-\frac{\mu\pi}{2}}}{2}(2\mu+0)+K. \tag{5}$$

Subtracting (4) from (5), we get

$$xwe^{-\frac{\mu\pi}{2}}=\frac{2\mu awe^{-\frac{\mu\pi}{2}}}{\mu^2+1}-\frac{aw(1-\mu^2)}{\mu^2+1}\Rightarrow x=\frac{2a\mu+(\mu^2-1)\,ae^{\frac{\mu\pi}{2}}}{1+\mu^2} \text{ (Proved).}$$

2. A uniform string rests in limiting equilibrium on a rough quadrant of a circle with one end at the highest point. Prove that the angle of friction λ satisfies the relation $\frac{1}{2}\pi\tan\lambda=\log\tan 2\lambda$.

Solution: Write the figure and up to (3) of example 1, i.e.,

$$Te^{-\mu\psi}=\frac{wae^{-\mu\psi}}{1+\mu^2}\left[2\mu\sin\psi+\left(1-\mu^2\right)\cos\psi\right]+k.$$

Putting $1=\tan\lambda$, i.e., $\sin 2\lambda=\frac{2\mu}{1+\mu^2}$ and $\cos 2\lambda=\frac{1-\mu^2}{1+\mu^2}$ in (3), we get

$$Te^{-\mu\psi}=wae^{-\mu\psi}[\sin 2\lambda\sin\psi+\cos 2\lambda\cos\psi]+k$$

$$\Rightarrow Te^{-\mu\psi}=wae^{-\mu\psi}\cos(\psi-2\lambda)+k. \tag{1}$$

At the highest point $A, \psi=0, T=0$, so from (1), we get

$$0=wa\cos 2\lambda+k. \tag{2}$$

At the lowest point $B, \psi=\frac{\pi}{2}, T=0$.

$$\therefore\ 0=wae^{-\frac{\mu\pi}{2}}\sin 2\lambda+k. \tag{3}$$

From (5) and (6), we get

$$wa\cos 2\lambda=wae^{-\frac{\mu\pi}{2}}\sin 2\lambda\Rightarrow\tan 2\lambda=e^{\frac{\mu\pi}{2}}\Rightarrow\log\tan 2\lambda=\frac{\mu\pi}{2}=\frac{\pi}{2}\tan\lambda \text{ (Proved).}$$

3. A heavy uniform chain rests on a rough cycloid whose axis is vertical and vertex upwards, one end of the chain being at the vertex and other at a cusp; if the equilibrium be limiting, show that $(1+\mu^2)\,e^{\frac{\mu\pi}{2}}=3$. [KU(H) 2007]

Solution: Let us consider the equilibrium of an element PQ of the chain. Resolving the forces along and perpendicular to the tangent, we get

$$(T+\delta T)\cos\delta\psi + w\delta s\sin\psi - T - \mu R\delta s = 0 \tag{1}$$

$$\text{and } (T+\delta T)\sin\delta\psi + w\delta s\cos\psi - R\delta s = 0, \tag{2}$$

where w is the weight per unit length of the string.

As $\delta\psi \to 0$, we have

$$\frac{dT}{ds} = \mu R - w\sin\psi \tag{3}$$

$$\text{and } T\frac{d\psi}{ds} = R - w\cos\psi. \tag{4}$$

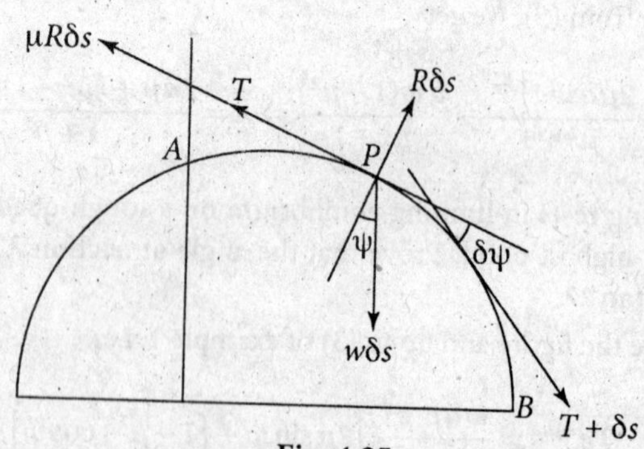

Fig. 4.25

Eliminating R between (3) and (4), we have

$$\frac{dT}{ds} = \mu\left(T\frac{d\psi}{ds} + w\cos\psi\right) - w\sin\psi$$

$$\Rightarrow \frac{dT}{d\psi}\cdot\frac{d\psi}{ds} - \mu T\frac{d\psi}{ds} - w(\mu\cos\psi - \sin\psi)$$

$$\Rightarrow \left(\frac{dT}{d\psi} - \mu T\right)\frac{d\psi}{ds} = w(\mu\cos\psi - \sin\psi)$$

$$\Rightarrow \frac{dT}{d\psi} - \mu T = w\rho(\rho\cos\psi - \sin\psi), \text{ as } \rho = \frac{ds}{d\psi}.$$

In a cycloid, $s = 4a\sin\psi$.

$$\therefore \rho = \frac{ds}{d\psi} = 4a\cos\psi.$$

$$\therefore \frac{dT}{d\psi} - \mu T = w4a\cos\psi(\mu\cos\psi - \sin\psi) = 4wa\left(\mu\cos^2\psi - \sin\psi\cos\psi\right)$$

$$\Rightarrow \frac{dT}{d\psi} - \mu T = 2wa[\mu(1+\cos 2\psi) - \sin 2\psi].$$

This is a linear differential equation and its IF $= e^{-\mu\int d\psi} = e^{-\mu\psi}$.

So, solution is

$$T(\text{IF}) = 2wa\int[\mu(1+\cos 2\psi)-\sin 2\psi](\text{IF})\,d\psi$$

$$\Rightarrow Te^{-\mu\psi} = 2wa\int[(1+\cos 2\psi)-\sin 2\psi]e^{-\mu\psi}\,d\psi$$

$$\Rightarrow Te^{-\mu\psi} = 2wa\left[-e^{-\mu\psi}+\frac{\mu e^{-\mu\psi}}{4+\mu^2}(-\mu\cos 2\psi+2\sin 2\psi)\right.$$
$$\left.-\frac{e^{-\mu\psi}}{4+\mu^2}(-\mu\sin 2\psi-2\cos 2\psi)\right]+k$$

$$\Rightarrow Te^{-\mu\psi} = -2wae^{-\mu\psi}+\frac{2wae^{-\mu\psi}}{4+\mu^2}\left[\left(2-\mu^2\right)\cos 2\psi+3\mu\sin 2\psi\right]+k. \tag{5}$$

At the vertex $A, \psi=0$ and $T=0$.
Also, at the cusp $B, \psi=\frac{\pi}{2}$ and $T=0$.
So from (5), we get

$$0=-2wa+\frac{2wa}{4+\mu^2}\left(2-\mu^2\right)+k \tag{6}$$

$$\text{and } 0=-2wae^{-\frac{\mu\pi}{2}}+\frac{2wae^{-\frac{\mu\pi}{2}}}{4+\mu^2}\left[-\left(2-\mu^2\right)\right]+k. \tag{7}$$

From (6) and (7), we get

$$-2wa+\frac{2wa}{4+\mu^2}\left(2-\mu^2\right)+k=-2wae^{-\frac{\mu\pi}{2}}-\frac{2wae^{-\frac{\mu\pi}{2}}}{4+\mu^2}\left(2-\mu^2\right)+k$$

$$\Rightarrow 1-\frac{2-\mu^2}{4+\mu^2}=e^{-\frac{\mu\pi}{2}}\left[1+\frac{2-\mu^2}{4+\mu^2}\right]\Rightarrow 2+2\mu^2=e^{-\frac{\mu\pi}{2}}6$$

$$\Rightarrow\left(1+\mu^2\right)=3e^{-\frac{\mu\pi}{2}}\Rightarrow\left(1+\mu^2\right)e^{\frac{\mu\pi}{2}}=3 \textbf{ (Proved).}$$

4. A string rests on a rough semi-circle being acted on by a constant attractive force towards one of its extremities and the friction is just sufficient to prevent the motion. Show that the coefficient of friction is given by $e^{\mu\pi}=\frac{3\mu}{1-2\mu^2}$. [VU(H) 2000]

Solution: Let F be the force per unit length of the string directed towards the extremity A. A frictional force $\mu R\delta s$ will act along the direction of T.

If $\angle COP=\theta$, $\angle AOP=\frac{\pi}{2}-\theta$, so that PA (or $F\delta s$) makes an angle $\frac{\pi}{4}-\frac{\theta}{2}$ with the direction of T.

We consider the equilibrium of an element PQ of the string.

Resolving forces along the tangent and normal at P, we have

$$(T+\delta T)\cos\delta\theta+F\delta s\cos\left(\frac{\pi}{4}-\frac{\theta}{2}\right)=T+\mu R\delta s \tag{1}$$

$$\text{and } (T+\delta T)\sin\delta\theta+F\delta s\sin\left(\frac{\pi}{4}-\frac{\theta}{2}\right)=R\delta s. \tag{2}$$

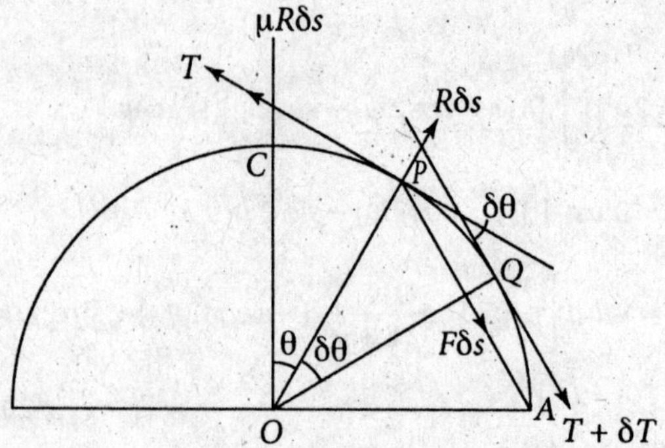

Fig. 4.26

In a circle, $s = a\theta$. $\therefore\ \delta s = a\delta\theta$.

$\therefore$ for small value of $\delta\theta$, the above equations in limit become

$$dT = \left[\mu R - F\cos\left(\frac{\pi}{4} - \frac{\theta}{2}\right)\right] a\,d\theta \qquad (3)$$

$$\text{and } T\,d\theta = \left[R - F\sin\left(\frac{\pi}{4} - \frac{\theta}{2}\right)\right] a\,d\theta. \qquad (4)$$

Eliminating R between (3) and (4), we have

$$dT - \mu T\,d\theta = a\,d\theta\left[-F\cos\left(\frac{\pi}{4} - \frac{\theta}{2}\right) + \mu F\sin\left(\frac{\pi}{4} - \frac{\theta}{2}\right)\right]. \qquad (5)$$

Let $\phi = \frac{\pi}{4} - \frac{\theta}{2}$.

$$\therefore\ d\phi = -\frac{1}{2}d\theta \Rightarrow d\theta = -2\,d\phi,$$

then (5) becomes

$$dT - \mu T(-2d\phi) = a(-2d\phi)F[-\cos\phi + \mu\sin\phi]$$

$$\Rightarrow dT + \mu T 2d\phi = 2aF(\cos\phi - \mu\sin\phi)\,d\phi$$

$$\Rightarrow \frac{dT}{d\phi} + 2\mu T = 2aF(\cos\phi - \mu\sin\phi),$$

which is a linear differential equation and its IF $= e^{2\mu\int d\phi} = e^{2\mu\phi}$.

So its solution is

$$T(\text{IF}) = 2aF\int(\cos\phi + \mu\sin\phi)(\text{IF})\,d\phi$$

$$\Rightarrow Te^{2\mu\phi} = 2aF\int(\cos\phi - \mu\sin\phi)e^{2\mu\phi}\,d\phi$$

$$\Rightarrow Te^{2\mu\phi} = 2aF\left[\frac{e^{2\mu\phi}(2\mu\cos\phi + \sin\phi)}{4\mu^2+1} - \frac{\mu e^{2\mu\phi}(2\mu\sin\phi - \cos\phi)}{4\mu^2+1}\right] + k$$

$$\Rightarrow Te^{2\mu\phi} = \frac{2aFe^{2\mu e^{\mu\phi}}}{1+4\mu^2}\left[3\mu\cos\phi + \left(1-2\mu^2\right)\sin\phi\right] + k. \qquad (6)$$

Tension is zero at the extremities of the diameter AB.

So, $T=0$, when $\theta=\pm\frac{\pi}{2}$, but we have $\phi=\frac{\pi}{4}-\frac{\theta}{2}$, so $T=0$, when $\phi=0$ or $\frac{\pi}{2}$.

Hence, using these conditions, we get

$$0=\frac{2aF\cdot 3\mu}{1+4\mu^2}+k \tag{7}$$

$$\text{and } 0=\frac{2aFe^{\mu\pi}}{1+4\mu^2}\left(1-2\mu^2\right)+k. \tag{8}$$

From (7) and (8), we get

$$\frac{2aF\cdot 3\mu}{1+4\mu^2}+k=\frac{2aFe^{\mu\pi}}{1+4\mu^2}\left(1-2\mu^2\right)+k\Rightarrow 3\mu=e^{\mu\pi}\left(1-2\mu^2\right)\Rightarrow e^{\mu\pi}=\frac{3\mu}{1-2\mu^2}.$$

5. A heavy string of variable density is hung up from two points. Prove that if T_1, T_2, T_3 are the tensions at point. A, B, C, where the inclinations of the tangents to the horizon are $\alpha-\beta$, α, $\alpha+\beta$, and w_1, w_2 the weights of the parts of the strings AB, BC respectively $\frac{1}{T_1}+\frac{1}{T_3}=\frac{2\cos\beta}{T_2}$, $\frac{T_1}{w_1}=\frac{T_3}{w_2}$.

Solution: For equilibrium of portions OA, AB, BC, where O is the lowest point, we have on resolving horizontally,

$$T_0=T_1\cos(\alpha-\beta) \text{ for } OA$$

$$T_1\cos(\alpha-\beta)=T_2\cos\alpha \text{ for } AB$$

$$T_2\cos\alpha=T_3\cos(\alpha+\beta) \text{ for } BC.$$

$$\therefore\ T_0=T_1\cos(\alpha-\beta)=T_2\cos\alpha=T_3\cos(\alpha+\beta). \tag{1}$$

$$\therefore\ \frac{T_2}{T_1}+\frac{T_2}{T_3}=\frac{\cos(\alpha-\beta)}{\cos\alpha}+\frac{\cos(\alpha+\beta)}{\cos\alpha}=\frac{\cos(\alpha-\beta)+\cos(\alpha+\beta)}{\cos\alpha}$$

$$=\frac{2\cos\alpha\cos\beta}{\cos\alpha}=2\cos\beta$$

$$\Rightarrow\frac{1}{T_1}+\frac{1}{T_3}=\frac{2\cos\beta}{T_2} \text{ (Proved).}$$

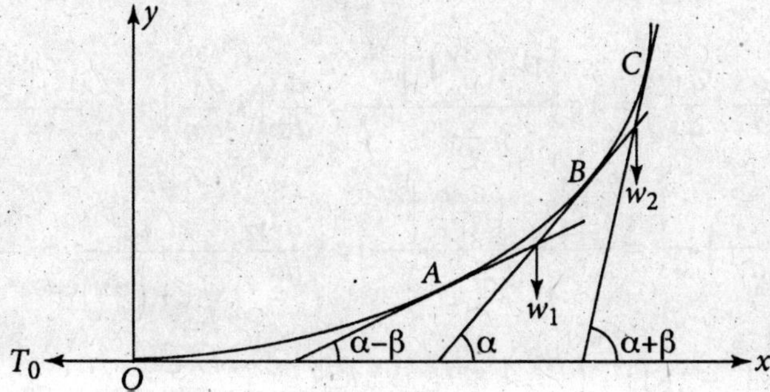

Fig. 4.27

Resolving vertically for equilibrium of AB and BC whose weights are w_1 and w_2.

$$T_2 \sin\alpha - T_1 \sin(\alpha-\beta) = w_1 \text{ for } AB \tag{2}$$

$$\text{and } T_3 \sin(\alpha+\beta) - T_2 \sin\alpha = w_2 \text{ for } BC. \tag{3}$$

Now,

$$\begin{aligned} w_1 T_3 - w_2 T_1 &= T_3\,[T_2 \sin\alpha - T_1 \sin(\alpha-\beta)] - T_1\,[T_3 \sin(\alpha+\beta) - T_2 \sin\alpha] \\ &= T_2 \sin\alpha (T_3 + T_1) - T_1 T_3 [\sin(\alpha-\beta) + \sin(\alpha+\beta)] \\ &= T_2 \sin\alpha \frac{2\cos\beta}{T_2} T_1 T_3 - T_1 T_3 2 \sin\alpha\cos\beta \left[\because \frac{1}{T_1} + \frac{1}{T_3} = \frac{2\cos\beta}{T_2}\right] \\ &= 2T_1 T_3 (\sin\alpha\cos\beta - \sin\alpha\cos\beta) = 0 \end{aligned}$$

$$\Rightarrow w_1 T_3 - w_2 T_1 = 0 \Rightarrow w_1 T_3 = w_2 T_1 \Rightarrow \frac{T_1}{w_1} = \frac{T_3}{w_2} \text{ (Proved).}$$

6. If the density at any point of a cord varies as the radius of curvature of the curve in which it hangs, show that the curve is the catenary of uniform strength. [IAS 2002]

Solution: Density varies as ρ (radius of curvature) $\Rightarrow m \propto \rho \Rightarrow m = \lambda\rho$, where λ is a constant.

Further, it is obvious that the cord hangs under gravity. Hence, the equations of the form of the string give

$$\frac{d}{ds}\left(T\frac{dx}{ds}\right) + m\cdot 0 = 0$$

$$\text{and } \frac{d}{ds}\left(T\frac{dy}{ds}\right) + m(-g) = 0 \text{ as } X = 0 \text{ and } Y = -g.$$

These equations can be written as

$$T\frac{dx}{ds} = \text{constant} = k \text{ (say)} \Rightarrow T = k\frac{ds}{dx} \tag{1}$$

and

$$\frac{d}{dx}\left(T\frac{dy}{ds}\right) = mg = \lambda\rho g. \tag{2}$$

Now putting the value of T from (1) in (2), we get

$$\frac{d}{ds}\left(k\frac{ds}{dx}\cdot\frac{dy}{ds}\right) = \lambda g\frac{\left[1+\left(\frac{dy}{dx}\right)^2\right]^{3/2}}{\frac{d^2y}{dx^2}} \Rightarrow \frac{d}{ds}\left(k\frac{dy}{dx}\right) = \lambda g\frac{\left[1+\left(\frac{dy}{dx}\right)^2\right]^{3/2}}{\frac{d^2y}{dx^2}}$$

$$\Rightarrow \frac{d}{dx}\left(k\frac{dy}{dx}\right)\frac{dx}{ds} = \frac{\lambda g\left[1+\left(\frac{dy}{dx}\right)^2\right]^{3/2}}{\frac{d^2y}{dx^2}} \Rightarrow k\frac{d^2y}{dx^2}\cdot\frac{1}{\sqrt{1+\left(\frac{dy}{dx}\right)^2}} = \frac{\lambda g\left[1+\left(\frac{dy}{dx}\right)^2\right]^{3/2}}{\frac{d^2y}{dx^2}}$$

$$\Rightarrow k\left(\frac{d^2y}{dx^2}\right)^2 = \lambda g\left[1+\left(\frac{dy}{dx}\right)^2\right]^2 \Rightarrow \frac{\frac{d^2y}{dx^2}}{1+\left(\frac{dy}{dx}\right)^2} = \sqrt{\frac{\lambda g}{k}}.$$

Integrating this, we get

$$\tan^{-1}\frac{dy}{dx} = \left(\sqrt{\frac{\lambda g}{k}}\right)x + A, \tag{3}$$

where A is the constant of integration.

But x and $\frac{dy}{dx}$ vanish together. $\therefore\ A = 0$.

Hence,

$$\tan^{-1}\frac{dy}{dx} = \sqrt{\frac{\lambda g}{k}}x \Rightarrow \frac{dy}{dx} = \tan\left(\sqrt{\frac{\lambda g}{k}}x\right).$$

Integrating this, we get

$$y = \sqrt{\frac{k}{\lambda g}}\log\sec\left(\sqrt{\frac{\lambda g}{k}}x\right).$$

The constant of integration vanishes as x and y vanish together.

Above is of the form $y = a\log\sec\frac{x}{a}$ and here the catenary is of uniform strength.

7. A non-uniform string hangs under gravity. Its cross section at any point is inversely proportional to the tension at that point. Prove that the curve in which the string hangs is an arc of a parabola with its axis vertical. [IAS 2004]

Solution: Equations of equilibrium arc

$$\frac{d}{ds}\left(T\frac{dx}{ds}\right) + mX = 0 \tag{1}$$

and

$$\frac{d}{ds}\left(T\frac{dy}{ds}\right) + mY = 0. \tag{2}$$

As the area is proportional to mass and mass is proportional to $\frac{1}{T}$, we have

$$m \propto \frac{1}{T} \Rightarrow m = \frac{\lambda}{T},$$

where λ is a constant and $X = 0, Y = -g$.

Hence, the (1) and (2) become

$$\frac{d}{ds}\left(T\frac{dx}{ds}\right) = 0. \qquad \therefore\ T\frac{dx}{ds} = c \tag{3}$$

and

$$\frac{d}{ds}\left(T\frac{dy}{ds}\right) = \frac{\lambda}{T}g. \tag{4}$$

Eliminating T from (3) and (4), we get

$$\frac{d}{ds}\left(c\frac{ds}{dx}\cdot\frac{dy}{ds}\right)=\frac{\lambda g}{c}\cdot\frac{dx}{ds}$$

$$\Rightarrow\frac{d}{dx}\left(c\frac{dy}{dx}\right)=\frac{\lambda g}{c}\cdot\frac{dx}{ds}$$

$$\Rightarrow\frac{d}{dx}\left(c\frac{dy}{dx}\right)\frac{dx}{ds}=\frac{\lambda g}{c}\cdot\frac{dx}{dx}$$

$$\Rightarrow c\frac{d^2y}{dx^2}=\frac{\lambda g}{c}\Rightarrow\frac{d^2y}{dx^2}=\frac{\lambda g}{c^2}=\frac{1}{2a}\text{ (say).}$$

Integrating, we have

$$\frac{dy}{dx}=\frac{x}{2a}.$$

The constant of integration vanishes as x and $\frac{dy}{dx}$ vanish together.
Again, integrating the last equation, we get

$$y=\frac{x^2}{4a}\text{ or, }x^2=4ay.$$

The constant of integration vanishes as x and y vanish together.
This equation represents a parabola.

EXERCISE 4B

1. A uniform string rests in limiting equilibrium on a rough vertical circle and subtends an angle β at the centre. If the string is on the point of slipping off, prove that the angular distance α of its upper end from the highest point of the circle is determined by the equation $\cos(\alpha+\beta-2\lambda)=e^{\beta\tan\lambda}\cos(\alpha-2\lambda)$, where λ is the angle of friction. **[KU(H) 2003]**

   [*Hints.* See worked-out Example 1, write up to (3). Here at $A, \psi=\alpha$ and $T=0$ and at $B, \psi=\alpha+\beta$ and $T=0$.]

2. A heavy uniform string is placed upon a rough catenary whose axis is vertical and vertex upwards, so that one extremity is at the vertex and the length of the string is equal to the parameter of the catenary. Prove that (a) μ is given by $e^{\mu\pi}=4$, (b) the string will just rest; provided that coefficient of friction is $\frac{2\log_e 2}{\pi}$.

3. A heavy string occupies a quadrant of the upper half of a rough vertical circle in a state bordering on motion. Prove that the radius through the lower end makes an angle α with the vertical given by $\tan(\alpha-2\lambda)=e^{-\frac{\mu\pi}{2}}$, where $\mu=\tan\lambda$ is the coefficient of friction.

Chapter 5

Stability of Equilibrium

5.1 Introduction

Consider the motion of a body on a smooth curve in a vertical plane as shown in Fig. 5.1. Obviously, the body can rest at the points A, B, C and D which are points of maxima or minima of the curve. If the body be slightly displaced from its position of rest at A or C (i.e., the points of minima), it will tend to return to its original position of rest, while if displaced from the position of rest at B or D (i.e., the points of maxima), it will tend to move still further away from its original position of rest. In the first case the body is said to be stable and in the second case body is said to be unstable.

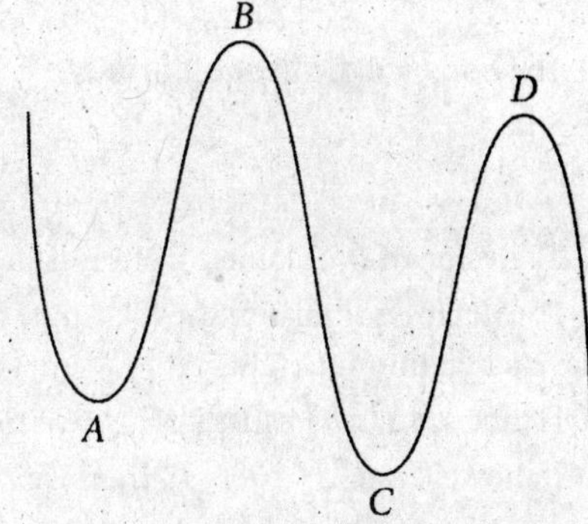

Fig. 5.1

5.2 Definition of Stable, Unstable and Neutral Equilibrium

[PU(H) 1999, LNMU(H) 1997, RU(H) 1999]

Stable Equilibrium

If a body is at rest in such a position that, if slightly disturbed it would tend to return to its original position, the body is said to be in stable equilibrium.

Therefore, for stable equilibrium the position of the CG of the body should be at its lowest height.

Example. (i) Egg on one side; (ii) A right cone resting on its base.

Unstable Equilibrium. If a body is at rest in such a position that if slightly disturbed, it would tend to move further from its original position, the body is said to be in unstable equilibrium.

Therefore, for unstable equilibrium the position of the CG of the body should be at its highest position.

Example. (i) Egg balanced on one end; (ii) Cone balanced on its vertex.

Neutral Equilibrium. If a body is at rest in such a position that if slightly disturbed, it shows no tendency to return to its original position or to move further from it, the body is said to be in neutral equilibrium.

Example. (i) Any body supported at its CG; (ii) Cone resting on its curved surface; (iii) Sphere lying on a table.

Conditions of stability for a body with one degree of freedom.

[PU(H) 1997, LNMU(H) 1993, RU(H) 1998, BVABU(H) 1995, CU(H) 2007]

If a body has more than one position of equilibrium, the one in which its CG is lowest will in general be the stable position and that in which CG is highest will be the unstable one. Hence, to determine the positions of equilibrium and the stability, we express the height of the CG of the body in terms of the variable θ, the angle through which the body is tilted from its position of equilibrium, above a fixed point or a fixed horizontal plane.

Let

$$z = z(\theta).$$

The body will be in equilibrium if z is either maximum or minimum.

$$\therefore \quad \frac{dz}{d\theta} = 0.$$

The value of θ obtained from the above equation will give the positions of equilibrium.

For stable equilibrium the CG must have a minimum and for unstable equilibrium it must have a maximum height.

Hence, $d^2z/d\theta^2$ must be positive for stable position and negative for unstable position.

If, however, $d^2z/d\theta^2 = 0$, then we consider

$$\frac{d^3z}{d\theta^3} \text{ and } \frac{d^4z}{d\theta^4}.$$

For position of rest, $d^3z/d\theta^3 = 0$. The equilibrium is stable or unstable according as $d^4z/d\theta^4$ is positive or negative.

Similarly, if z be the depth of CG below a fixed horizontal plane, equilibrium is stable or unstable according as z is maximum or minimum.

If there be more than one position of equilibrium, one position will be stable and the next one unstable, i.e., stability and instability will occur alternately as maxima and minima occur alternately.

5.3 Work Function Test for the Nature of Stability of Equilibrium

[PU(H) 2000]

We suppose that a rigid body is so constrained that only one geometrical quantity is required to determine its position. In this case the body is said to possess only one degree of freedom. Let θ be this geometrical quantity. We shall assume that the work done in bringing the body from some standard position to its present position is given by $w = f(\theta)$.

Let R be the resultant force acting on the body and let it receive a small displacement δS in the direction of R. Then

$$\delta W = R\delta S \text{ or, } f'(\theta)\delta\theta = R\delta S.$$

Proceeding to the limit, we have

$$f'(\theta)\frac{d\theta}{dS}=R. \tag{1}$$

But when the body is in equilibrium, $R=0$. Therefore,

$$f'(\theta)\frac{d\theta}{dS}=0.$$

But $d\theta/dS$ is not zero, since there are always a number of quantities each of which can be used to fix the position of the body with one degree of freedom. We choose θ to be that which continuously increases or decreases as the body moves in the same direction. Hence, $f'(\theta)=0$.

Let θ_0 be a root of this equation. θ_0 defines the position of equilibrium, θ_0 changes by a small quantity $\delta\theta_0$ and a small force δR is brought into play. For stable equilibrium, this force acts in an opposite direction to the displacement so that $dR/d\theta$ is negative, that is

$$\left\{f''(\theta_0)\left(\frac{d\theta}{dS}\right)^2+f'(\theta_0)\left(\frac{d^2\theta}{dS^2}\right)\right\}$$

must be negative.

But $f'(\theta_0)=0$ and $(d\theta/dS)^2$ is positive.

Hence, $f''(\theta_0)$ must be negative.

To sum up, if $W=f(\theta)$, then for stable equilibrium at $\theta=\theta_0$.

(i) $f'(\theta_0)=0$; (ii) $f''(\theta_0)$ must be negative.

5.4 Energy Test for Stability

[MU(H) 2006, PU(H) 2011, LNMU(H) 1998, TMBU(H) 1995, RU(H) 2000, KU(H) 2007]

To prove that for a system in equilibrium under conservative forces, the PE is minimum for stable and maximum for unstable equilibrium.

For any dynamical system, the sum of two energies, namely kinetic and potential, is constant. In the position of equilibrium, KE is zero, therefore, the PE is constant, i.e., PE is either maximum or minimum.

Let the PE be maximum. Now let the system be slightly displaced from the position of maximum PE and then set free. In the new position PE will decrease. Consequently the KE will increase. In other words, the KE will be positive as in the position of equilibrium it was zero. Hence, the system moves further away from the position of maximum PE and therefore, the equilibrium in the position of maximum PE is an unstable one.

Again, let the PE be minimum. Now let the system be slightly displaced from the position of minimum PE and then set free. In the new position the PE should increase. Hence, KE should decrease. But in the position of minimum PE, KE was zero. Hence, KE should become negative meaning thereby when released it will have the tendency to move in the direction opposite to the displacement. Hence, it will have the tendency to return to its original position. Thus, in this case, equilibrium is stable.

Alternative Method

The PE of a body, acted upon by a conservative system of forces, is defined as its capacity to do work by virtue of the position it has acquired. It is measured by the amount of work it can do in passing from the present position to some standard position.

If W be the work function of the body in any position referred to some standard position and V be the PE of the body in that position; then $V = -W$. If V_A and V_B are the values of the PE at the two positions A and B, then $V_A - V_B$ is the work done by the forces in displacing the body from A to B.

Let A be the position of equilibrium of a rigid body under the action of a given system of forces and V be the PE of the body in this position A. Suppose the body undergoes a small displacement and takes a position B near the position of equilibrium A, then the PE of the body in the position B will be $V + dV$. Therefore, the work done by the forces in displacing the body from the equilibrium position A to the nearby position B is $V-(V+dV)$, i.e., $-dV$. Since the body is in equilibrium in the position A, therefore, by the principle of virtual work, we have

$$-dV = 0 \Rightarrow dV = 0.$$

Hence, the PE V is stationary (maximum or minimum) in the position of equilibrium.

First suppose that V is minimum at the equilibrium position. Imagine that the body is slightly displaced to a position B and let V' be the PE there. Since V is minimum at A, therefore, $V' > V$, so that $V - V'$ is negative. It means that in displacing the body from A to B the work done by the forces acting on the body is negative, i.e., the work is done against the forces and so the forces will have a tendency to bring the body back to the original position of equilibrium at A. Hence, the equilibrium at A is stable.

Thus, we see that in the position of stable equilibrium, the PE of the body is minimum.

Next suppose that V is maximum at the equilibrium position A. If V' is the value of the PE in a slightly displaced position B of the body, then in this case $V' < V$, so that $V - V'$ is positive. It means that in displacing the body from A to B the work done by the force is positive, i.e., the work is done by the forces and so the forces will tend to move the body further away from the position of equilibrium. Hence, the equilibrium at A is unstable.

Thus, in the positions of equilibrium of the body the PE V is either maximum or minimum. If it is minimum, the equilibrium is stable and if it is maximum, the equilibrium is unstable.

» **Example 1.** *A body rests in equilibrium upon another fixed body, the portions of the two bodies in contact being spheres of radii r and R respectively, and the straight line joining the centres of spheres being vertical; if the first body be slightly displaced, find when the equilibrium is stable or unstable, the bodies being rough enough to prevent sliding.*

[MU(H) 2011, PU(H) 2009, LNMU(H) 2011, TMBU(H) 1976, CU(H) 2008]

Solution: Let O be the centre of the spherical surface of the lower body and O_1 that of the upper body. The figure is a section on the bodies by a vertical plane through the CG G_1 of the upper body. Let A_1 be the point of contact and let $A_1G_1 = h$. Let the upper body be slightly displaced by rolling so that the new position of the centre of the upper body is O_2, the new point of contact is A_2, the new position of the CG is G_2 and the new position of the A_1 is C. But $A_1G_1 = h$, therefore, $CG_2 = h$. Also let $\angle A_1OA_2 = \theta$ and $\angle CO_2A_2 = \phi$.

$\therefore\ \angle G_2O_2H = \theta + \phi.$

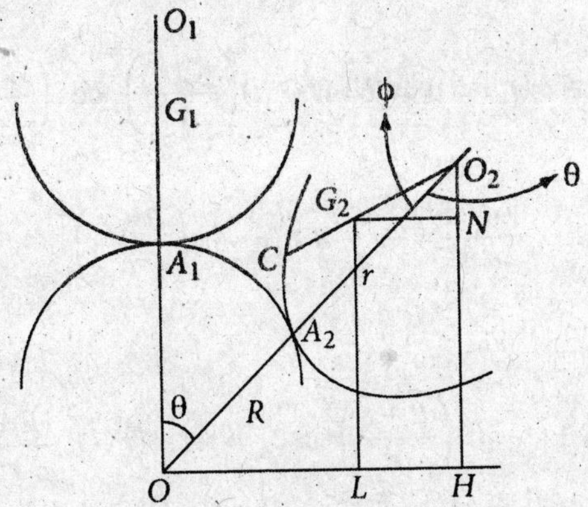

Fig. 5.2

Since the upper body has rolled into its new position, the arc A_1A_2 of the lower sphere (radius R) = the arc CA_2 of the upper sphere (radius r) so that $R\theta = r\phi$.

Now in order to find the nature of equilibrium we should find the height z of CG; G_2 in the new position above O.

Now

$$z = G_2L = NH = O_2H - O_2N. \tag{1}$$

In ΔOO_2H,

$$\cos\theta = \frac{O_2H}{OO_2} = \frac{O_2H}{r+R}.$$

$$OO_2 = OA_2 + A_2O_2 = R + r;\ O_2H = (r+R)\cos\theta.$$

In ΔO_2G_2N,

$$\cos(\theta+\phi) = \frac{O_2N}{OG_2} = \frac{O_2N}{r-h}$$

$$OG_2 = OC - CG_2 = r - h, O_2N = (r-h)\cos(\theta+\phi).$$

So from (1),

$$z = (r+R)\cos\theta - (r-h)\cos(\theta+\phi)$$

$$= (r+R)\cos\theta - (r-h)\cos\left(\theta + \frac{R\theta}{r}\right) \quad [\because R\theta = r\phi]$$

$$\text{or, } z = (r+R)\cos\theta - (r-h)\cos\left(\frac{R+r}{r}\right)\theta$$

$$\therefore \frac{dz}{d\theta} = -(r+R)\sin\theta + (r-h)\frac{(R+r)}{r}\sin\left(\frac{R+r}{r}\right)\theta.$$

For equilibrium, we have

$$\frac{dz}{d\theta} = 0, \text{ i.e., } -(R+r)\sin\theta + (r-h)\frac{(R+r)}{r}\sin\left(\frac{R+r}{r}\right)\theta = 0.$$

This is satisfied by $\theta = 0$.

Now

$$\frac{d^2z}{d\theta^2} = -(r+R)\cos\theta + (r-h)\left(\frac{R+r}{r}\right)^2 \cos\left(\frac{R+r}{r}\right)\theta. \quad (1)$$

At $\theta = 0$,

$$\frac{d^2z}{d\theta^2} = -r - R + (r-h)\left(\frac{R+r}{r}\right)^2.$$

So z will be minimum, if $\frac{d^2z}{d\theta^2} > 0$,

$$\Rightarrow -r - R + (r-h)\left(\frac{R+r}{r}\right)^2 > 0, \Rightarrow -1 + (r-h)\frac{(R+r)}{r^2} > 0,$$

$$\Rightarrow -r^2 + rR + r^2 - hR - hr > 0, \Rightarrow rR > h(r+R), \Rightarrow \frac{1}{h} > \frac{1}{R} + \frac{1}{r}$$

and z will be maximum if $\frac{d^2z}{d\theta^2} < 0$,

$$\text{i.e., } \frac{1}{h} < \frac{1}{R} + \frac{1}{r}.$$

Hence, the equilibrium is stable when $\frac{1}{h} > \frac{1}{r} + \frac{1}{R}$ and unstable when $\frac{1}{h} < \frac{1}{r} + \frac{1}{R}$.
Now third case arises, when

$$\frac{1}{h} = \frac{1}{r} + \frac{1}{R}.$$

In this case,

$$\frac{d^2z}{d\theta^2} = -(r+R) + (r-h)\left(\frac{r+R}{r}\right)^2.$$

Now

$$\frac{1}{h} = \frac{r+R}{rR}, \; h = \frac{rR}{r+R}$$

$$\therefore \; r - h = r - \frac{rR}{r+R} = \frac{r^2}{r+R}. \quad (2)$$

Hence,

$$\frac{d^2z}{d\theta^2} = -(r+R) + \frac{(r+R)^2}{r^2} \cdot \frac{r^2}{r+R} = -(r+R) + (r+R) = 0.$$

Hence, we find

$$\frac{d^3z}{d\theta^3} \text{ and } \frac{d^4z}{d\theta^4}.$$

Then, from (1)

$$\frac{d^3z}{d\theta^3} = (r+R)\sin\theta - (r-h)\left(\frac{r+R}{r}\right)^3 \sin\left\{\frac{\theta(r+R)}{r}\right\}$$

$$\frac{d^4z}{d\theta^4} = (r+R)\cos\theta - (r-h)\left(\frac{r+R}{r}\right)^4 \cos\left\{\frac{(r+R)\theta}{r}\right\}.$$

At $\theta=0$; $\frac{d^3z}{d\theta^3}=0$.

$$\frac{d^4z.}{d\theta^4}=(r+R)-(r-h)\left(\frac{r+R}{r}\right)^4=(r+R)-\frac{r^2}{r+R}\frac{(R+r)^4}{r^4}\text{ [using (2)]}$$

$$=(r+R)-\frac{(R+r)^3}{r^2}=(r+R)\left[1-\left(\frac{r+R}{r}\right)^2\right]=(r+R)\left[1-\left(1+\frac{R}{r}\right)^2\right]<0.$$

This shows that z is maximum and as such the equilibrium is unstable.

» **Example 2.** *A body rests in equilibrium upon another fixed body, the portions of the two bodies in contact have radii of curvatures ρ_1 and ρ_2 respectively. The CG of the first body is at a height h above the point of contact and the common normal makes an angle α with the vertical. If the upper body be slightly displaced, to find whether the equilibrium is stable or unstable, the bodies being rough enough to prevent sliding.*

Solution: Let O and O_1 be the centres of curvature of the lower and upper bodies in the position of rest and A_1 be their point of contact. In this position of equilibrium the common normal OA_1O_1 makes an angle α with the vertical OY.

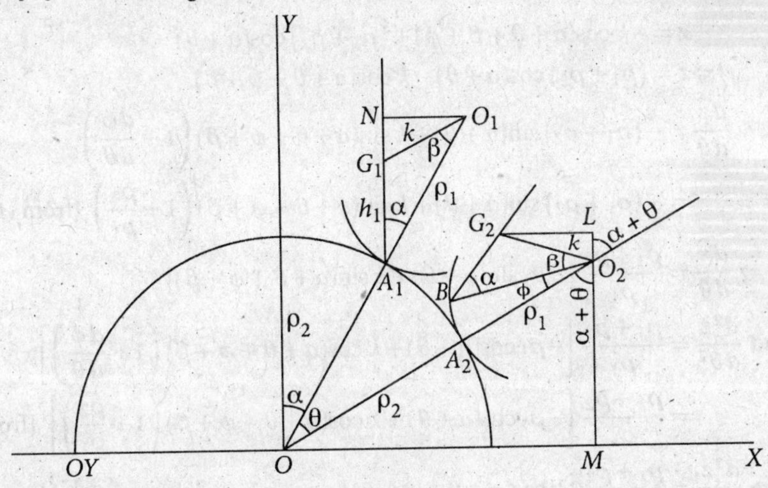

Fig. 5.3

If G_1 is the CG of the upper body, then for equilibrium the line A_1G_1 must be vertical. It is given that $A_1G_1=h$.

Let $O_1G_1=k$ and $\angle OO_1G_1=\beta$.

Suppose the upper body is slightly displaced by pure rolling over the lower body which is fixed. Let A_2 be the new point of contact. O_2 is the new position of O_1 and the point A_1 of the upper body rolls up to the position B so that O_2, B is the new position of the original normal O_1A_1. Also G_2 is the new position of G_1 so that $O_2G_2=O_1G_1=k$.

Suppose the common normal at A_2 makes angles θ and ϕ with the original OA_1 and O_2B.

We have $O_1A_1=\rho_1$ and $OA_1=\rho_2$.

Also $O_2A_2=\rho_1$ and $OA_2=\rho_2$.

Since the upper body rolls on the lower body without slipping, therefore, arcA_1A_2 = arc$A_2B \Rightarrow \rho_2\theta = \rho_1\phi$.

$$\rho_1\phi = \rho_2\theta. \qquad \therefore \rho_1\frac{d\phi}{d\theta} = \rho_2 \Rightarrow \frac{d\phi}{d\theta} = \frac{\rho_2}{\rho_1}. \tag{1}$$

Now,

$$z = LM = LO_2 + O_2M \tag{2}$$

In ΔO_2G_2L,

$$\cos G_2O_2L = \frac{LO_2}{O_2G_2} = \frac{LO_2}{k} \Rightarrow LO_2 = k\cos\{\pi - (\alpha+\theta+\beta+\phi)\}$$
$$\Rightarrow LO_2 = -k\cos(\alpha+\theta+\beta+\phi).$$

In ΔOO_2M,

$$\cos(\alpha+\theta) = \frac{O_2M}{O_2O} = \frac{O_2M}{\rho_1+\rho_2} \Rightarrow O_2M = (\rho_1+\rho_2)\cos(\alpha+\theta).$$

So from (2),

$$z = -k\cos(\alpha+\theta+\beta+\phi) + (\rho_1+\rho_2)\cos(\alpha+\theta)$$
$$\Rightarrow z = (\rho_1+\rho_2)\cos(\alpha+\theta) - k\cos(\alpha+\theta+\phi+\beta).$$
$$\therefore \frac{dz}{d\theta} = -(\rho_1+\rho_2)\sin(\alpha+\theta) + k\sin(\alpha+\theta+\phi+\beta)\left(1+\frac{d\phi}{d\theta}\right)$$
$$= -(\rho_1+\rho_2)\sin(\alpha+\theta) + k\sin(\alpha+\theta+\phi+\beta)\left(1+\frac{\rho_2}{\rho_1}\right) \quad \text{[from (1)]}$$
$$\Rightarrow \frac{dz}{d\theta} = \frac{\rho_1+\rho_2}{\rho_1}[-\rho_1\sin(\alpha+\theta) + k\sin(\alpha+\theta+\phi+\beta)]$$
$$\text{and } \frac{d^2z}{d\theta^2} = \frac{\rho_1+\rho_2}{\rho_1}\left[-\rho_1\cos(\alpha+\theta) + k\cos(\alpha+\theta+\phi+\beta)\left(1+\frac{d\phi}{d\theta}\right)\right]$$
$$= \frac{\rho_1+\rho_2}{\rho_1}\left[-\rho_1\cos(\alpha+\theta) + k\cos(\alpha+\theta+\phi+\beta)\left(1+\frac{\rho_2}{\rho_1}\right)\right] \quad \text{[from (1)]}$$
$$\Rightarrow \frac{d^2z}{d\theta^2} = \frac{\rho_1+\rho_2}{\rho_1^2}\left[k\cos(\alpha+\theta+\phi+\beta)(\rho_1+\rho_2) - \rho_1^2\cos(\alpha+\theta)\right].$$

Now for the position of equilibrium, we have

$$\frac{dz}{d\theta} = 0 \Rightarrow \frac{\rho_1+\rho_2}{\rho_1}[k\sin(\alpha+\theta+\phi+\beta) - \rho_1\sin(\alpha+\theta)] = 0$$
$$\Rightarrow \frac{\rho_1}{k} = \frac{\sin(\alpha+\beta+\theta+\phi)}{\sin(\alpha+\theta)}. \tag{3}$$

Also,

$$h = A_1G = A_1N - NG_1 \Rightarrow h = \rho_1\cos\alpha - k\cos(\alpha+\beta)$$
$$\Rightarrow \frac{h}{k} = \frac{\rho_1}{k}\cos\alpha - \cos(\alpha+\beta). \tag{4}$$

Also

$$\frac{h}{\sin\beta}=\frac{k}{\sin\alpha}\Rightarrow\frac{h}{k}=\frac{\sin\beta}{\sin\alpha}.$$

So from (4), we have

$$\frac{\sin\beta}{\sin\alpha}+\cos(\alpha+\beta)=\frac{\rho_1}{k}\cos\alpha$$
$$\Rightarrow\frac{\sin\beta+\sin\alpha(\cos\alpha\cos\beta-\sin\alpha\sin\beta)}{\sin\alpha}=\frac{\rho_1\cos\alpha}{k}$$
$$\Rightarrow\frac{\sin\beta\left(1-\sin^2\alpha\right)+\sin\alpha\cos\alpha\cos\beta}{\sin\alpha}=\frac{\rho_1}{k}\cos\alpha$$
$$\Rightarrow\frac{\sin\beta\cos\alpha+\sin\alpha\cos\beta}{\sin\alpha}=\frac{\rho_1}{k}\Rightarrow\frac{\sin(\alpha+\beta)}{\sin\alpha}=\frac{\rho_1}{k}. \tag{5}$$

Comparing (3) and (5), we get $\theta=\phi=0$.

Again,

$$\left(\frac{d^2z}{d\theta^2}\right)_{\theta=\phi=0}=\frac{\rho_1+\rho_2}{\rho_1^2}\left[k\cos(\alpha+\beta)(\rho_1+\rho_2)-\rho_1^2\cos\alpha\right]$$
$$=\frac{\rho_1+\rho_2}{\rho_1^2}\left[(\rho_1\cos\alpha-h)(\rho_1+\rho_2)-\rho_1^2\cos\alpha\right]$$
$$=\frac{\rho_1+\rho_2}{\rho_1^2}\left[\rho_1\rho_2\cos\alpha-h\rho_1-h\rho_2\right]$$
$$=\frac{\rho_1+\rho_2}{\rho_1^2}\left[\rho_1\rho_2\cos\alpha-h(\rho_1+\rho_2)\right].$$

Now,

$$\frac{d^2z}{d\theta^2}>0\Rightarrow\frac{\rho_1+\rho_2}{\rho_1^2}\left[\rho_1\rho_2\cos\alpha-h(\rho_1+\rho_2)\right]>0$$
$$\Rightarrow\rho_1\rho_2\cos\alpha-h(\rho_1+\rho_2)>0\Rightarrow\rho_1\rho_2\cos\alpha>h(\rho_1+\rho_2)$$
$$\Rightarrow h<\frac{\rho_1\rho_2\cos\alpha}{\rho_1+\rho_2}$$

and

$$\frac{d^2z}{d\theta^2}<0,\text{ if } h>\frac{\rho_1\rho_2\cos\alpha}{\rho_1+\rho_2}.$$

Hence, the equilibrium will be stable, if $h<\frac{\rho_1\rho_2\cos\alpha}{\rho_1+\rho_2}$ and unstable, if $h>\frac{\rho_1\rho_2\cos\alpha}{\rho_1+\rho_2}$.

Corollary. If $\alpha=0$, the above conditions give that the equilibrium is stable or unstable according as

$$h<\text{ or, }>\frac{\rho_1\rho_2}{\rho_1+\rho_2}\Rightarrow\frac{1}{h}>\text{ or, }<\frac{\rho_1+\rho_2}{\rho_1\rho_2}\Rightarrow\frac{1}{h}>\text{ or, }<\frac{1}{\rho_1}+\frac{1}{\rho_2}.$$

Thus, suppose that a body rests in equilibrium upon another body which is fixed and the portion of the two bodies in contact have radii of curvatures ρ_1 and ρ_2 respectively. The CG of the first body is at a height h above the point of contact and the common normal coincides with the vertical. Then the equilibrium is stable or unstable according as $\frac{1}{h}>$ or $<\frac{1}{\rho_1}+\frac{1}{\rho_2}$.

If the portions of the bodies in contact are spheres of radii r_1 and r_2, then in the above condition, we put $\rho_1 = r_1$ and $\rho_2 = r_2$.

Thus, the equilibrium is stable or unstable according as $\frac{1}{h} >$ or $< \frac{1}{r_1} + \frac{1}{r_2}$.

5.5 A Body Resting Inside Other Fixed Concave Body

Let O be the centre of the spherical surface of the lower body and O_1 that of the upper body. The figure is a section of the bodies by a vertical plane through G_1, where G_1 is the CG of the body resting inside the concave body of radius R.

Let $A_1G_1 = h$, A_2 is the new point of contact after displacement (by rolling). A_1 moves to the position C in the displaced position.

Let $\angle A_1OA_2 = \theta = \angle A_2O_2N$ and $\angle CO_2A_2 = \phi$ so that $\angle CO_2N = \phi - \theta$.

Now arc A_1A_2 of sphere of radius R = arc A_2C of sphere of radius r

$$\Rightarrow R\theta = r\phi \Rightarrow \phi = \frac{R\theta}{r}. \qquad (1)$$

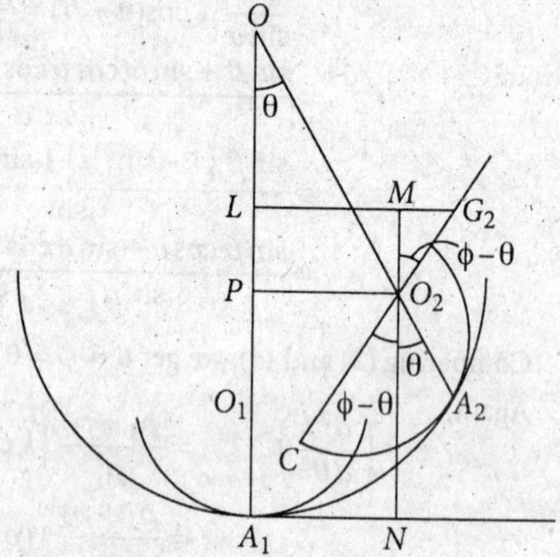

Fig. 5.4

Now, in order to find the nature of equilibrium, we should find the depth z of CG G_2 below O.

$$\therefore\ z = OL = OP - PL = OP - O_2M. \qquad (2)$$

In ΔOO_2P,

$$\cos\theta = \frac{OP}{OO_2} = \frac{OP}{OA_2 - A_2O_2} = \frac{OP}{R-r} \Rightarrow OP = (R-r)\cos\theta.$$

In ΔO_2G_2M,

$$\cos(\phi-\theta) = \frac{O_2M}{O_2G_2} = \frac{O_2M}{CG_2 - CO_2} = \frac{O_2M}{A_1G_1 - r} = \frac{O_2M}{h-r} \Rightarrow O_2M = (h-r)\cos(\phi-\theta).$$

So from (2),

$$z = (R-r)\cos\theta - (h-r)\cos(\phi-\theta)$$

$$\Rightarrow z = (R-r)\cos\theta - (h-r)\cos\left(\frac{R\theta}{r} - \theta\right)$$

$$\Rightarrow z = (R-r)\cos\theta - (h-r)\cos\left(\frac{R}{r} - 1\right)\theta.$$

$$\therefore\ \frac{dz}{d\theta} = -(R-r)\sin\theta + (h-r)\sin\left(\frac{R-r}{r}\right)\theta\left(\frac{R-r}{r}\right)$$

$$\frac{d^2z}{d\theta^2} = -(R-r)\cos\theta + (h-r)\cos\left(\frac{R-r}{r}\right)\theta\left(\frac{R-r}{r}\right)^2$$

$$\frac{d^3z}{d\theta^3}=(R-r)\sin\theta-(h-r)\sin\left(\frac{R-r}{r}\right)\theta\left(\frac{R-r}{r}\right)^3$$

$$\text{and } \frac{d^4z}{d\theta^4}=(R-r)\cos\theta-(h-r)\cos\left(\frac{R-r}{r}\right)\theta\left(\frac{R-r}{r}\right)^4.$$

For stable equilibrium height of CG should be minimum and hence, the depth of CG should be maximum. For stable equilibrium $\frac{d^2z}{d\theta^2}$ should be negative and for unstable equilibrium $\frac{d^2z}{d\theta^2}$ should be positive.

Now for the position of equilibrium,

$$\frac{dz}{d\theta}=0\Rightarrow-(R-r)\sin\theta+(h-r)\sin\left(\frac{R-r}{r}\right)\theta\left(\frac{R-r}{r}\right)=0\Rightarrow\sin\theta=0\Rightarrow\theta=0.$$

Now, $$\left(\frac{d^2z}{d\theta^2}\right)_{\theta=0}=-(R-r)+(h-r)\left(\frac{R-r}{r}\right)^2=\left(\frac{R-r}{r}\right)^2\left[h-r-\frac{r^2}{R-r}\right]$$

$$=\left(\frac{R-r}{r}\right)^2\left[h-\frac{rR}{R-r}\right].$$

So the equilibrium will be stable, if

$$\left(\frac{d^2z}{d\theta^2}\right)_{\theta=0}<0\Rightarrow\left(\frac{R-r}{r}\right)^2\left[h-\frac{rR}{R-r}\right]<0\Rightarrow h-\frac{rR}{R-r}<0$$

$$\Rightarrow h<\frac{rR}{R-r}\Rightarrow\frac{1}{h}>\frac{R-r}{rR}\Rightarrow\frac{1}{h}>\frac{1}{r}-\frac{1}{R}$$

and the equilibrium will be unstable, if $\frac{1}{h}<\frac{1}{r}-\frac{1}{R}$, when $\frac{1}{h}=\frac{1}{r}-\frac{1}{R}$, then $\frac{d^2z}{d\theta^2}=0$.

Also,

$$\left(\frac{d^3z}{d\theta^3}\right)_{\theta=0}=0$$

and

$$\left(\frac{d^4z}{d\theta^4}\right)_{\theta=0}=(R-r)-(h-r)\left(\frac{R-r}{r}\right)^4=(R-r)\left[1-\frac{(R-r)^3}{r^4}(h-r)\right]$$

$$=(R-r)\left[1-\frac{(R-r)^3}{r^4}\left(\frac{rR}{R-r}-r\right)\right]\left(\because h=\frac{rR}{R-r}\right)$$

$$=(R-r)\left[1-\frac{(R-r)^3}{r^4}\cdot\frac{r^2}{R-r}\right]=(R-r)\left[1-\left(\frac{R-r}{r}\right)^2\right].$$

So the equilibrium will be stable or unstable according as

$$1-\left(\frac{R-r}{r}\right)^2<0\text{ or, }>0\Rightarrow\frac{R-r}{r}<-1;\frac{R-r}{r}>1\text{ or, }-1<\frac{R-r}{r}<1$$

$$\Rightarrow R-r<-r, R-r>r\text{ or, }\left|\frac{R-r}{r}\right|<1\Rightarrow R<0, R>2r\text{ or, }R-r<r$$

$$\Rightarrow R>2r(\because R\nless 0)\text{ or, }R<2r.$$

Hence, in the case when $\frac{1}{h}=\frac{1}{r}-\frac{1}{R}$, the equilibrium is stable, if $R>2r$ and unstable, if $R<2r$.

Rules for making sums

1. If the lower body of radius R near the point of contact be a plane, then $R = \infty$. $\therefore \frac{1}{R} = 0$. Thus, equilibrium is stable, if $\frac{1}{h} > \frac{1}{r} + \frac{1}{R}$ or, $\frac{1}{h} > \frac{1}{r}$ or, $h < r$; and $h > r$ for unstable.
2. If the upper body of radius r near the point of contact be a plane, then $r = \infty$, $\frac{1}{r} = 0$. Hence, the equilibrium is stable, if
$$\frac{1}{h} > \frac{1}{r} + \frac{1}{R} \text{ or, } \frac{1}{h} > \frac{1}{R} \text{ or, } h < R; \text{ and } h > R \text{ for unstable.}$$
3. If a body of radius r resting inside another concave body of radius R and h is the height of CG of the body resting above the point of contact,
$\frac{1}{h} > \frac{1}{r} - \frac{1}{R}$ for stable and $\frac{1}{h} < \frac{1}{r} - \frac{1}{R}$ for unstable.
If, however, $\frac{1}{h} = \frac{1}{r} - \frac{1}{R}$, then $R > 2r$ for stable and $R < 2r$ for unstable.

WORKED-OUT EXAMPLES

1. A hemisphere rests in equilibrium on a sphere of equal radius, show that the equilibrium is unstable when the curved surface of the hemisphere and stable when the flat surface of the hemisphere rests on the sphere.

[LNMU(H) 2011, TMBU(H) 1976; NBU(H) 2006]

Solution:

Case I. *When the curved surface of the hemisphere rests on the sphere:* A hemisphere of centre O_1 rests on a sphere of centre O with its curved surface in contact with the sphere. The point of contact is A and $OA = O_1A = a$.

If G is the CG, of the hemisphere, then G lies on O_1A and $O_1G = 3a/8$.

Here $r =$ the radius of upper body $= a$.

Also $R = a$.

And

$$h = AG = O_1A - O_1G = a - \frac{3a}{8} = \frac{5}{8}a.$$

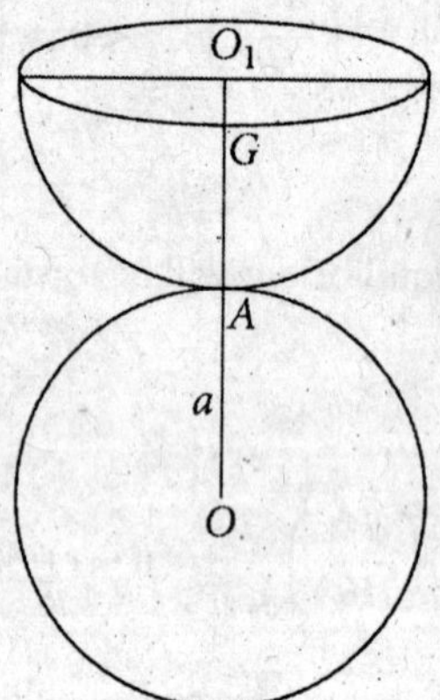

Fig. 5.5

Now

$$\frac{1}{h} = \frac{8}{5a} \text{ and } \frac{1}{r} + \frac{1}{R} = \frac{1}{a} + \frac{1}{a} = \frac{1}{2a}.$$

Thus, $\frac{1}{h} < \frac{1}{r} + \frac{1}{R}$. Hence, the equilibrium is unstable in this case.

Case II. *When the flat surface of the hemisphere rests on the sphere:*

Here the upper body has a plane face in contact with the lower. $\therefore$ $r = \infty$.

$R =$ the radius of the lower body $= a$.

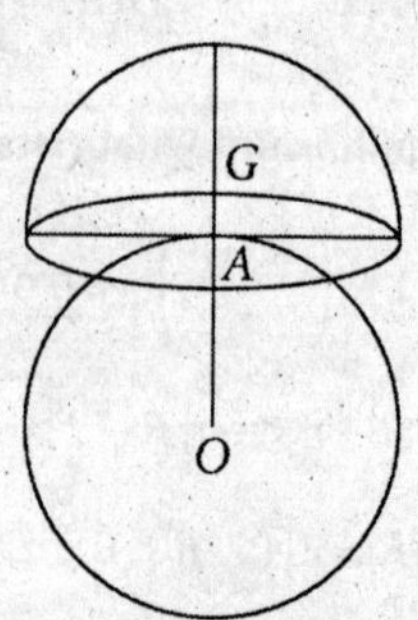

Fig. 5.6

Also $h=3a/8$.

Now

$$\frac{1}{R}+\frac{1}{r}=\frac{1}{a}+0=\frac{1}{a} \text{ and } \frac{1}{h}=\frac{8}{3a}.$$

Thus, $\frac{1}{h}>\frac{1}{r}+\frac{1}{R}$.

Hence, the equilibrium is stable.

2. A body, consisting of a cone and a hemisphere on the same base, rests on a rough horizontal table, the hemisphere being in contact with the table. Show that the greatest height of the cone, so that equilibrium may be stable, is $\sqrt{3}$ times the radius of the hemisphere. [RU(H) 1997, BVABU(H) 1996, KU(H) 2007, CU(H) 2005, PU(H) 2013]

Solution: AB is the common base of the hemisphere and the cone and COD is their common axis which must be vertical for their equilibrium. The hemisphere touches the table at C.

Let H be the height OD of the cone and r be the radius OA, OB or OC of the hemisphere.

Let G_1 and G_2 be the CG of the hemisphere and the cone respectively. Then $OG_1=3r/8$ and $OG_2=H/4$.

If h be the height of the CG of the combined body composed of the hemisphere and the cone above the point of contact C, then using the formula

$$\bar{x}=\frac{w_1x_1+w_2x_2}{w_1+w_2}. \qquad (5.1)$$

w_1 = weight of the hemisphere = $\frac{2}{3}\pi r^3\rho g$.

x_1 = distance of the CG of the hemisphere from the table = CG_1 = $CO-OG_1=r-3r/8=5r/8$.

w_2 = weight of the cone $=\frac{1}{3}\pi r^2 H\rho g$.

x_2 = distance of the CG of the cone from the table $=CG_2=CO+OG_2=r+H/4$.

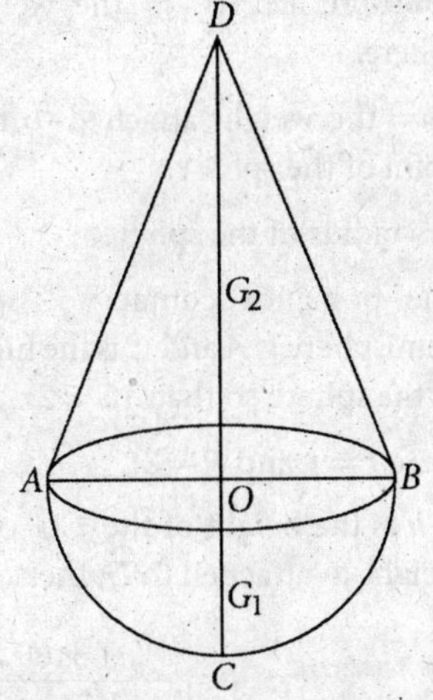

Fig. 5.7

So from (1),

$$h=\frac{\frac{2}{3}\pi r^3\rho g\cdot\frac{5r}{8}+\frac{1}{3}\pi r^2H\rho g\left(r+\frac{H}{4}\right)}{\frac{2}{3}\pi r^3\rho g+\frac{1}{3}\pi r^2H\rho g}=\frac{H\left(r+\frac{H}{4}\right)+2r\frac{5r}{8}}{2r+H}=\frac{H\left(r+\frac{H}{4}\right)+\frac{5}{4}r^2}{2r+H}.$$

Since the lower surface near the point of contact is a plane, therefore, $R=\infty$.

Also $r=r$.

The equilibrium will be stable, if

$$\frac{1}{h} > \frac{1}{r} + \frac{1}{R} \text{ or, } \frac{2r+H}{\frac{5}{4}r^2 + Hr + \frac{H^2}{4}} > \frac{1}{r} + \frac{1}{\infty}, \text{ i.e., } 2r^2 + rH > \frac{5}{4}r^2 + rH + \frac{H^2}{4}$$

$$\text{or, } 8r^2 + 4rH > 5r^2 + 4rH + H^2 \text{ or, } 3r^2 > H^2$$

$$\Rightarrow H^2 < 3r^2 \Rightarrow H^2 - 3r^2 < 0 \Rightarrow -\sqrt{3}r < H < \sqrt{3}r \Rightarrow |H| < \sqrt{3}r.$$

Hence, the greatest height of the cone, so that the equilibrium may be stable, is $\sqrt{3}$ times the radius of the hemisphere.

3. A solid sphere rests inside a fixed rough hemisphere bowl of twice its radius. Show that however large a weight is attached to the highest point of the sphere, the equilibrium is stable.

[PU(H) 1999, LNMU(H) 1999, RU(H) 1992, BVABU(H) 1996, NBU(H) 2004, PU(H) 2009, MU(H) 2007]

Solution: Let W = the weight of the sphere.

w = the weight attached to the highest point of the sphere.

r = radius of the sphere.

The point of contact of sphere and hemisphere is A and C is the highest point of the sphere so that $AC = 2r$.

Here $r = r$ and $R = 2r$.

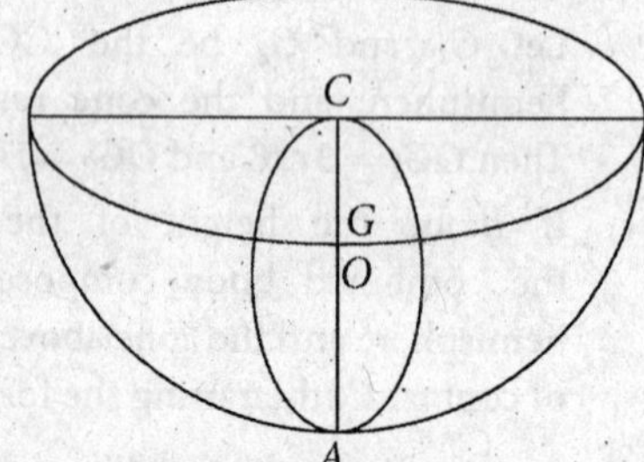

Fig. 5.8

If h is the height of the CG of the combined body consisting of the sphere and the weight w attached to O, then

$$h = \frac{W \cdot r + w \cdot 2r}{W + w} \quad \left(\text{by the formula } \bar{x} = \frac{w_1 x_1 + w_2 x_2}{w_1 + w_2}\right)$$

$$\Rightarrow \frac{1}{h} = \frac{w + W}{r(W + 2w)}.$$

The equilibrium will be stable, provided that

$$\frac{1}{h} > \frac{1}{r} - \frac{1}{R} \Rightarrow \frac{W+w}{r(W+2w)} > \frac{1}{r} - \frac{1}{2r} \Rightarrow \frac{W+w}{r(W+2w)} > \frac{1}{2r}$$

$$\text{or, } 2W + 2w > W + 2w, \text{ or, } 2W > W \text{ or, } 2 > 1,$$

which is true for all values of W.

Hence, the equilibrium is stable, however large w may be.

4. A solid frustum of a paraboloid of revolution of height h and latus rectum $4a$ rests with its vertex on the vertex of a paraboloid of revolution, whose latus rectum is $4b$. Show that the equilibrium is stable, if $h < 3ab/(a+b)$.

[IAS 1997, RU(H) 1998, NBU(H) 2006, CU(H) 2008, VU(H) 2002]

Solution: The point of contact of the two bodies is A and $AB = h$.

Let the equation of the generating parabola of the upper paraboloid be $y^2 = 4ax$. Differentiating $y^2 = 4ax$ w.r.t. x, we get

$$2y\frac{dy}{dx} = 4a.$$

$$y_1 = \frac{dy}{dx} = \frac{2a}{y} = \frac{2a}{2\sqrt{ax}} = \sqrt{\frac{a}{x}}.$$

$$y_2 = \frac{d^2y}{dx^2} = -\frac{1}{2}\sqrt{a}\, x^{-3/2}.$$

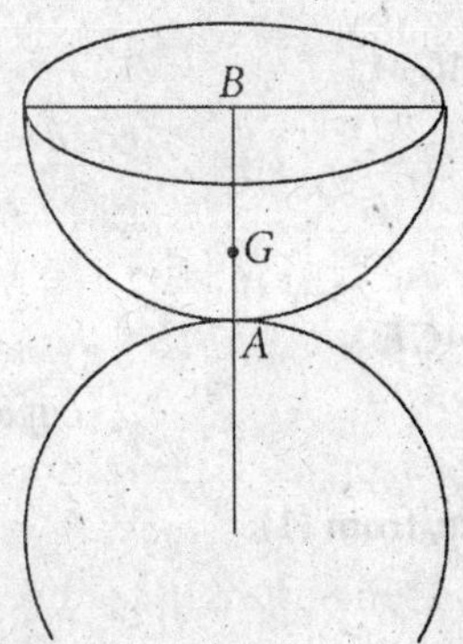

Fig. 5.9

Let G be the CG of the upper body and H be the height of the CG of the upper body above the point of contact A, then

$$H = AG = x = \frac{\int_0^h x\pi y^2 dx}{\int_0^h \pi y^2 dx} = \frac{\int_0^h x(4ax)\, dx}{\int_0^h 4ax dx} = \frac{\int_0^h x^2 dx}{\int_0^h x\, dx} = \frac{h^3/3}{h^2/2} = \frac{2}{3}h.$$

Now

$$\rho_1 = \frac{\left(1+y_1^2\right)^{3/2}}{y_2} = \frac{\left(1+\frac{a}{x}\right)^{3/2}}{\frac{-\sqrt{a}}{2x^{3/2}}} = \frac{2(x+a)^{3/2}}{-\sqrt{a}} = \frac{2(x+a)^{3/2}}{\sqrt{a}}$$

(in magnitude).

At A, $\rho_1 = \frac{2a^{3/2}}{\sqrt{a}} = 2a$.

Similarly, radius of curvature ρ_2 of the lower paraboloid at its vertex is $2b$, $\rho_2 = 2b$.

Now the equilibrium will be stable, if

$$\frac{1}{x} > \frac{1}{\rho_1} + \frac{1}{\rho_2}, \text{ i.e., if } \frac{3}{2h} > \frac{1}{2a} + \frac{1}{2b} \text{ or, } \frac{3}{h} > \frac{a+b}{ab} \text{ or, } h < \frac{3ab}{a+b}.$$

5. A heavy uniform rod rests with one end against a smooth vertical wall and with a point in its length resting on a smooth peg. Find the position of equilibrium and show that it is stable. [IAS 1985]

Solution: Let AB be the rod with its end A resting against a vertical wall and a point of its length resting against a peg at P. Let G be the CG of the rod. Let $PC = a$ be the distance of the wall from the peg P. Let $AB = 2l$ be the length of the rod. Let $z = GH$ be the height of CG above the peg P. Then

$$z = GH = GM - MH = GM - AC. \quad (1)$$

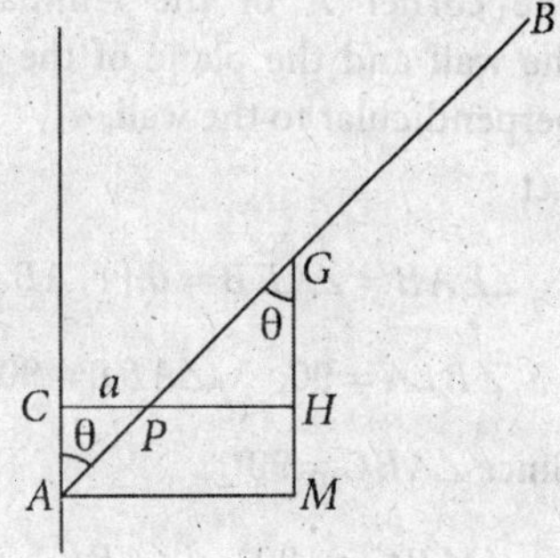

Fig. 5.10

In $\triangle AGM$,

$$\cos\theta = \frac{GM}{AG} = \frac{GM}{l}\ [\because AG = GB = l];$$
$$GM = l\cos\theta.$$

In $\triangle ACP$,

$$\cot\theta = \frac{AC}{CP} = \frac{AC}{a};\ AC = a\cot\theta.$$

Hence, from (1),

$$z = l\cos\theta - a\cot\theta$$
$$\therefore\ \frac{dz}{d\theta} = -l\sin\theta + a\,\text{cosec}^2\,\theta$$
$$\text{and}\ \frac{d^2z}{d\theta^2} = -l\cos\theta - 2a\,\text{cosec}^2\,\theta\cot\theta.$$

For equilibrium, $\frac{dz}{d\theta} = 0$,

$$\Rightarrow -l\sin\theta + a\,\text{cosec}^2\,\theta = 0 \Rightarrow l\sin\theta = a\,\text{cosec}^2\,\theta = \frac{a}{\sin^2\theta} \Rightarrow \sin^3\theta = \frac{a}{l}.$$

$$\therefore\ \sin\theta = \left(\frac{a}{l}\right)^{1/3}.$$

This gives the position of equilibrium.

Again, $d^2z/d\theta^2$ is negative, for all acute values of θ.

Hence, the equilibrium is stable.

6. A square lamina rests with its plane perpendicular to a smooth wall, one corner being attached to a point of the wall by a fine string of length a equal to a side of the square. Find the position of equilibrium and show that it is stable.

[PU(H) 1995, MU(H) 2008, CU(H) 2000, VU(H) 2005, KU(H) 2008]

Solution: Let $ABCD$ be a square lamina of side $2a$. It is suspended from the point E in the wall by a fine string EB of length $2a$.

The corner A of the lamina touches the wall and the plane of the lamina is perpendicular to the wall.

Let

$$\angle EAB = \angle AEB = \theta\ (\because AB = EB)$$

$$\angle BLA = 90,\ \therefore\ \angle ABL = 90° - \theta.$$

Since $\angle ABC = 90°$,

$$\therefore\ \angle MBC = 180° - (\angle ABL + \angle ABC)$$
$$= 180° - (90° - \theta + 90°) = \theta.$$

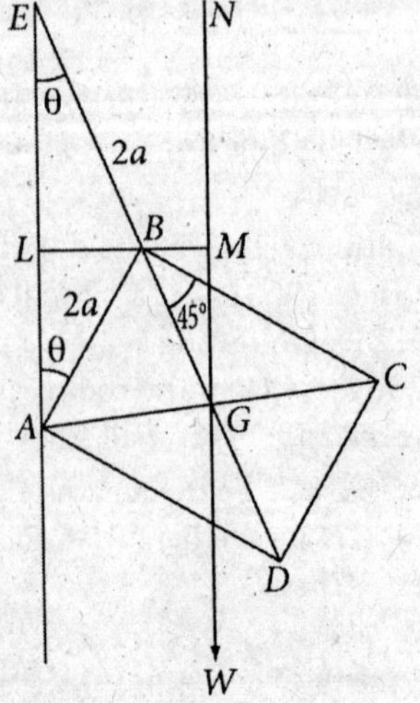

Fig. 5.11

$$\angle MBG' = \theta + 45°.$$

In ΔBGM,

$$\sin(45° + \theta) = \frac{GM}{BG}$$
$$BG = \tfrac{1}{2}BD = \tfrac{1}{2}2a\sqrt{2} = a\sqrt{2}$$
$$\therefore\ GM = a\sqrt{2}\sin(45° + \theta) = a\sqrt{2}(\sin 45° \cos\theta + \cos 45° \sin\theta)$$
$$= \frac{a\sqrt{2}}{\sqrt{2}}(\cos\theta + \sin\theta) = a(\cos\theta + \sin\theta).$$

Now z = depth of G below $E = GN = GM + MN = GM + EL$.
In ΔELB,

$$\cos\theta = \frac{EL}{EB} = \frac{EL}{2a}.$$
$$\therefore\ EL = 2a\cos\theta.$$

$$z = a(\cos\theta + \sin\theta) + 2a\cos\theta = a(3\cos\theta + \sin\theta).$$

$$\therefore\ \frac{dz}{d\theta} = a(-3\sin\theta + \cos\theta)$$
$$\frac{d^2z}{d\theta^2} = a(-3\cos\theta - \sin\theta).$$

For equilibrium, $dz/d\theta = 0$.

$$-3a\sin\theta + a\cos\theta = 0,\ 3\sin\theta = \cos\theta,\ \tan\theta = \frac{1}{3}.$$

This gives position of equilibrium.
Now

$$\frac{d^2z}{d\theta^2} = -a\left[\frac{3}{\sec\theta} + \frac{1}{\operatorname{cosec}\theta}\right] = -a\left[\frac{3}{\sqrt{1+\tan^2\theta}} + \frac{1}{\sqrt{1+\cot^2\theta}}\right]$$
$$= -a\left[\frac{3}{\sqrt{1+\frac{1}{9}}} + \frac{1}{\sqrt{1+9}}\right] = -a\left[\frac{9}{\sqrt{10}} + \frac{1}{\sqrt{10}}\right] = \frac{-a10}{\sqrt{10}} = -a\sqrt{10} < 0.$$

Therefore, the depth of the CG is maximum or its height above the horizontal is minimum. Hence, the equilibrium is stable.

7. A uniform cubical box of edge A is placed on the top of a fixed sphere, the centre of the face of the cube being in contact with the highest point of the sphere. What is the least radius of the sphere for which the equilibrium will be stable?

[MU(H) 1991, CU(H) 2003]

Solution: A uniform cubical box of edge a is placed on the top of a fixed sphere of centre O. The point of contact is A. If G is the CG of the box, then for equilibrium the line OAG must be vertical.

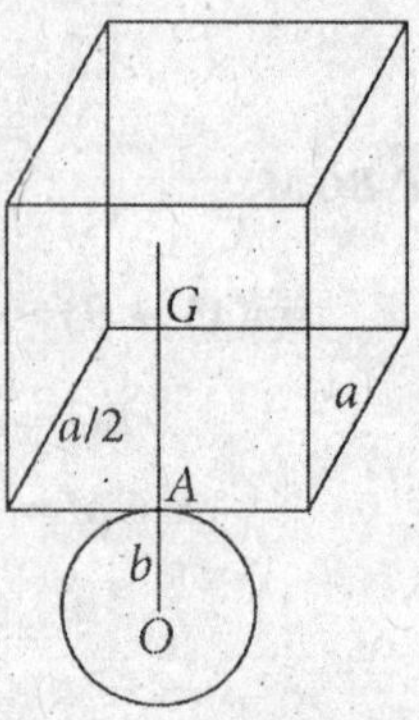

Fig. 5.12

Let the radius of the sphere be b.

Here $r=\infty$, $R=b$.

h = the height of the CG of the box above the point of contact A = half the edge of the box = $a/2$.

The equilibrium will be stable, if $\frac{1}{h}>\frac{1}{r}+\frac{1}{R}$

$$\Rightarrow \frac{2}{a}>\frac{1}{\infty}+\frac{1}{b}\Rightarrow \frac{2}{a}>\frac{1}{b}\Rightarrow \frac{a}{2}<b\Rightarrow b>\frac{a}{2}.$$

Hence, the least value of b for the equilibrium to be stable is $a/2$.

8. A solid homogeneous hemisphere of radius r has a solid right circular cone of the same substance constructed on the base, the hemisphere rests on the convex side of the fixed sphere of radius R. Show that the axis of the cone consistent with stability for a small rolling displacement is

$$\frac{r}{R+r}\left[\sqrt{\{(3R+r)(R-r)\}}-2r\right]_V$$

Solution: Let G_1 and G_2 be the centre of gravities of the hemisphere and the cone respectively, and G be their common CG, such that $AG=h$. Let H be the height of the cone.

$$\therefore\ h=\frac{\frac{1}{3}\pi r^2H\rho g\left(r+\frac{H}{4}\right)+\frac{2}{3}\pi r^3\rho g\cdot\frac{5r}{8}}{\frac{1}{3}\pi r^2H\rho g+\frac{2}{3}\pi r^3\rho g}\Rightarrow h=\frac{H\left(r+\frac{H}{4}\right)+\frac{5}{4}r^2}{H+2r}$$

$$\Rightarrow h=\frac{H(4r+H)+5r}{4(H+2r)}. \quad (1)$$

The equilibrium will be stable of

$$\frac{1}{h}>\frac{1}{r}+\frac{1}{R}$$

$$\Rightarrow\frac{4(H+2r)}{H(4r+H)+5r^2}>\frac{r+R}{rR}$$

$$\Rightarrow(4H+8r)rR>(r+R)\left(5r^2+4rH+H^2\right)$$

$$\Rightarrow 4rRH+8r^2R>5r^3+4r^2H+rH^2+5Rr^2+4RrH+H^2R$$

$$\Rightarrow 0>5r^2+4r^2H+rH^2-3Rr^2+H^2R$$

$$\Rightarrow H^2(R+r)+4r^2H+5r^3-3Rr^2=0.$$

The corresponding equation is

$$H^2(R+r)+4r^2H+5r^3-3Rr^2=0.$$

$$\therefore\ H=\frac{-4r^2\pm\sqrt{16r^4-4(R+r)r^2(5r-3R)}}{2(R+r)}$$

$$=\frac{-2r^2\pm\sqrt{4r^4-r^2(R+r)(5r-3R)}}{(R+r)}$$

$$=\frac{-2r^2\pm\sqrt{r^2\{4r^2-(5rR+5r^2-3r^2-3Rr)\}}}{R+r}$$

$$=\frac{-2r^2\pm\sqrt{r^2(3R^2-2rR-r^2)}}{R+r}-2r^2$$

$$\Rightarrow H=\frac{-2r^2\pm r\sqrt{(3R+r)(R-r)}}{R+r}$$

$$\Rightarrow H=\frac{r}{R+r}\left[-2r\pm\sqrt{(3R+r)(R-r)}\right].$$

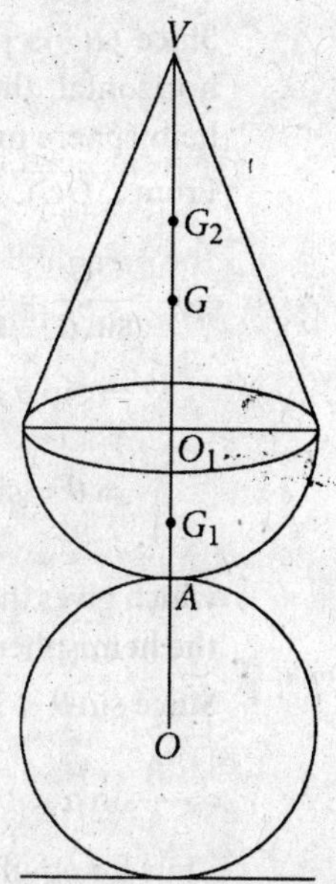

Fig. 5.13

Let the two values of H be α and β, where

$$\alpha=\frac{r}{R+r}\left[-2r-\sqrt{(3R+r)(R-r)}\right]<0$$

and

$$\beta=\frac{r}{R+r}\left[-2r+\sqrt{(3R+r)(R-r)}\right].$$

Also, $R+r>0$.

$\therefore\ \alpha<H<\beta$ and also $H>0$.

But $\alpha<0$.

So, we must have $H<\beta$, i.e.,

$$H<\frac{r}{R+r}\left[\sqrt{(3R+r)(R-r)}-2r\right].$$

So for stability,

$$H<\frac{r}{R+r}\left[\sqrt{(3R+r)(R-r)}-2r\right]$$

or the greatest value of H consistent with the stability of the equilibrium is

$$\frac{r}{R+r}\left[\sqrt{(3R+r)(R-r)}-2r\right].$$

9. A solid hemisphere rests on a plane inclined to the horizon at an angle $\alpha<\sin^{-1}\frac{3}{8}$ and the plane is rough enough to prevent any sliding. Find the position of equilibrium and show that it is stable. **[PU(H) 1991]**

Solution: Let O be the centre of the base of the hemisphere and r be its radius. If C is the point of contact of the hemisphere and the inclined plane, then $OC=r$.

Let G be the CG of the hemisphere. Then $OG=3r/8$. In the position of equilibrium the line CG must be vertical.

Since OC is perpendicular to the inclined plane and CG is perpendicular to the horizontal, therefore, $\angle OCG = \alpha$. Let in the equilibrium position, the axis of the hemisphere make an angle θ with the vertical.

From ΔOGC, we have

$$\frac{OG}{\sin\alpha} = \frac{OC}{\sin\theta} \Rightarrow \frac{3r/8}{\sin\alpha} = \frac{r}{\sin\theta}$$

$$\Rightarrow \sin\theta = \frac{8}{3}\sin\alpha$$

$$\Rightarrow \theta = \sin^{-1}\left(\frac{8}{3}\sin\alpha\right), \qquad (1)$$

which gives the position of equilibrium of the hemisphere.

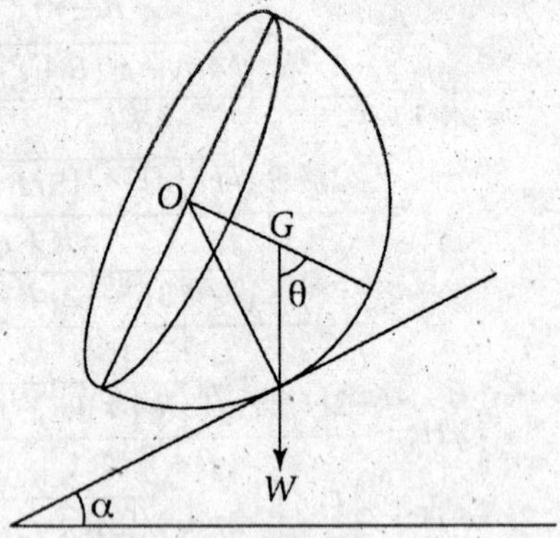

Fig. 5.14

Since $\sin\theta < 1$

$$\Rightarrow \frac{8}{3}\sin\alpha < 1 \Rightarrow \sin\alpha < \frac{3}{8} \Rightarrow \alpha < \sin^{-1}\frac{3}{8}.$$

Thus for equilibrium to exist, we must have $\alpha < \sin^{-1} 3/8$.

Let CG $= h$. Then

$$\frac{h}{\sin(\theta-\alpha)} = \frac{3r/8}{\sin\alpha} \Rightarrow h = \frac{3r\sin(\theta-\alpha)}{8\sin\alpha}.$$

Here $\rho_1 = r$ and $\rho_2 = \infty$. So the equilibrium will be stable, if

$$h < \frac{\rho_1\rho_2\cos\alpha}{\rho_1+\rho_2} \Rightarrow \frac{1}{h} > \frac{\rho_1+\rho_2}{\rho_1\rho_2}\sec\alpha \Rightarrow \frac{1}{h} > \left(\frac{1}{\rho_1}+\frac{1}{\rho_2}\right)\sec\alpha \Rightarrow \frac{1}{h} > \frac{1}{r}\sec\alpha$$

$$\Rightarrow h < r\cos\alpha \Rightarrow \frac{3r\sin(\theta-\alpha)}{8\sin\alpha} < r\cos\alpha \Rightarrow 3\sin(\theta-\alpha) < 8\sin\alpha\cos\alpha$$

$$\Rightarrow 3\sin\theta\cos\alpha - 3\cos\theta\sin\alpha < 8\sin\alpha\cos\alpha$$

$$\Rightarrow 8\sin\alpha\cos\alpha - 3\sin\alpha\sqrt{1-\sin^2\theta} < 8\sin\theta\cos\alpha \text{ [from (1)]}$$

$$\Rightarrow 8\sin\alpha\cos\alpha - 3\sin\alpha\sqrt{1-\frac{64}{9}\sin^2\alpha} < 8\sin\alpha\cos\alpha \quad \left[\because \sin\theta = \frac{8}{3}\sin\alpha\right]$$

$$\Rightarrow -\sin\alpha\sqrt{9-64\sin^2\alpha} < 0 \Rightarrow \sin\alpha\sqrt{9-64\sin^2\alpha} > 0$$

$$\Rightarrow \sqrt{9-64\sin^2\alpha} > 0 \; (\because \sin\alpha > 0)$$

$$\Rightarrow 9-64\sin^2\alpha > 0 \Rightarrow 9 > 64\sin^2\alpha \Rightarrow \sin^2\alpha < \frac{9}{64} \Rightarrow -\frac{3}{8} < \sin\alpha < \frac{3}{8} \Rightarrow |\sin\alpha| < \frac{3}{8},$$

which is true.

Hence, the equilibrium is stable.

10. An isosceles triangle of angle 2α rests between two smooth pegs at the same level, distance $2c$ apart, prove that if h be the distance of the CG from the vertex and if $2c\sec\alpha < h < 2c\sec\alpha\operatorname{cosec}\alpha$, the oblique positions of equilibrium exists, which are unstable. Discuss the stability of the vertical position in case $h = 2c\sec\alpha\operatorname{cosec}\alpha$.

[VU(H) 2008]

Solution: Let ABC be an isosceles triangle, A its vertex, P, Q the two pegs and G the CG. Let z be the height of G above PQ.

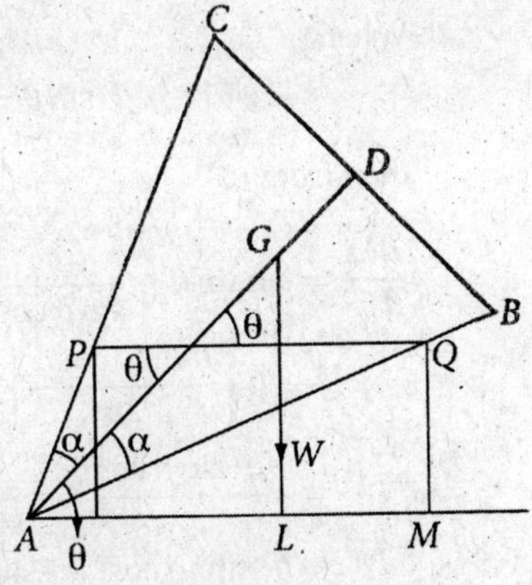

Fig. 5.15

$$z = GL - QM$$
$$= AG\sin\theta - AQ\sin(\theta-\alpha). \qquad (1)$$

In $\triangle APQ$, $PQ = 2c$ and

$$\frac{AQ}{\sin\{\pi-(\theta+\alpha)\}} = \frac{2c}{\sin 2\alpha}$$
$$\Rightarrow AQ = \frac{2c\sin(\theta+\alpha)}{\sin 2\alpha}.$$

So from (1),

$$z = h\sin\theta - \frac{2c\sin(\theta+\alpha)\sin(\theta-\alpha)}{\sin 2\alpha} = h\sin\theta - 2c\operatorname{cosec}2\alpha(\sin^2\theta - \sin^2\alpha).$$

$$\therefore \frac{dz}{d\theta} = h\cos\theta - 4c\operatorname{cosec}2\alpha\sin\theta\cos\theta \qquad (2)$$

and

$$\frac{d^2z}{d\theta^2} = -h\sin\theta - 4c\operatorname{cosec}2\alpha\cos 2\theta. \qquad (3)$$

For equilibrium $\frac{dz}{d\theta} = 0$

$$\Rightarrow \cos\theta(h - 4c\operatorname{cosec}2\alpha\sin\theta) = 0$$
$$\Rightarrow \text{either } \cos\theta = 0$$
$$\text{or, } \sin\theta = \frac{h\sin 2\alpha}{4c} = \frac{h\sin\alpha\cos\alpha}{2c}.$$

When

$$h\sin\alpha\cos\alpha < 2c \text{ or, } h < \frac{2c}{\sin\alpha\cos\alpha},$$

there are three positions of equilibrium, one $\theta = \pi/2$ and the other two given by

$$\sin\theta = \frac{h\sin\alpha\cos\alpha}{2c} = \sin(\pi-\theta).$$

If, however, $h\sin\alpha\cos\alpha > 2c$, then there is only one position of equilibrium given by $\cos\theta = 0$, i.e., $\theta = \pi/2$, because the other equation

$$\sin\theta = \frac{h\sin\alpha\cos\alpha}{2c}$$

will not hold good.

When

$$\sin\theta = \frac{h\sin\alpha\cos\alpha}{2c} \quad \text{and} \quad h < \frac{2c}{\sin\alpha\cos\alpha},$$

then from (3),

$$\frac{d^2z}{d\theta^2} = -h\sin\theta - \frac{2c}{\sin\alpha\cos\alpha}(1-2\sin^2\theta)$$

$$= -h\left\{\frac{h\sin\alpha\cos\alpha}{2c}\right\} - \frac{2c}{\sin\alpha\cos\alpha}\left\{1 - \frac{2h^2\sin^2\alpha\cos^2\alpha}{4c^2}\right\}$$

$$= -\frac{h^2\sin\alpha\cos\alpha}{2c} - \frac{2c}{\sin\alpha\cos\alpha} + \frac{2h^2\sin\alpha\cos\alpha}{2c} = \frac{h^2\sin\alpha\cos\alpha}{2c} - \frac{2c}{\sin\alpha\cos\alpha}$$

$$= \frac{h^2\sin^2\alpha\cos^2\alpha - 4c^2}{2c\sin\alpha\cos\alpha} = \frac{(h\sin\alpha\cos\alpha - 2c)(h\sin\alpha\cos\alpha + 2c)}{2c\sin\alpha\cos\alpha} < 0$$

as $h < \dfrac{2c}{\sin\alpha\cos\alpha}$.

Also

$$\alpha < \theta \Rightarrow \sin\alpha < \sin\theta = \frac{h\sin\alpha\cos\alpha}{2c} \Rightarrow 2c\sec\alpha < h.$$

Hence, from (4), z is maximum which shows that equilibrium is unstable under the condition

$$2c\sec\alpha < h < \frac{2c}{\sin\alpha\cos\alpha}.$$

Again,

$$\left(\frac{d^2z}{d\theta^2}\right)_{\theta=\pi/2} = -h + \frac{2c}{\sin\alpha\cos\alpha} = \frac{-2c}{\sin\alpha\cos\alpha} + \frac{2c}{\sin\alpha\cos\alpha} = 0, \text{ if } h = \frac{2c}{\sin\alpha\cos\alpha}.$$

$$\therefore \left(\frac{d^3z}{d\theta^3}\right)_{\theta=\pi/2} = [-h\cos\theta + 8c\,\text{cosec}\,2\alpha 2\alpha\sin 2\theta]_{\theta=\frac{\pi}{2}} = 0.$$

$$\left(\frac{d^4z}{d\theta^4}\right)_{\theta=\pi/2} = [h\sin\theta + 16c\,\text{cosec}\,2\alpha\cos 2\theta]_{\theta=\frac{\pi}{2}} = h - 16c\,\text{cosec}\,2\alpha$$

$$= \frac{2c}{\sin\alpha\cos\alpha} - \frac{8c}{\sin\alpha\cos\alpha}, \text{ if } h = \frac{2c}{\sin\alpha\cos\alpha}$$

$$= \frac{-6c}{\sin\alpha\cos\alpha} < 0.$$

Hence, z is maximum, i.e., equilibrium is unstable in the vertical position under the condition

$$h = \frac{2c}{\sin\alpha\cos\alpha}.$$

11. A heavy hemispherical shell of radius r has a particle attached to a point on the rim and rests with the curved surface in contact with a rough sphere of radius R at the highest point. Prove that if $\frac{R}{r} > \sqrt{5} - 1$, the equilibrium is stable, whatever be the weight of the particle. [RU(H) 2000]

Solution: Let O' be the centre of the base of the hemispherical shell of radius r. Let a weight be attached to the rim of the hemispherical shell at A. The CG G_1 of the hemispherical shell is on its symmetrical radius $O'D$ and

$$O'G_1 = \tfrac{1}{2}O'G = \tfrac{1}{2}r.$$

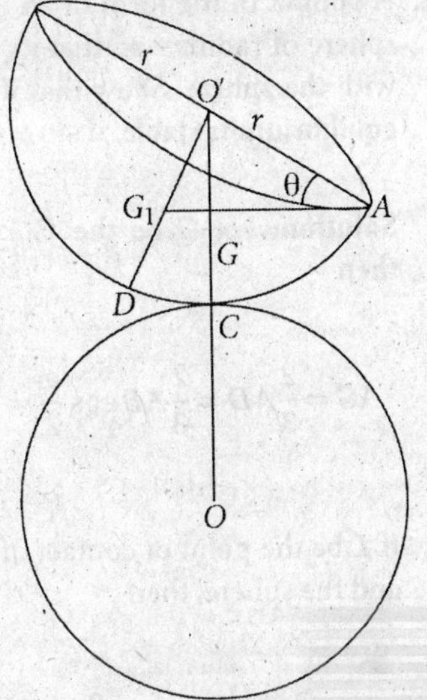

Fig. 5.16

Let G be the CG of the combined body consisting of the hemispherical shell and the weight at A. Then G lies on the line AG_1.

The hemispherical shell rests with its curved surface in contact with a rough sphere of radius R and centre O at the highest point C. For equilibrium the line $OCGO'$ must be vertical but AG_1 need not be horizontal.

Let $CG = h$. The equilibrium will be stable, if

$$\frac{1}{h} > \frac{1}{r} + \frac{1}{R} = \frac{R+r}{rR}, \text{ i.e., } h < \frac{rR}{r+R}. \tag{5.2}$$

The value of h depends on the weight of the particle attached at A. So the equilibrium will be stable, whatever be the weight of the particle attached at A, if the relation (1) holds even for the maximum value of h.

Now h will be maximum if $O'G$ is minimum, i.e., if $O'G$ is perpendicular to AG_1 or if $\Delta AO'G$ is right-angled.

Let $\angle O'AG = \theta$. Then from right-angled $\Delta AO'G_1$,

$$\tan\theta = \frac{O'G_1}{O'A} = \frac{r/2}{r} = \frac{1}{2}. \therefore \sin\theta = \frac{1}{\sqrt{5}}.$$

Hence, the minimum value of

$$O'G = O'A\sin\theta = \frac{r}{\sqrt{5}}.$$

So, the maximum value of $h = r-$ the minimum value of $O'G$

$$= r - \frac{r}{\sqrt{5}} = \frac{r\left(\sqrt{5}-1\right)}{\sqrt{5}}.$$

Hence, the equilibrium will be stable, whatever be the weight of the particle at A, if

$$\frac{r\left(\sqrt{5}-1\right)}{\sqrt{5}} < \frac{rR}{R+r} \Rightarrow \frac{\sqrt{5}-1}{\sqrt{5}} < \frac{R}{R+r} \Rightarrow \left(\sqrt{5}-1\right)R + \left(\sqrt{5}-1\right)r < R\sqrt{5}$$

$$\Rightarrow \left(\sqrt{5}-1\right)r < R \Rightarrow \frac{R}{r} > \sqrt{5}-1 \textbf{ (Proved).}$$

12. **A lamina in the form of an isosceles triangle whose vertical angle is α, placed on a sphere of radius r so that its plane is vertical and one of its equal sides is in contact with the sphere. Show that if the triangle be slightly displaced in its own plane, the equilibrium is stable, if $\sin\alpha < \frac{3r}{a}$, where a is one of the equal sides of the triangle.**

[CU(H) 1999]

Solution: Let G be the CG of ΔABC, then

$$AG = \frac{2}{3}AD = \frac{2}{3}AB\cos\frac{\alpha}{2} = \frac{2}{3}a\cos\frac{\alpha}{2}.$$

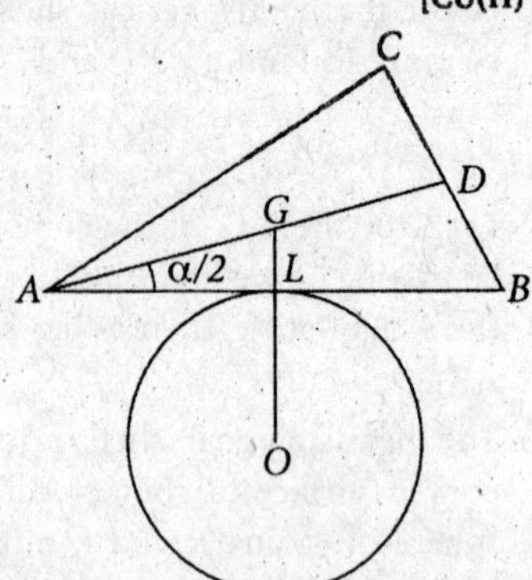

Fig. 5.17

If L be the point of contact of the lamina and the sphere, then

$$h = GL = AG\sin\frac{\alpha}{2} = \frac{2}{3}a\cos\frac{\alpha}{2}\sin\frac{\alpha}{2} = \frac{a}{3}\sin\alpha.$$

Hence, the equilibrium will be stable, if

$$\frac{1}{h} > \frac{1}{r} + \frac{1}{R} \Rightarrow \frac{3}{a\sin\alpha} > \frac{1}{r} + \frac{1}{\infty}.$$

Here $R = \infty$, as the upper body has a plane face in contact with the sphere

$$\Rightarrow \frac{3}{a\sin\alpha} > \frac{1}{r} \Rightarrow \frac{a\sin\alpha}{3} < r \Rightarrow \sin\alpha < \frac{3r}{a}.$$

13. **A uniform rod of length $2l$ has its lower end attached by alight string of length r to a point O and it is constrained to pass through a fixed point A at a distance c vertically above O. Show that the rod is table in the vertical position of $l < \frac{(c+r)^2}{r}$.** **[MU(H) 2009]**

Solution: Let AC be the given uniform rod of length $2l$. Let W be the weight of the rod which acts at its CG G vertically downwards. The lower end C of the rod is connected by a light string CO.

Then $CO = r$. Also $OL = c$.

Let CL and CO make angle θ and ϕ with the vertical.

Let z be the height of G above O.

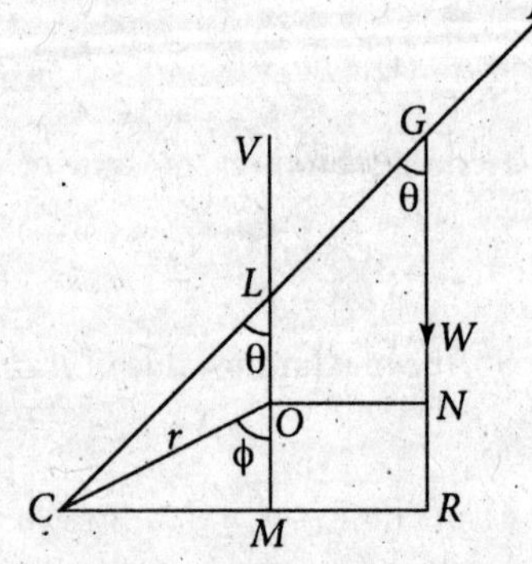

Fig. 5.18

Then,
$$z = GN = GR - NR = CG\cos\theta - OM = l\cos\theta - r\cos\phi$$

$$\Rightarrow z = l\cos\theta - r\cos\phi. \tag{1}$$

In ΔCLM,

$$\tan\theta = \frac{CM}{LM} = \frac{OC\sin\phi}{LO+OM} = \frac{r\sin\phi}{c+r\cos\phi}. \tag{2}$$

$\therefore$ the potential energy V is given by

$$V = Wz = W(l\cos\theta - r\cos\phi).$$

$$\therefore \frac{dV}{d\theta} = W\left[-\sin\theta\frac{d\theta}{d\phi} + r\sin\phi\right]$$

and

$$\frac{d^2V}{d\phi^2} = W\left[-l\cos\theta\left(\frac{d\theta}{d\phi}\right)^2 - l\sin\theta\frac{d^2\theta}{d\phi^2} + r\cos\phi\right]. \tag{3}$$

Also, from (2),

$$\tan\theta = \frac{r\sin\phi}{r\cos\phi + c}.$$

$$\therefore \sec^2\theta\frac{d\theta}{d\phi} = \frac{r\left[\cos\phi(r\cos\phi+c)+r\sin^2\phi\right]}{(c+r\cos\phi)^2} \Rightarrow \sec^2\theta\frac{d\theta}{d\phi} = \frac{r[c\cos\phi+r]}{(c+r\cos\phi)^2}.$$

Now, in the vertical position, $\theta = \phi = 0$.

$$\therefore \left(\frac{d\theta}{d\phi}\right)_{\theta=\phi=0} = \frac{r(c+r)}{(c+r)^2} = \frac{r}{c+r}. \tag{4}$$

So from (3),

$$\left(\frac{d^2V}{d\phi^2}\right)_{\theta=\phi=0} = W\left[-l\cos\theta\left(\frac{d\theta}{d\phi}\right)^2 - l\sin\theta\frac{d^2\theta}{d\phi^2} + r\cos\phi\right]_{\theta=\phi=0}$$

$$= W\left[-l\left(\frac{d\theta}{d\phi}\right)^2_{\theta=\phi=0} + r\right] = W\left[-l\left(\frac{r}{c+r}\right)^2 + r\right] \text{ [from (4)]}$$

$$= Wr\left[1 - \frac{lr}{(c+r)^2}\right].$$

So the equilibrium will be stable, if

$$\left(\frac{d^2V}{d\phi}\right)_{\theta=\phi=0} > 0 \Rightarrow Wr\left[1-\frac{lr}{(c+r)^2}\right] > 0 \Rightarrow 1 - \frac{lr}{(c+r)^2} > 0$$

$$\Rightarrow 1 > \frac{lr}{(c+r)^2} \Rightarrow \frac{(c+r)^2}{r} > l \Rightarrow l < \frac{(c+r)^2}{r}.$$

14. A smooth beam AB of weight W, rests with one end A on a smooth horizontal plane AC and the other end B against a smooth vertical wall BC. The end A is connected by a string which passes over a smooth pulley at C and is attached to a weight W, A, B, C being in one vertical plane. Find the position of equilibrium and show that it is unstable. [CU(H) 1996]

Solution: Let the rod AB be inclined at an angle θ with the horizontal.

Let G be the CG of the beam where its weight W acts.

Also, $AG = GB = a$ (say).

From G, we draw GL perpendicular to the horizontal plane AC.

Then height of G above AC = GL = $a\sin\theta$.

Let the length of the string $ACW' = l$.

Then $CW' = l - AC = l - 2a\cos\theta$.

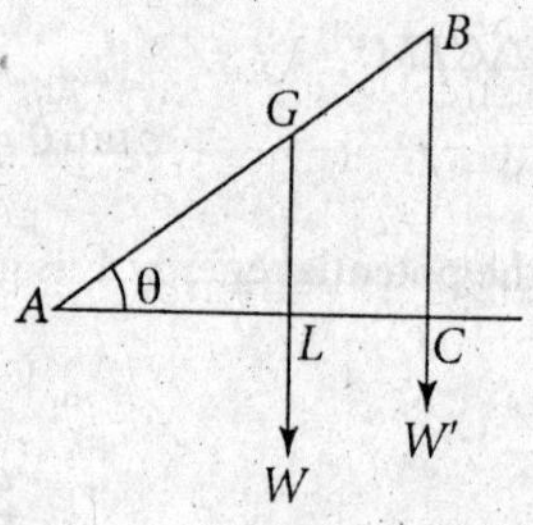

Fig. 5.19

$\therefore$ depth of W' below $AC = CW' = l - 2a\cos\theta$ or height of W' above AC

$$= -(l - 2a\cos\theta)$$

Now, if z be the height of CG of the combined body (rod AB and weight W') above AC, then

$$z = \frac{W(GL) + W'(\text{length } CW')}{W + W'} \Rightarrow z = \frac{Wa\sin\theta - W'(l - 2a\cos\theta)}{W + W'}$$

$$\Rightarrow z = \frac{1}{W + W'}\left[-W'l + Wa\sin\theta + W'2a\cos\theta\right].$$

$$\therefore \frac{dz}{d\theta} = \frac{1}{W + W'}(Wa\cos\theta - 2aW'\sin\theta)$$

and

$$\frac{d^2z}{d\theta^2} = \frac{1}{W + W'}\left[-Wa\sin\theta - 2aW'\cos\theta\right].$$

For equilibrium,

$$\frac{dz}{d\theta} = 0 \Rightarrow Wa\cos\theta - 2aW'\sin\theta = 0 \Rightarrow \tan\theta = \frac{W}{2W'}.$$

For this value of θ,

$$\frac{d^2z}{d\theta^2} < 0$$

So z is maximum for

$$\tan\theta = \frac{W}{2W'} \text{ or, } \theta = \tan^{-1}\frac{W}{2W'}.$$

15. One end A of a uniform rod AB of weight W and length l is smoothly hinged at a fixed point, while B is tied to a light string which passes over a small smooth pulley at a distance a vertically above A and carries a weight $W/4$. If $l < a < 2l$, show that the system is in stable equilibrium when AB is vertically upwards, and that there is also a configuration of equilibrium in which the rod is at a certain angle to the vertical.

[VU(H) 2006]

Solution: Let the end A of the rod is fixed at A, and the other end is tied to the string which passes over a small pulley at a point C vertically above A.

And this string sustains a weight $W/4$. Let the total length of the string BCW' be l' so that the length CW' becomes $(l'-CB)=(l'-r)$, where $CB=r$. Obviously, the height of W' from A is

$$AW'=AC-CW'=a-(l'-r)=a-l'+r.$$

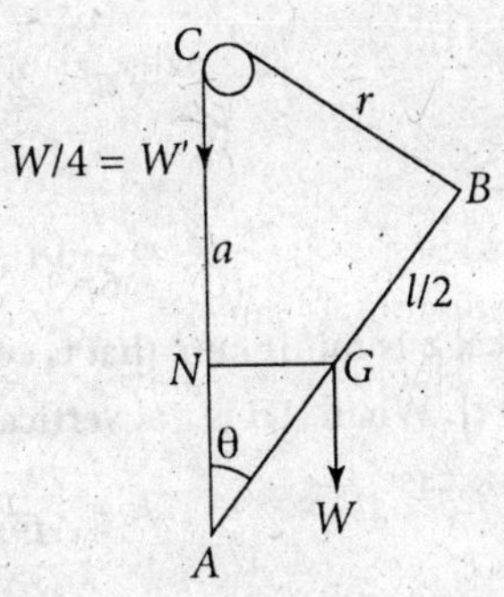

Fig. 5.20

Also the height of CG G of the rod AB above A is

$$AN=AG\cos\theta=\frac{l}{2}\cos\theta.$$

Now, if z be the height of CG of the combined body (rod AB and weight $W/4$) above A, then

$$z=\frac{W\cdot AN+\frac{W}{4}(AW')}{W+\frac{W}{4}}\Rightarrow z=\frac{W\cdot\frac{l}{2}\cos\theta+\frac{W}{4}(a-l'+r)}{\frac{5W}{4}}\Rightarrow z=\frac{2l\cos\theta+(a-l'+r)}{5}.$$

$$\therefore\ \frac{dz}{d\theta}=\frac{1}{5}\left[-2l\sin\theta+\frac{dr}{d\theta}\right]. \tag{1}$$

In $\triangle ABC$,

$$r^2=l^2+a^2-2al\cos\theta.$$

$$\therefore\ 2r\frac{dr}{d\theta}=2al\sin\theta\Rightarrow\frac{dr}{d\theta}=\frac{al}{r}\sin\theta.$$

So from (1),

$$\frac{dz}{d\theta}=\frac{1}{5}\left[-2l\sin\theta+\frac{al}{r}\sin\theta\right]$$

and

$$\frac{d^2z}{d\theta^2}=\frac{1}{5}\left[-2l\cos\theta+\frac{al}{r^2}\left\{r\cos\theta-\sin\theta\frac{dr}{d\theta}\right\}\right]$$
$$=\frac{1}{5}\left[-2l\cos\theta+\frac{al}{r^2}\left\{r\cos\theta-\sin^2\theta\frac{al}{r}\right\}\right]$$
$$=\frac{1}{5}\left[\frac{l\cos\theta}{r}(a-2r)-\frac{a^2l^2}{r^3\sin^2\theta}\right].$$

For the position of equilibrium,

$$\frac{dz}{d\theta}=0\Rightarrow\frac{1}{5}\left[-2l+\frac{al}{r}\right]\sin\theta=0\Rightarrow\theta=0 \text{ or, } a=2r.$$

Now, we shall find the nature of equilibrium in the two cases ($\theta=0$ and $a=2r$).

Case I. When the rod AB is vertical be $\theta=0$.

In this case,

$$\frac{d^2z}{d\theta^2}=\frac{1}{5}\left[\frac{l}{r}(a-2r)\right]=\frac{l}{5r}(a-2r)$$

$$\therefore\ r^2 = l^2 + a^2 - 2al\cos\theta \Rightarrow r^2 = l^2 + a^2 - 2al\ (\because\ \theta = 0).$$

$$\Rightarrow r = a - l$$

$$\therefore\ \frac{d^2z}{d\theta^2} = \frac{l}{5r}[a - 2(a-l)] = \frac{l}{5r}(2l - a) > 0\ (\because\ 2l > a)$$

Hence, z is minimum, that is equilibrium is stable.

Case II. When AB is not vertical, i.e., $a = 2r$.

In this case,

$$\frac{d^2z}{d\theta^2} = -\frac{1}{5}\cdot\frac{a^2l^2}{r^3}\sin^2\theta < 0.$$

Hence, z is maximum, that is, equilibrium is unstable.

16. A uniform rod, of length $2a$, is attached by smooth rings at both ends of a parabolic wire, fixed with its axis vertical and vertex downwards, and of latus rectum $4b$. Show that the angle which the rod makes with the horizontal in a slanting position of equilibrium is given by $\cos^2\theta = \frac{2b}{a}$ and that, of these position exist, they are stable. Show also that the positions in which the rod is horizontal are stable or unstable according as the rod is below or above the focus. [CU(H) 2001]

Solution: Let the equation of the parabola be $x^2 = 4by$. AB is the rod of length $2l$. Let the coordinates of A and B be $\left(2bt_1, bt_1^2\right)$ and $\left(2bt_2, bt_2^2\right)$ respectively. If G be the CG of the rod, then

$$z = \text{height of } G \text{ above } OX = \tfrac{1}{2}(AL + BN)$$
$$= \tfrac{1}{2}\left(bt_1^2 + bt_2^2\right) = \tfrac{b}{2}\left(t_1^2 + t_2^2\right)$$

Fig. 5.21

and

$$BM = BN - MN = BN - AL = bt_2^2 - bt_1^2$$
$$= b\left(t_2^2 - t_1^2\right). \qquad (1)$$

In ΔABM,

$$\sin\theta = \frac{BM}{AB} = \frac{BM}{2a} \Rightarrow BM = 2a\sin\theta. \qquad (2)$$

So from (1),

$$2a\sin\theta = b\left(t_2^2 - t_1^2\right). \qquad (3)$$

Also in ΔABM,

$$\cos\theta = \frac{AM}{AB} = \frac{AM}{2a} \Rightarrow AM = 2a\cos\theta \Rightarrow LN = 2a\cos\theta \Rightarrow LO + ON = 2a\cos\theta$$

$$\Rightarrow 2bt_1 + 2bt_2 = 2a\cos\theta \Rightarrow 2b(t_1 + t_2) = 2a\cos\theta \Rightarrow t_1 + t_2 = \frac{a}{b}\cos\theta. \qquad (3)$$

Now from (2),

$$b(t_2 + t_1)(t_2 - t_1) = 2a\sin\theta \Rightarrow a\cos\theta\,(t_2 - t_1) = 2a\sin\theta \text{ [from (3)]}$$

$$\Rightarrow t_2 - t_1 = 2\tan\theta. \qquad (4)$$

Squaring and adding (3) and (4), we get

$$(t_2+t_1)^2+(t_2-t_1)^2=\frac{a^2}{b^2}\cos^2\theta+4\tan^2\theta$$

$$\Rightarrow 2\left(t_2^2+t_1^2\right)=\frac{a^2}{b^2}\cos^2\theta+4\tan^2\theta$$

$$\Rightarrow t_2^2+t_1^2=\frac{1}{2}\left(\frac{a^2}{b^2}\cos^2\theta+4\tan^2\theta\right).$$

Now

$$z=\frac{b}{2}\left(t_1^2+t_2^2\right)=\frac{b}{2}\cdot\frac{1}{2}\left(\frac{a^2}{b^2}\cos^2\theta+4\tan^2\theta\right)$$

$$\Rightarrow z=\frac{b}{4b^2}\left(a^2\cos^2\theta+4b^2\tan^2\theta\right)$$

$$\Rightarrow z=\frac{1}{4b}\left(a^2\cos^2\theta+4b^2\tan^2\theta\right).$$

$$\therefore\ \frac{dz}{d\theta}=\frac{1}{4b}\left[-2a^2\cos\theta\sin\theta+8b^2\tan\theta\sec^2\theta\right]$$

$$\Rightarrow \frac{dz}{d\theta}=\frac{1}{4b}\left[-a^2\sin 2\theta+8b^2\tan\theta\sec^2\theta\right].$$

$$\therefore\ \frac{d^2z}{d\theta^2}=\frac{1}{4b^2}\left[-2a^2\cos 2\theta+8b^2\left\{\sec^4\theta+2\sec^2\theta\tan^2\theta\right\}\right].$$

For equilibrium,

$$\frac{dz}{d\theta}=0\Rightarrow\frac{1}{4b}\left[-2a^2\cos\theta\sin\theta+8b^2\tan\theta\sec^2\theta\right]=0$$

$$\Rightarrow -2a^2\cos\theta\sin\theta+8b^2\tan\theta\sec^2\theta=0$$

$$\Rightarrow -2a^2\cos\theta\sin\theta+\frac{8b^2\sin\theta}{\cos^3\theta}=0\Rightarrow 2\sin\theta\left[\frac{4b^2}{\cos^3\theta}-a^2\cos\theta\right]$$

$$\Rightarrow\frac{\sin\theta}{\cos^3\theta}\left[4b^2-a^2\cos^4\theta\right]=0\Rightarrow\sin\theta\left[4b^2-a^2\cos^4\theta\right]=0$$

$$\Rightarrow\sin\theta=0\text{ or, }4b^2-a^2\cos^4\theta=0\Rightarrow\theta=0\text{ or, }\cos^4\theta=\frac{4b^2}{a^2}$$

$$\Rightarrow\theta=0\text{ or, }\cos^2\theta=\frac{2b}{a}.$$

When $\cos^2\theta=\frac{2b}{a}$.

$$\frac{d^2z}{d\theta^2}=\frac{1}{4b}\left[-2a^2\cos 2\theta+8b^2\sec^2\theta\left(\sec^2\theta+2\tan^2\theta\right)\right]$$

$$=\frac{1}{2b}\left[-a^2\cos 2\theta+4b^2\sec^2\theta\left\{\sec^2\theta+2\left(\sec^2\theta-1\right)\right\}\right]$$

$$=\frac{1}{2b}\left[-a^2\cos^2\theta+4b^2\sec^2\theta\left(3\sec^2\theta-2\right)\right]$$

$$=\frac{1}{2b}\left[-a^2\left(2\cos^2\theta-1\right)+4b^2\cdot\frac{a}{2b}\left\{\frac{3\times a}{2b}-2\right\}\right]$$

$$=\frac{1}{2b}\left[-a^2\left(\frac{4b}{a}-1\right)+\frac{2ba}{2b}(3a-4b)\right]$$

$$= \frac{1}{2b}[-a(4b-a)+a(3a-4b)] = \frac{1}{2b}\left[4a^2-8ab\right] = \frac{4a^2}{2b}\left(1-\frac{2b}{a}\right)$$

$$= \frac{2a^2}{b}\left(1-\frac{2b}{a}\right) = \frac{2a^2}{b}\left(1-\cos^2\theta\right) = \frac{2a^2}{b}\sin^2\theta > 0 \text{ as } \sin^2\theta < 1.$$

Hence, the equilibrium is stable.

When the rod is horizontal, i.e., when $\theta = 0$,

$$\frac{d^2z}{d\theta^2} = \frac{1}{4b}\left[-2a^2+8b^2\right] = \frac{4b^2-a^2}{2b}.$$

The equilibrium in this case is stable or unstable according as $\frac{d^2z}{d\theta^2}$ is positive or negative

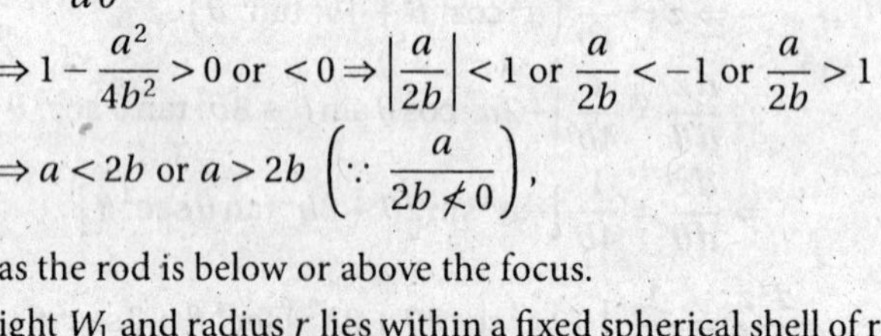

$$\text{i.e., } \frac{d^2z}{d\theta^2} > 0 \text{ or } < 0 \Rightarrow 4b^2 - a^2 > 0 \text{ or } < 0$$

$$\Rightarrow 1 - \frac{a^2}{4b^2} > 0 \text{ or } < 0 \Rightarrow \left|\frac{a}{2b}\right| < 1 \text{ or } \frac{a}{2b} < -1 \text{ or } \frac{a}{2b} > 1$$

$$\Rightarrow a < 2b \text{ or } a > 2b \left(\because \frac{a}{2b \nless 0}\right),$$

i.e., according as the rod is below or above the focus.

17. A sphere of weight W_1 and radius r lies within a fixed spherical shell of radius $R(> 2r)$ and a particle of weight W_2 is attached to its highest point. Show that the equilibrium is stable, if $W_1 \geq \frac{(R-2r)}{r} W_2$. [IAS 2003; CU(H) 2008, VU(H) 2003]

Solution: Let A be the point of contact of the sphere and the spherical shell, O is the centre of the sphere, AA' is the vertical diameter of the sphere and O' is the centre of the spherical shell.

We have $OA = r$ and $O'A = R$.

The weight W_1 of the sphere acts at O and a particle of weight W_2 is a attached to A'. If h is the height of the CG of the combined body consisting of the sphere and the weight W_2 attached at A', then

$$h = \frac{W_2 \cdot 2r + W_1 \cdot r}{W_2 + W_1} = \frac{(2W_2 + W_1)\, r}{W_2 + W_1}.$$

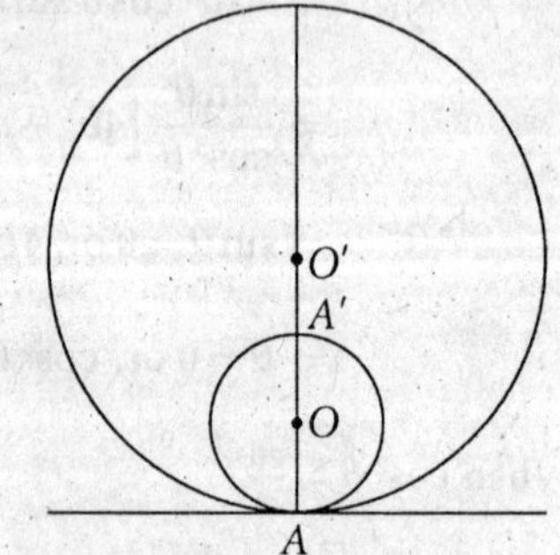

Fig. 5.22

As $R > 2r$, the equilibrium will be stable, if $\frac{1}{h} \geq \frac{1}{r} - \frac{1}{R}$

$$\Rightarrow \frac{W_1 + W_2}{(W_1 + 2W_2)\, r} \geq \frac{R-r}{rR}$$

$$\Rightarrow R(W_1 + W_2) \geq (R-r)(W_1 + 2W_2)$$

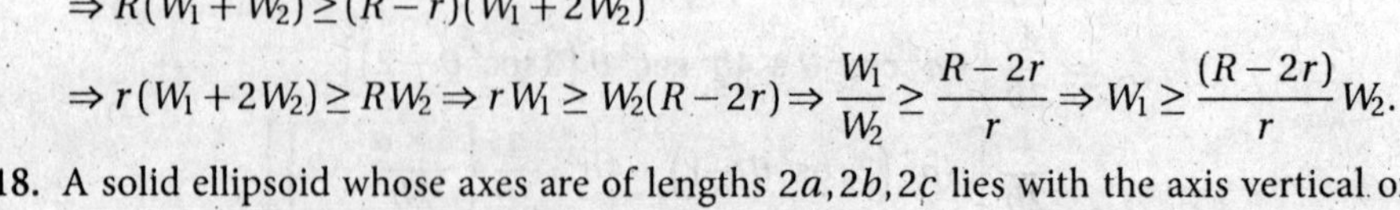

$$\Rightarrow r(W_1 + 2W_2) \geq RW_2 \Rightarrow rW_1 \geq W_2(R-2r) \Rightarrow \frac{W_1}{W_2} \geq \frac{R-2r}{r} \Rightarrow W_1 \geq \frac{(R-2r)}{r} W_2.$$

18. A solid ellipsoid whose axes are of lengths $2a, 2b, 2c$ lies with the axis vertical on a rough horizontal plane. The CG is on the vertical axis at a distance h from the bottom vertex. Show that the equilibrium is stable, if h is less than $\frac{a^2}{c}$ and $\frac{b^2}{c}$ both.

Solution: The figure shows the section of the ellipsoid $\frac{x^2}{a^2}+\frac{y^2}{b^2}+\frac{z^2}{c^2}=1$ by the plane $y=0$.

Putting $y=0$ in the above equation, we get

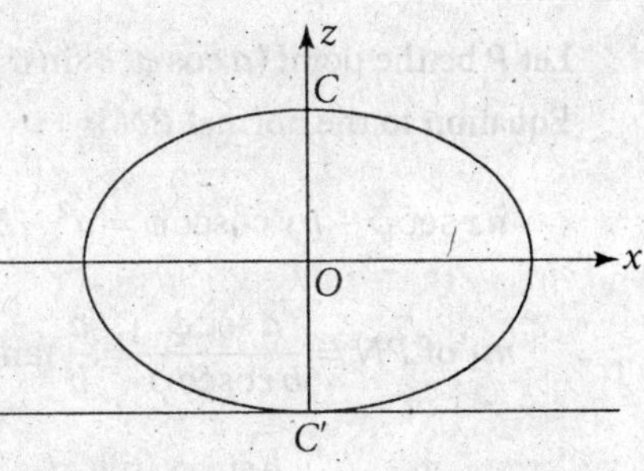

Fig. 5.23

$$\frac{x^2}{a^2}+\frac{z^2}{c^2}=1 \Rightarrow \frac{z^2}{c^2}=1-\frac{x^2}{a^2}=\frac{a^2-x^2}{a^2}$$

$$\Rightarrow z=\frac{c}{a}\sqrt{a^2-x^2}.$$

$$\therefore \frac{dz}{dx}=-\frac{c}{a}x\left(a^2-x^2\right)^{-1/2}$$

and
$$\frac{d^2z}{dx^2}=-\frac{c}{a}\left[\left(a^2-x^2\right)^{-1/2}-\frac{1}{2}\left(a^2-x^2\right)^{-3/2}(-2x)x\right]$$

$$=-\frac{c}{a}\left(a^2-x^2\right)^{-3/2}\left[a^2-x^2+x^2\right]=-ca\left(a^2-x^2\right)^{-3/2}.$$

Now,

$\rho'=$ the radius of curvature of the upper section at B

$$=\frac{\left[1+\left(\frac{dz}{dx}\right)^2\right]^{3/2}}{\frac{d^2z}{dx^2}}=\frac{\left[1+\frac{c^2x^2}{a^2\left(a^2-x^2\right)}\right]}{-\frac{ca}{\left(a^2-x^2\right)^{3/2}}}.$$

$\therefore$ at the point of contact C', $x=0$ and so $\rho'=-\frac{a^2}{c}$ or $\frac{a^2}{c}$ in magnitude.

Hence, the equilibrium is stable, if

$$\frac{1}{h}>\frac{1}{\rho}+\frac{1}{\rho'}\Rightarrow\frac{1}{h}>\frac{1}{\infty}+\frac{1}{a^2/c},$$

as the lower surface at the point of contact is a horizontal plane, hence $\rho=\infty$

$$\Rightarrow\frac{1}{h}>\frac{c}{a^2}\Rightarrow h<\frac{a^2}{c}.$$

In a similar way by taking section of the ellipsoid by the plane $x=0$, we get the condition for stability as $h<\frac{b^2}{c}$.

Hence, the equilibrium is stable, if h is less than $\frac{a^2}{c}$ and $\frac{b^2}{c}$ both.

19. An elliptic cylinder is placed with its axis horizontal on a rough plane inclined to the horizon at angles less than the angle of friction, prove that the cylinder cannot rest if the inclination of the plane exceeds $\sin^{-1}\left(\frac{a^2-b^2}{a^2+b^2}\right)$, and if the inclination is equal to $\sin^{-1}\left(\frac{a^2-b^2}{a^2+b^2}\right)$, the equilibrium is neutral to a first approximation. [NBU(H) 2006]

Solution: C is the centre of the ellipse (which is the cross section of the cylinder) and P the point of contact. PN is the normal to the ellipse at P, CP is vertical and let $\angle CPN=\alpha$.

Let P be the point $(a\cos\phi, b\sin\phi)$.

Equation to the normal PN is

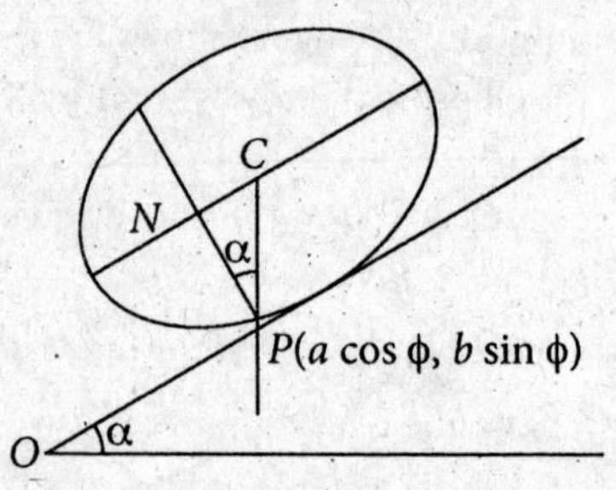

Fig. 5.24

$$ax\sec\phi - by\,\mathrm{cosec}\,\phi = a^2 - b^2.$$

$$m_1 \text{ of } PN = \frac{a\sec\phi}{b\,\mathrm{cosec}\,\phi} = \frac{a}{b}\tan\phi.$$

$$m_2 \text{ of } CP = \frac{b\sin\phi}{a\cos\phi} = \frac{b}{a}\tan\phi.$$

$$\therefore \tan\alpha = \frac{m_1 - m_2}{1 + m_1 m_2} = \frac{\left(\frac{a}{b}\right)\tan\phi - \left(\frac{b}{a}\right)\tan\phi}{1 + \frac{a}{b}\tan\phi\cdot\frac{b}{a}\tan\phi} = \frac{(a^2-b^2)}{ab}\cdot\frac{\tan\phi}{1+\tan^2\phi}$$
$$= \frac{a^2-b^2}{2ab}\sin 2\phi.$$

Now α is maximum if $\sin 2\phi$ is maximum, i.e., if $\sin 2\phi = 1$, i.e., if $2\phi = \frac{\pi}{2}$, i.e., if $\phi = \frac{\pi}{4}$ and in that case we have

$$\tan\alpha = \frac{a^2-b^2}{2ab}.$$

$$\therefore \sin\alpha = \frac{1}{\mathrm{cosec}\,\alpha} = \frac{1}{\sqrt{1+\cot^2\theta}} = \frac{1}{\sqrt{1+\frac{4a^2b^2}{(a^2-b^2)^2}}} = \frac{a^2-b^2}{\sqrt{(a^2-b^2)+4a^2b^2}}$$
$$= \frac{a^2-b^2}{\sqrt{(a^2+b^2)^2}} = \frac{a^2-b^2}{a^2+b^2}.$$

So the maximum value of $\alpha = \sin^{-1}\left[\frac{a^2-b^2}{a^2+b^2}\right]$.

Now, if

$$\alpha = \sin^{-1}\left[\frac{a^2-b^2}{a^2+b^2}\right], \text{ i.e., } \phi = \frac{\pi}{4},$$

then

$$h = \mathrm{CP} = \sqrt{a^2\cos^2\phi + b^2\sin^2\phi} = \sqrt{\frac{a^2+b^2}{2}}, \qquad (1)$$

where $\phi = \pi/4$.

Also, for the ellipse $x = a\cos\phi, y = b\sin\phi$, on differentiating, we get

$$x' = -a\sin\phi, \quad y' = b\cos\phi$$
$$x'' = -a\cos\phi, \quad y'' = -b\sin\phi.$$

$$\therefore \rho' = \frac{(x'^2+y'^2)}{x'y''-x''y'} = \frac{\left(a^2\sin^2\phi + b^2\cos^2\phi\right)^{3/2}}{ab\sin^2\phi + ab\cos^2\phi} = \frac{\left(a^2\sin^2\phi + b^2\cos^2\phi\right)^{3/2}}{ab}.$$

When $\phi = \pi/4$,

$$\rho' = \frac{\left(\frac{a^2}{2}+\frac{b^2}{2}\right)^{3/2}}{ab} = \frac{(a^2+b^2)^{3/2}}{2\sqrt{2}ab}.$$

Also, here $\rho = \infty$ and $\beta = \alpha$.

Now, the equilibrium is neutral, if

$$h = \left(\frac{\rho\rho'}{\rho+\rho'}\right)\cos\beta = \frac{\cos\beta}{\frac{1}{\rho}+\frac{1}{\rho'}} = \frac{\cos\alpha}{\frac{1}{\infty}+\frac{1}{\rho'}} = \rho'\cos\alpha = \frac{(a^2+b^2)^{3/2}}{2\sqrt{2}ab}\sqrt{\sin^2\alpha}$$

$$= \frac{(a^2+b^2)^{3/2}}{2\sqrt{2}ab}\sqrt{1-\left(\frac{a^2-b^2}{a^2+b^2}\right)^2} = \frac{(a^2+b^2)^{3/2}}{2\sqrt{2}ab}\cdot\frac{\sqrt{(a^2+b^2)^2-(a^2-b^2)^2}}{a^2+b^2}$$

$$= \frac{\sqrt{a^2+b^2}\,\sqrt{4a^2b^2}}{2\sqrt{2}ab} = \frac{\sqrt{a^2+b^2}\,2ab}{2\sqrt{2}ab} = \frac{\sqrt{a^2+b^2}}{\sqrt{2}} = h \text{ [from (1)]}$$

Hence, the equilibrium is neutral to the first approximation.

20. A uniform square of mass M is supported in a vertical plane on two smooth pegs at the same horizontal level. The distance between the pegs is a and the diagonal of the square is $D > 4a$. If one diagonal is vertical and a mass m is attached to its lower end, prove that the equilibrium is stable, if $4am > M(D-4a)$. **[NB(H) 2005]**

Solution: Let $ABCD$ be the square and P, Q the pegs.

Then $PQ = a$ and $AC = D$.

Let the diagonal AC be inclined at an angle θ to the horizontal AX.

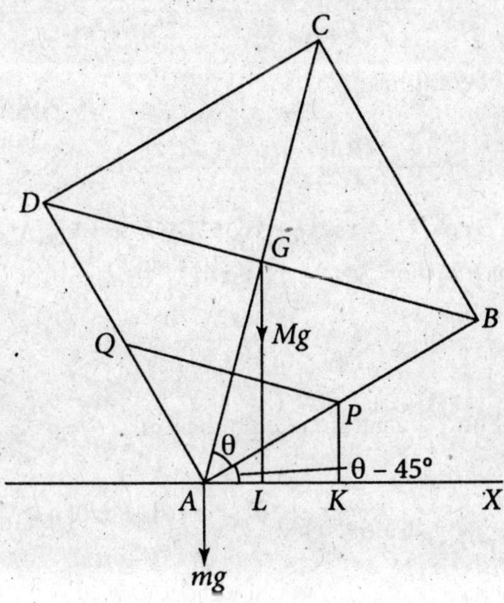

Fig. 5.25

Now, the height of CG G above PQ is given by

$$=GN=GL-NL=GL-PK=AG\sin\theta-AP\sin(\theta-45^\circ)$$
$$=\frac{D}{2}\sin\theta-(PQ\cos QPA)\sin(\theta-45^\circ)=\frac{D}{2}\sin\theta-a\cos(\theta-45^\circ)\sin(\theta-45^\circ)$$
$$=\frac{D}{2}\sin\theta-\frac{a}{2}\sin(2\theta-90^\circ)=\frac{D}{2}\sin\theta+\frac{a}{2}\cos 2\theta.$$

And the depth of CG of mass m below PQ

$$PK=AP\sin(\theta-45^\circ)=PQ\cos QPA\sin(\theta-45^\circ)=a\cos(\theta-45^\circ)\sin(\theta-45^\circ)$$
$$=\frac{a}{2}\sin(2\theta-90^\circ)=-\frac{a}{2}\cos 2\theta.$$

So the height of CG of mass m above $PQ=\frac{a}{2}\cos 2\theta$.

Now, the height of the CG of the combined body (square and mass m at A) above PQ is

$$z=\frac{M\left(\frac{D}{2}\sin\theta+\frac{a}{2}\cos 2\theta\right)+m\cdot\frac{a}{2}\cos 2\theta}{M+m}$$
$$\Rightarrow z=\frac{MD\sin\theta+a\cos 2\theta(M+m)}{2(M+m)}.$$
$$\therefore\ \frac{dz}{d\theta}=\frac{MD\cos\theta-2a\sin 2\theta(M+m)}{2(M+m)},$$

and

$$\frac{d^2z}{d\theta^2}=\frac{-MD\sin\theta-4a\cos 2\theta(M+m)}{2(M+m)}.$$

For position of equilibrium,

$$\frac{dz}{d\theta}=0\Rightarrow\frac{MD\cos\theta-2a\sin 2\theta(M+m)}{2(M+m)}=0$$
$$\Rightarrow MD\cos\theta-4a\sin\theta\cos\theta(M+m)=0\quad[MD-4a\sin\theta(M+m)]\cos\theta=$$
$$\Rightarrow\cos\theta=0\text{ or, }MD-4a\sin\theta(M+m)=0$$
$$\Rightarrow\theta=\frac{\pi}{2}\text{ or, }\sin\theta=\frac{MD}{4a(M+m)}.$$

Now, we shall find the nature of equilibrium at $\theta=\pi/2$.

$$\therefore\ \left(\frac{d^2z}{d\theta^2}\right)_{\theta=\pi/2}=\frac{1}{2(M+m)}[-MD+4a(M+m)]>0,\text{if }-MD+4a(M+m)>0$$
$$\text{or, }4am>M(D-4a).$$

From above, the height of CG above a fixed horizontal is minimum; hence the equilibriu is stable.

EXERCISE 5

1. An open cylindrical can whose height is equal to its diameter is poised on the top of a sphere. Find the least diameter of the sphere consistent with stability. [LNMU(H) 1999]
2. A heavy right circular cone rests with its base on a fixed rough sphere of given radius a. Find the greatest height of the cone if it is an stable equilibrium. [PU(H) 2000, TMBU(H) 1994]
3. A uniform hemisphere rests in equilibrium with its base upwards on the top of a sphere double of its radius. Show that the greatest weight which can be placed at the centre of the plane face without rendering the equilibrium unstable is one eighth of the weight of the hemisphere.
4. A uniform beam of thickness $2b$ rests symmetrically on a perfectly rough horizontal cylinder of radius a. Show that the equilibrium of the beam is stable or unstable according as $b < a$ or, $b > a$. [PU(H) 1996, MU(H) 2004]
5. A thin hemispherical bowl of radius b and weight w rests in equilibrium on the highest point of a fixed sphere of radius a, which is rough enough to prevent any sliding. Inside the bowl is placed a smooth sphere of weight W_1. Show that the equilibrium is not stable unless $W_1 < \frac{W(a-b)}{2b}$. [CU(H) 2001, VU(H) 2006]
6. A heavy uniform cube balances on the highest point of a sphere whose radius is r. If the sphere be rough enough to prevent sliding and if the side of the cube is $\frac{\pi r}{2}$, show that the cube can rock through a right-angle without falling. [CU(H) 2006]
7. A solid circular cone whose height is h and semi-vertical angle α is placed keeping downwards in a smooth circular hole whose radius is a and which is cut in a horizontal table. Investigate the stability of possible positions of equilibrium. [PU(H) 1992, CU(H) 1997]
8. A uniform rod AB of length $2a$ is hinged at A, a string attached to the middle point G of the rod passes over a smooth pulley at C at a height a vertically above A and supports a weight W hanging freely.
   Find the positions of equilibrium and determine their nature. [IAS 1984]
9. Two equal rods are freely jointed at one end so that the angle between them is α and they rest in a vertical plane on a smooth sphere of radius r. Show that they are in a stable or unstable equilibrium accordingly as the length of the rod is greater than or less than $4r \operatorname{cosec} \alpha$. [IAS 1988, PU(H) 2003]
10. A uniform rectangular heavy plate is suspended from a smooth peg fixed in a vertical wall by means of a string of length $2a$, the extremities of which are attached to two fixed points, distance $2c$ apart, placed symmetrically on one of the edges of the plate. The dimensions of the plate are known, show that under certain circumstances there are three positions of equilibrium of which two are stable.
11. A uniform beam of length l rests with its ends on two smooth planes which intersect in a horizontal line. If the inclinations of the plane to the horizontal are α and β $(\alpha > \beta)$, show that the inclination θ of the beam to the horizontal in one of the equilibrium positions is given by $\tan\theta = \frac{1}{2}(\cot\beta - \cot\alpha)$ and show that the beam is unstable in this position. [IAS 2007, VU(H) 2004, CU(H) 2006]
12. A uniform rod is hung from a smooth peg by means of a light inextensible string attached to its ends and passing over the peg. Show that the rod is unstable when it is horizontal. When is the rod stable? [VU(H) 2007]

Answers

1. The least diameter consistent with stability is $\frac{4}{5}$ of the diameter of the can.
2. The height of the cone is four times the radius of the sphere.
7. Equilibrium is stable when $16a > 3h\sin 2\alpha$ and unstable when $16a < 3h\sin 2\alpha$ when $16a = 3h\sin 2\alpha$, it gives inclined position.
8. The equilibrium is stable when $P < 2W$ and unstable when $P \geq 2W$.
10. If the depth of the picture is less than $\frac{2c^2}{\sqrt{a^2-c^2}}$, then there are three positions of equilibriu of which the symmetrical one is unstable. If the depth exceeds the above value, the symmetrical position of equilibrium is the only one and is stable.
12. The rod is stable when it is vertical.

Chapter 6

Forces in Three Dimensions

6.1 Introduction

When forces act on a rigid body in different directions in the space, the body is said to be acted upon by forces in three dimensions. In case the rigid body under the action of different forces is not in equilibrium, then the displacement is of two mutually independent types, i.e., translation and rotation. The translation is the displacement of each particle of the rigid body in the same direction and of equal magnitude, while in rotation the body rotates about some point itself. Thus, we conclude that for the equilibrium of a rigid body under the action of different forces, both of these displacements, i.e., translation and rotation, should be zero.

» **Example 1.** *To find the resultant of any given number of forces acting on a particle.*

Let $\vec{F}_1, \vec{F}_2, \ldots; \vec{F}_n$ be the forces acting on a particle at O and OX, OY, OZ be any three mutually perpendicular axes in space through O.

If $\vec{R}$ is the resultant of the above given system of forces, then by a repeated application of the parallelogram law of forces, we have

$$\vec{R} = \vec{F}_1 + \vec{F}_2 + \cdots + \vec{F}_n. \tag{1}$$

Let $\vec{i}, \vec{j}, \vec{k}$ be the unit vectors along the three axes OX, OY, OZ respectively, then $\vec{R}\cdot\vec{i}$, $\vec{R}\cdot\vec{j}$, $\vec{R}\cdot\vec{k}$ are the resolved parts of the resultant $\vec{R}$ along OX, OY, OZ respectively.

From (1), we get

$$\left.\begin{aligned} \vec{R}\cdot\vec{i} &= \vec{F}_1\cdot\vec{i} + \vec{F}_2\cdot\vec{i} + \cdots + \vec{F}_n\cdot\vec{i} = X \\ \vec{R}\cdot\vec{j} &= \vec{F}_1\cdot\vec{j} + \vec{F}_2\cdot\vec{j} + \cdots + \vec{F}_n\cdot\vec{j} = Y \\ \text{and } \vec{R}\cdot\vec{k} &= \vec{F}_1\cdot\vec{k} + \vec{F}_2\cdot\vec{k} + \cdots + \vec{F}_n\cdot\vec{k} = Z \end{aligned}\right\}. \tag{2}$$

From the relation (2), we conclude that *the resolved part of the resultant $\vec{R}$ along any axis is equal to the sum of the resolved parts of the forces $\vec{F}_1, \vec{F}_2, \ldots, \vec{F}_n$ along that axis.*

If X, Y, Z are the algebraic sums of the resolved parts of the given forces along the three axes respectively, then

$$\vec{R} = X\vec{i} + Y\vec{j} + Z\vec{k}.$$

Note. Unit vectors along coordinate axes are denoted by $\vec{i}, \vec{j}, \vec{k}$ universally.

The magnitude R of the resultant $\vec{R}$ is given by

$$R=\sqrt{X^2+Y^2+Z^2}.$$

Also,

$$\hat{R}=\frac{\vec{R}}{|\vec{R}|}=\frac{X}{R}\vec{i}+\frac{Y}{R}\vec{j}+\frac{Z}{R}\vec{k}.$$

$\therefore \frac{X}{R}, \frac{Y}{R}, \frac{Z}{R}$ are the direction cosines of the line of action of the resultant force $\vec{R}$.

» **Example 2.** *Necessary and sufficient conditions of equilibrium of a particle under the action of a system of forces.*

The necessary and sufficient conditions of equilibrium of a particle under the action of a system of forces are that the algebraic sums of the resolved parts of the forces along any three mutually perpendicular directions vanish separately.

Proof. Let a particle at O be acted upon by a system of forces $\vec{F}_1, \vec{F}_2, \ldots, \vec{F}_n$. Then, the resultant $\vec{R}$ of this system of forces is given by

$$\vec{R}=X\vec{i}+Y\vec{j}+Z\vec{k},$$

where X, Y, Z are the sums of the resolved parts of the given forces along any three mutually perpendicular axes OX, OY, OZ through O and $\vec{i}$, $\vec{j}$, $\vec{k}$ are unit vectors along these axes respectively.

The conditions are necessary: If the particle at O is in equilibrium, then the resultant force

$$\vec{R}=X\vec{i}+Y\vec{j}+Z\vec{k}=\vec{O},$$

which gives $X=0$, $Y=0$ and $Z=0$.

Thus, if the particle is in equilibrium, then the sums of the resolved parts of the forces along OX, OY, OZ vanish separately.

The conditions are sufficient: If the sums of the resolved parts of the forces along three mutually perpendicular axes OX, OY, OZ vanish separately, i.e., if $X=0$, $Y=0$, $Z=0$, then the resultant

$$\vec{R}=X\vec{i}+Y\vec{j}+Z\vec{k}=\vec{O}.$$

Hence, the particle is in equilibrium.

» **Example 3.** *To find the resultant of any given system of forces acting at given points of a rigid body.*

[RU(H) 1994, TMBU(H) 2005]

Classical Method. Let an arbitrarily chosen point O be taken as origin and the three mutually perpendicular lines Ox, Oy, Oz as rectangular axes of coordinates.

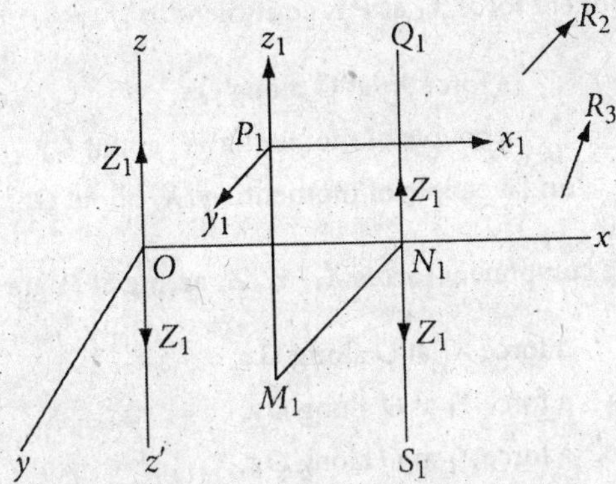

Fig. 6.1

Let the given system of forces R_1, R_2, R_3, ..., act respectively at the points $P_1(x_1, y_1, z_1)$, $P_2(x_2, y_2, z_2)$, $P_3(x_3, y_3, z_3)$ whose components parallel to the axes are respectively X_1, Y_1, Z_1; X_2, Y_2, Z_2; X_3, Y_3, Z_3.

From P_1 we draw P_1M_1 perpendicular to xy plane. From M_1 we draw M_1N_1 perpendicular to Ox and we draw $Q_1N_1S_1$ parallel to Oz.

Then,

$$P_1M_1 = z_1, M_1N_1 = y_1, ON_1 = x_1.$$

At O, we introduce two equal and opposite forces each equal to Z_1 acting in the direction Oz and the opposite direction Oz'. At N_1, we introduce, similarly, two equal and opposite forces, each equal to Z_1, acting in the direction N_1Q_1 and the opposite direction N_1S_1. These forces, being in equilibrium among themselves, do not alter the effect of the given forces. Thus, we have five forces (parallel to the z-axis) each equal to Z_1 as shown in the figure.

Forces Z_1 along P_1z_1 and N_1S_1 form a couple of moment $Z_1 \cdot M_1N_1$, i.e., Z_1y_1 is a plane at right angles to Ox and in the positive direction about Ox, they are, therefore, equivalent to a couple whose axis is along Ox and is positive.

Similarly, the forces Z_1 along N_1Q_1 and Z_1 along Oz form a couple whose moment is $Z_1 \cdot ON_1$, i.e., Z_1x_1 is a plane perpendicular to Oy and in the negative direction about Oy.

Hence, the component force Z_1 at P_1 is equivalent to

a force Z_1 at O along Oz,
a couple of moment y_1Z_1 about Ox
and a couple of moment $-x_1Z_1$ about Oy.

Similarly, the component force Y_1 at P_1 is equivalent to

a force Y_1 at O along Oy,
a couple of moment x_1Y_1 about Oz
and a couple of moment $-z_1Y_1$ about Ox.

Similarly, the component force X_1 at P is equivalent to

$$\begin{aligned}&\text{a force } X_1 \text{ at O along } Ox,\\ &\text{a couple of moment } z_1X_1 \text{ about } Oy\\ \text{and }&\text{a couple of moment } -y_1X_1 \text{ about } Oz.\end{aligned}$$

In this way the three component forces X_1, Y_1, Z_1 acting at P_1 are equivalent to

$$\begin{aligned}&\text{a force } X_1 \text{ at } O \text{ along } Ox,\\ &\text{a force } Y_1 \text{ at } O \text{ along } Oy,\\ &\text{a force } Z_1 \text{ at } O \text{ along } Oz,\\ &\text{a couple of moment } (y_1Z_1 - z_1Y_1) \text{ about } Ox,\\ &\text{a couple of moment } (z_1X_1 - x_1Z_1) \text{ about } Oy\\ \text{and }&\text{a couple of moment } (x_1Y_1 - y_1X_1) \text{ about } Oz.\end{aligned}$$

Similar results hold good for other forces of the system acting at different points of a rigid body.

Hence, the whole system of forces is equivalent to

$$\begin{aligned}&\text{a force } X_1 + X_2 + \ldots, \text{i.e., } \Sigma X_1 = X \text{ along } Ox,\\ &\text{a force } Y_1 + Y_2 + \ldots, \text{i.e., } \Sigma Y_1 = Y \text{ along } Oy,\\ &\text{a force } Z_1 + Z_2 + \ldots, \text{i.e., } \Sigma Z_1 = Z \text{ along } Oz,\\ &\text{a couple about } Ox = \Sigma(y_1Z_1 - z_1Y_1) = L,\\ &\text{a couple about } Oy = \Sigma(z_1X_1 - x_1Z_1) = M,\\ &\text{a couple about } Oz = \Sigma(x_1Y_1 - y_1X_1) = N.\end{aligned}$$

These three forces are equivalent to a single force R acting through O such that $R^2 = X^2 + Y^2 + Z^2$ along a line whose direction cosines are $\frac{X}{R}, \frac{Y}{R}, \frac{Z}{R}$.

These three component couples are equivalent to a couple of moment G, such that

$$G^2 = L^2 + M^2 + N^2,$$

whose axis is along a line whose direction cosine are $\frac{L}{G}, \frac{M}{G}, \frac{N}{G}$.

Hence, the system of forces has been reduced to single force acting through an arbitrarily chosen point O and a couple whose axis passes through O.

Vector Method

Let there be a system of forces $\vec{F}_1, \vec{F}_2, \vec{F}_3, \ldots, \vec{F}_n$ acting at the points $A_1, A_2, \ldots, A_n$ respectively, of a rigid body.

Let O be any arbitrarily chosen point and $\vec{a}_1, \vec{a}_2, \ldots, \vec{a}_n$ the position vectors of the points $A_1, A_2, \ldots, A_n$ respectively, w.r.t. O.

Let us first consider the force $\vec{F}_r$ acting at A_r, where $\overrightarrow{OA_r} = \vec{a}_r$ and apply two forces $\vec{F}_r$, $-F_r$ at O parallel to the force $\vec{F}_r$ at A_r, in opposite directions as shown in the figure. Since two equal and opposite forces are applied at the same point of the body, therefore, they will neutralise each other and will have no extra effect on the body. Thus, the force $\vec{F}_r$ at A_r is equivalent to force $\vec{F}_r$ at A_r and the force $\vec{F}_r$ and $-\vec{F}_r$ at O. The force $\vec{F}_r$ at $\vec{A}_r$ and $-\vec{F}_r$ at O will form a couple of moment $\vec{a}_r \times \vec{F}_r$.

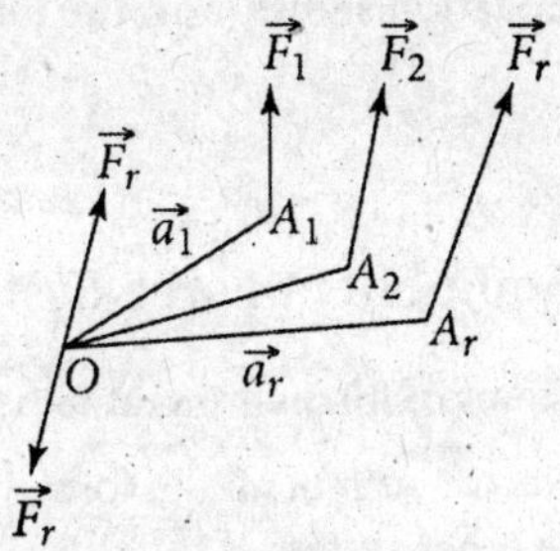

Fig. 6.2

Thus, the force $\vec{F}_r$ acting at the point A_r of a rigid body is equivalent to an equal force $\vec{F}_r$ at O together with a couple of moment $\vec{a}_r \times \vec{F}_r$.

Similarly, all the forces $\vec{F}_1, \vec{F}_2, \vec{F}_3, \ldots, \vec{F}_n$ acting at the points $A_1, A_2, A_3, \ldots, A_n$ of the rigid body are equivalent to forces $\vec{F}_1, \vec{F}_2, \vec{F}_3, \ldots, \vec{F}_n$ at O together with the couples of moments $\vec{a}_1 \times \vec{F}_1, \vec{a}_2 \times \vec{F}_2, \vec{a}_3 \times \vec{F}_3, \ldots, \vec{a}_n \times \vec{F}_n$.

Consequently, the given system of forces is equivalent to resultant force $\vec{R}$ acting at O together with a couple of moment $\vec{G}$ whose axis passes through O such that

$$\vec{R} = \vec{F}_1 + \vec{F}_2 + \vec{F}_3 \cdots + \vec{F}_n$$
$$\vec{G} = \vec{a}_1 \times \vec{F}_1 + \vec{a}_2 \times \vec{F}_2 + \vec{a}_3 \times \vec{F}_3 + \cdots + \vec{a}_n \times \vec{F}_n.$$

Such a combination of $\vec{R}$, $\vec{G}$ is called the *Dyname* and the point O is known as the *base point of the force system.*

6.2 Change of Base Point

From the above discussion it is clear that $\vec{R}$ does not depend upon the choice of base point O. On the other hand, $\vec{G}$ does depend on this choice. Let us see how $\vec{G}$ alters, when the base point is changed from O to O_1, where $\overrightarrow{OO_1} = \vec{a}$.

Let $\vec{a}_1, \vec{a}_2, \ldots, \vec{a}_n$ be the position vectors of the particles relative to O, then we have

$$\vec{a}_r = \vec{a} + \vec{a}'_r.$$

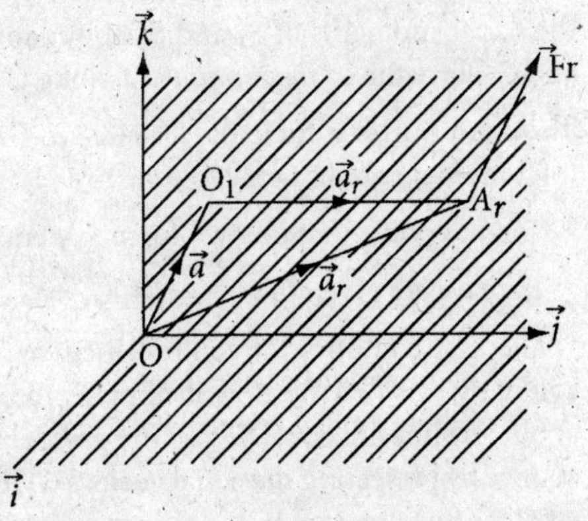

Fig. 6.3

Now if $\vec{G}'$ is the total moment about O_1, we have

$$\vec{G}' = \sum_{r=1}^{n} \vec{a}'_r \times \vec{F}_r = \sum_{r=1}^{n} (\vec{a}_r - \vec{a}) \times \vec{F}_r = \sum_{r=1}^{n} \vec{a}_r \times \vec{F}_r - \sum_{r=1}^{n} \vec{a} \times \vec{F}_r = \vec{G} - \vec{a} \times \sum_{r=1}^{n} \vec{F}_r$$
$$\Rightarrow \vec{G}' = \vec{G} - \vec{a} \times \vec{R}.$$

This equation shows, how the total moment changes with the change of base point.

[PU(H) 1995]

6.3 Equipollent Force-Systems

Two systems of forces are said to be equipollent when the following conditions are satisfied:

1. The vector sum of all the forces of one system is equal to the vector sum of all the forces of the other system.
2. The sum of the moments of all the forces of one system about an arbitrary line is equal to the sum of the moments of all the forces of the other system about that line.

6.4 Necessary and Sufficient Condition for a System to Equipollent

[PU(H) 2000]

The necessary and sufficient condition for two force-systems to be equipollent is that they have the same total force and the same total moment about an arbitrary base point, the same for both systems.

[MU(H) 1991]

Proof. Let S_1 and S_2 be two force-systems and O any base point. The total forces of the two systems will be denoted by $\vec{R}_1$, $\vec{R}_2$ and their total moments about O by G_1, G_2 respectively.

The condition is necessary. Here we assume that the two force-systems S_1 and S_2 are equipollent.

To prove that S_1 and S_2 have same total force and same total moment about O.

When S_1 and S_2 are equipollent, then $\vec{R}_1 = \vec{R}_2$ from the definition of equipollence. Further, the scalar moments of S_1 and S_2 about any line through O are equal and so the vectors $\vec{G}_1$ and $\vec{G}_2$ have the same component along any line through O implying $\vec{G}_1 = \vec{G}_2$.

The condition is sufficient. Here we are given that S_1 and S_2 have same total force and same total moment about O.

Then to prove that the two force-systems S_1 and S_2 are equipollent.

Here we are given that $\vec{R}_1 = \vec{R}_2, \vec{G}_1 = \vec{G}_2$.

The first condition of equipollence, namely, the equality of two forces, viz., $\vec{R}_1 = \vec{R}_2$ is evidently satisfied. So now it remains to prove only that the scalar moments of S_1 and S_2 about any line are equal. If the line passes through O, the equality of scalar moments follows at once by projecting the equal vectors $\vec{G}_1, \vec{G}_2$ on the line. If the line does not pass through O, let O' be any point on it, the total moment $\vec{G}$ of S_1 about O can be expressed by the formula, $\vec{G}' = \vec{G} - \vec{a} \times \vec{R}_1$ and similarly, $\vec{G}_2'$ of S_2 about O' can be determined by the same formula $\vec{G}_2' = \vec{G}_2 - \vec{a} \times \vec{R}_2$. But $\vec{R}_1 = \vec{R}_2$ and $\vec{G}_1 = \vec{G}_2$. So $\vec{G}_1' = \vec{G}_2'$. Hence, the scalar moments are also equal. This completes the proof of the theorem.

Note. If the total force $\vec{R}$ of a system is zero and also the total moment G about any base point, it follows that $\vec{R}$ and $\vec{G}$ are zero for all base points and then we say that the system is equipollent to zero.

6.5 Moment of a Force about a Line

[RU(H) 2000]

Let a force F act along a line

$$\frac{x-x_1}{\cos\alpha}=\frac{y-y_1}{\cos\beta}=\frac{z-z_1}{\cos\gamma}. \tag{1}$$

We are to find the moment of this force F about any other line, say,

$$\frac{x-0}{0}=\frac{y-0}{0}=\frac{z-0}{1}, \text{ i.e., } z \text{ axis.} \tag{2}$$

Now moment of F about z-axis

$$= N_1 = x_1Y_1 - y_1X_1 = F\left(x_1\cos\beta - y_1\cos\alpha\right)$$

$$\text{or, } N_1 = \left(x_1\cos\beta - y_1\cos\alpha\right)F. \tag{3}$$

The shortest distance r between the lines (1) and (2) is given by

$$r=\frac{1}{\sin\theta}\begin{vmatrix} x_1-0 & y_1-0 & z_1-0\\ \cos\alpha & \cos\beta & \cos\gamma\\ 0 & 0 & 1\end{vmatrix}=\frac{x_1\cos\beta-y_1\cos\alpha}{\sin\theta}$$

$$\Rightarrow x_1\cos\beta - y_1\cos\alpha = r\sin\theta$$

so (3) becomes

$$N_1 = Fr\sin\theta, \tag{4}$$

where θ is the angle between the lines (1) and (2).

Now (4) can be written as

$$N_1 = \text{force} \times \text{shortest distance} \times \text{sine of the included angle.}$$

We can generalise this result thus:

The moment of a force about a line is equal to the product of the force, shortest distance and the sine of the included angle.

From above it is obvious that if the lines are parallel (i.e., $\theta = 0$), then moment of the force F vanishes.

Theorem 1. *To find the necessary and sufficient conditions for equilibrium of a rigid body.*

Classical Method

The necessary and sufficient conditions of equilibrium of a rigid body under the action of a system of forces acting at any point of it are that the sums of the resolved parts of the forces along any three mutually perpendicular axes and the sums of the moments of the forces about these axes must vanish separately.

We know that the system of forces (X, Y, Z, L, M, N) acting on a rigid body can be reduced to a single force R acting at an arbitrarily chosen point O and a couple of moment G.

Since a force and a couple cannot balance each other, therefore, for equilibrium the force R and the couple G must separately vanish, i.e., $R=0$ and $G=0$.

But $R^2 = X^2 + Y^2 + Z^2$ and $G^2 = L^2 + M^2 + N^2$

$$\therefore\ X^2 + Y^2 + Z^2 = 0 \text{ and } L^2 + M^2 + N^2 = 0.$$

Therefore, for equilibrium, we must have $X = 0$, $Y = 0$, $Z = 0$; $L = 0$, $M = 0$ and $N = 0$.

Again, if $X = 0 = Y = Z$ and $L = 0 = M = N$, then $R = 0$ and $G = 0$, and consequently, the body will remain in equilibrium. Thus, condition is sufficient.

Vector Method

Let a system of forces $\vec{F}_1, \vec{F}_2, \ldots, \vec{F}_n$ act at the points $A_1, A_2, \ldots, A_n$ respectively, of a rigid body. Let $\vec{a}_1, \vec{a}_2, \ldots, \vec{a}_n$ be the position vectors of the points $A_1, A_2, \ldots, A_n$ respectively, referred to an arbitrary point O as origin. Then, we know that the system is reducible to a single force $\vec{R}$ acting at an arbitrarily chosen point O and a couple of moment $\vec{G}$ whose axis passes through O such that

$$\vec{R} = \sum_{r=1}^{n} \vec{F}_r \tag{1}$$

$$\text{and } \vec{G} = \sum_{r=1}^{n} \vec{a}_r \times \vec{F}_r, \tag{2}$$

$\vec{a}_r$ being the position vector of the point A_r.

Let $\vec{i}, \vec{j}, \vec{k}$ be the unit vectors along three mutually perpendicular axes OX, OY, OZ through O and let (x_r, y_r, z_r) be the coordinates of the point A_r w.r.t. these axes and X_r, Y_r, Z_r be the components in the direction of these axes of the forces. $\vec{F}_r$ is acting at the point A_r. Then, we have

$$\vec{F}_r = X_r\vec{i} + Y_r\vec{j} + Z_r\vec{k}$$

$$\text{and } \vec{a}_r = x_r\vec{i} + y_r\vec{j} + z_r\vec{k}$$

$$\therefore\ \vec{R} = \sum_{r=1}^{n} \vec{F}_r = \sum_{r=1}^{n} \left(X_r\vec{i} + Y_r\vec{j} + Z_r\vec{k}\right)$$

$$\text{or, } \vec{R} = \vec{i}\sum_{r=1}^{n} X_r + \vec{j}\sum_{r=1}^{n} Y_r + \vec{k}\sum_{r=1}^{n} Z_r - X\vec{i} + Y\vec{j} + Z\vec{k}$$

where

$$X = \sum_{r=1}^{n} X_r,\ Y = \sum_{r=1}^{n} Y_r, Z = \sum_{r=1}^{n} Z_r,$$

are the sums of the components of the given forces along the axes OX, OY, OZ respectively.

Also,

$$\vec{G} = \sum_{r=1}^{n} \vec{a}_r \times \vec{F}_r = \sum_{r=1}^{n} \left(x_r\vec{i} + y_r\vec{j} + z_r\vec{k}\right) \times \left(X_r\vec{i} + Y_r\vec{j} + Z_r\vec{k}\right)$$

$$\text{or, } \vec{G} = \vec{i}\sum_{r=1}^{n} (y_r Z_r - z_r Y_r) + \vec{j}\sum_{r=1}^{n} (z_r X_r - x_r Z_r) + \vec{k}\sum_{r=1}^{n} (x_r Y_r - y_r X_r)$$

$$\text{or, } \vec{G} = L\vec{i} + M\vec{j} + N\vec{k},$$

where

$$L=\sum_{r=1}^{n}(y_r Z_r - z_r Y_r),\ M=\sum_{r=1}^{n}(z_r X_r - x_r Z_r),\ N=\sum_{r=1}^{n}(x_r Y_r - y_r X_r)$$

are the sums of the moments of the forces about OX, OY, OZ respectively.

The conditions are necessary: If the body is in equilibrium, then there is neither the motion of translation nor the motion of rotation. Hence, for the equilibrium of the body it is necessary that $\vec{R}=0$ and $\vec{G}=0$.

Now

$$\vec{R}=0 \Rightarrow X\vec{i}+Y\vec{j}+Z\vec{k}=0 \Rightarrow X=0, Y=0, Z=0$$

$$\text{and } \vec{G}=0 \Rightarrow L\vec{i}+M\vec{j}+N\vec{k}=0 \Rightarrow L=0, M=0, N=0.$$

Hence, for a body acted upon by a number of forces at any point of it to be in equilibrium, it is necessary that

$$X=0=Y=Z$$

$$\text{and } L=0=M=N.$$

The conditions are sufficient: If $X=0=Y=Z$ and $L=0=M=N$ then

$$\vec{R}=X\vec{i}+Y\vec{j}+Z\vec{k}=\vec{O}$$

$$\text{and } \vec{G}=L\vec{i}+M\vec{j}+N\vec{k}=0.$$

Hence, the body is in equilibrium.

Theorem 2. *Any system of forces acting on a rigid body can be reduced to a single force together with a couple whose axis is along the direction of the single force.*

[MU(H) 2000, RU(H) 1992, LNMU(H) 1994, TMBU(H) 2011]

We know that a system of forces acting on a rigid body can be reduced to a single force R acting at O along OA and a couple of moment G whose axis is the line OD passing through O. Let $\angle AOD = \theta$. We draw OB perpendicular to OA so that OA, OB, OD — all lie in one plane. Again, we draw a line OC perpendicular to the plane AOB.

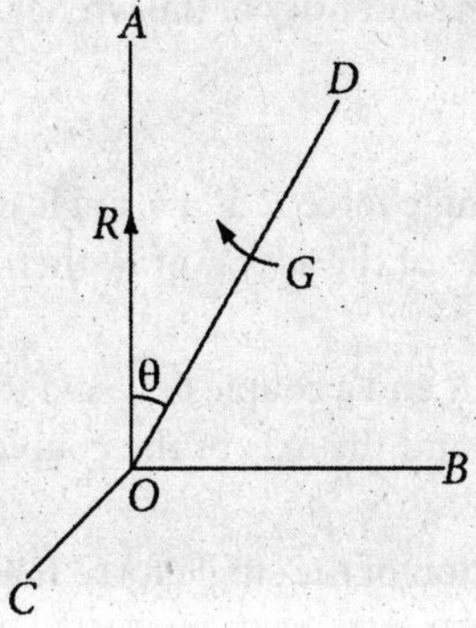

Fig. 6.4

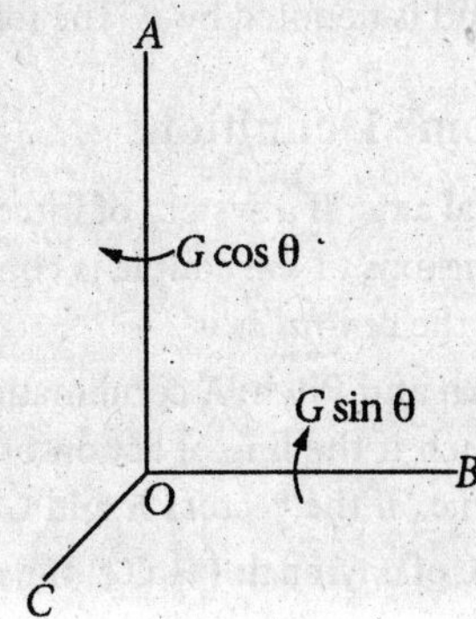

Fig. 6.5

The couple of moment G about OD as its axis is equivalent to two couples of moments $G\cos\theta$ and $G\sin\theta$ whose axes are OA and OB respectively (See Fig. 6.5).

The couple $G\sin\theta$ about OB as axis acts in the plane AOC, since its axis OB is perpendicular to the plane AOC. It follows that the couple of moment $G\sin\theta$ can be replaced by two equal and unlike parallel forces R, $-R$ of moment $G\sin\theta$. Let R act along $\overrightarrow{AO}$ and $-R$ act at some point O' on OC in the direction $\overrightarrow{O'A'}$ such that

$$R\cdot OO' = G\sin\theta \text{ or, } OO' = \frac{G\sin\theta}{R} \text{ (See Fig. 6.6).}$$

The forces R and $-R$ both at O balance each other. We are now left with a force R at O' and a couple of moment $G\cos\theta$ about OA as axis. The axis of the couple can also be transformed to a parallel axis. Consequently, we can select $O'A'$ as the axis of the couple $G\cos\theta$ (in Fig. 6.7).

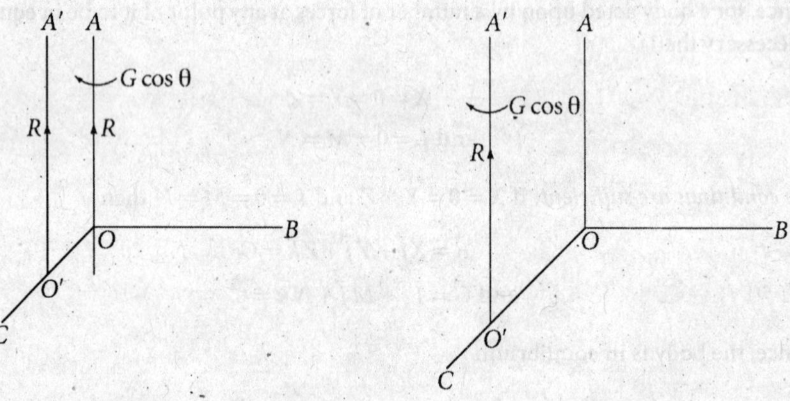

Fig. 6.6

Fig. 6.7

Thus, we have a force R acting at O' in the direction of $O'A'$ and a couple of moment $G\cos\theta$ about $O'A'$ as axis, showing thereby the axis of the couple is coincident with the axis of the force.

The axis $O'A'$, the axis of the single couple and also the line of action of single force R to which system of force is reduced, is called *Poisson's central axis* or only *central axis*. Such a system is called a *Wrench*. $G\cos\theta$ is denoted by k. The ratio k/R is called the *pitch of the wrench* and is denoted by p. The force R is said to be the intensity of the wrench.

6.6 Some Definitions

1. **Central axis:** If a system of forces is reduced to a single force R and a single couple such that the axis of the couple is coincident with the line of the action of R, then that line is called the *central axis*. [IAS 1992]
2. **Wrench and Pitch:** A combination of a single force $\vec{R}$ and a couple $\vec{G}$ is said to constitute a wrench if the line of action of the single force $\vec{R}$ and the axis of the couple $\vec{G}$ are the same, i.e., if the vectors $\vec{R}$ and $\vec{G}$ are parallel.

   In case of a wrench $(\vec{R}, \vec{G})$, the common line of action of the single force $\vec{R}$ and the axis of the couple $\vec{G}$ is said to be the axis of the wrench. If $R = |\vec{R}|$, then R is called the *intensity of the wrench.* Also, if $\vec{G} = p\vec{R}$, then p is called the *pitch of the wrench.*

3. **Screw:** The straight line along which the single force acts, when considered together with the pitch, is called *screw.* Hence, screw is a definite straight line associated with a definite pitch.

Theorem 1. *To determine the condition in order that a given system of forces should compound into a single force.* [LMMU(H) 1998, TMBU(H) 1994, RU(H) 2000]

Classical Method

We know that any system of forces acting at given points of a rigid body can be reduced to a single force R acting through an arbitrarily chosen point O along OA and a couple G whose axis OB passes through O. Let $\angle AOB = \theta$.

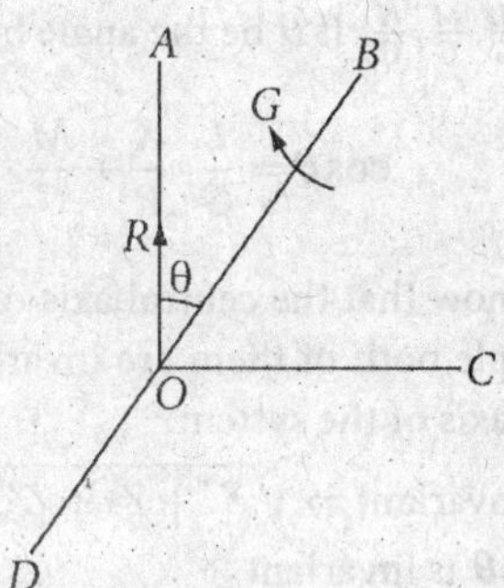

Fig. 6.8

Now, the resolved part of R along OB is $R\cos\theta$ and perpendicular to OB is $R\sin\theta$.

We know that a single force and a couple acting in the same plane upon a rigid body cannot produce equilibrium but are equivalent to the single force acting in a direction parallel to its original direction.

Hence, $R\sin\theta$ and the parallel forces of G are equivalent to a parallel force $R\sin\theta$ which does not pass through O and, therefore, cannot, in general, compound with $R\cos\theta$ into a single force, since only concurrent forces can be compounded into a single force through that point.

But if $R\cos\theta = 0$, i.e., if $\cos\theta = 0$, i.e., if $\theta = \pi/2$, then we are left with a single force $R\sin\theta$ which does not pass through O. Therefore, the angle between OA, whose direction cosines are $\frac{X}{R}, \frac{Y}{R}, \frac{Z}{R}$ and OB whose direction cosines are $\frac{L}{G}, \frac{M}{G}, \frac{N}{G}$ is a right angle.

$$\therefore \quad \frac{X}{R}\cdot\frac{L}{G} + \frac{Y}{R}\cdot\frac{M}{G} + \frac{Z}{R}\cdot\frac{N}{G} = 0,$$

$\therefore$ $XL + YM + ZN = 0$, which is the required condition.

Vector Method

We suppose that at some point O' as the base point the system $(\vec{R}, \vec{G})$ reduces to a single force R' and a couple $\vec{G}'$. Then, $\vec{R}' = \vec{R}$.

If at O' the system reduces to a single resultant force, then we must have $\vec{R}' \neq \vec{O}$ and $\vec{G}' \neq \vec{O}$

$$\Rightarrow \vec{R} \neq \vec{O} \text{ and } \vec{R}\cdot\vec{G}' = 0$$

$$\Rightarrow \vec{R} \neq \vec{O} \text{ and } \vec{R}\cdot\vec{G} = 0 \; [\because \vec{R}\cdot\vec{G} \text{ is invariant}]$$

$$\Rightarrow X^2 + Y^2 + Z^2 \neq 0 \text{ and } LX + MY + NZ = 0$$

$$\left[\because \vec{R} = X\vec{i} + Y\vec{j} + Z\vec{k}; \vec{G} = L\vec{i} + M\vec{j} + N\vec{k}\right]$$

6.7 Invariants

Whatever origin and axis are chosen, to show that the quantities $(LX + MY + NZ)$ and $(X^2 + Y^2 + Z^2)$ are invariants for any given system of forces.

Classical Proof: We know that

$$R^2 = X^2 + Y^2 + Z^2,$$
$$G^2 = L^2 + M^2 + N^2.$$

The direction of R has the direction cosine $\frac{X}{R}, \frac{Y}{R}, \frac{Z}{R}$. The axis of the couple has direction cosine $\frac{L}{G}, \frac{M}{G}, \frac{N}{G}$. If θ be the angle between direction of R and axis of G, then

$$\cos\theta = \frac{L}{G}\cdot\frac{X}{R} + \frac{M}{G}\cdot\frac{Y}{R} + \frac{N}{G}\cdot\frac{Z}{R} \text{ or, } G\cos\theta = \frac{LX+MY+NZ}{R}. \tag{1}$$

We know that the central axis of the system is unique. Hence, R, the force, and $G\cos\theta$, the couple both of them are invariable, because R and $G\cos\theta$ both are defined along the central axis of the system.

R is invariant $\Rightarrow \sqrt{X^2+Y^2+Z^2}$ is invariant $\Rightarrow X^2+Y^2+Z^2$ is invariant.

$G\cos\theta$ is invariant

$$\Rightarrow \frac{LX+MY+NZ}{R} \text{ is invariant [from (1)].}$$

Also, R is invariant

$$\Rightarrow LX+MY+NZ \text{ is invariant.}$$

Then,

$$\text{Pitch} = p = \frac{K}{R} = \frac{G\cos\theta}{R} = \frac{LX+MY+NZ}{R^2} = \frac{LX+MY+NZ}{X^2+Y^2+Z^2}.$$

Vector Proof: Let O be the origin and OX, OY, OZ be the rectangular coordinate axes, $\vec{i}$, $\vec{j}$, $\vec{k}$ be the unit vectors along these coordinate axes. Let a system of forces $\vec{F}_1, \vec{F}_2, \ldots, \vec{F}_n$ acting at different points of a rigid body reduce to a single force $\vec{R}$ acting at O and a couple $\vec{G}$ where

$$\vec{R} = X\vec{i} + Y\vec{j} + Z\vec{k}$$
$$\text{and } \vec{G} = L\vec{i} + M\vec{j} + N\vec{k}.$$

Again, let $\vec{i}', \vec{j}', \vec{k}'$ be the unit vectors along another system of rectangular coordinate axes $O'X'$, $O'Y'$, $O'Z'$ with O' as the origin. We suppose that the same system of forces $\vec{F}_1, \vec{F}_2, \ldots, \vec{F}_n$ reduce to a single force $\vec{R}'$ acting at O' and a couple G' where

$$\vec{R}' = X'\vec{i} + Y'\vec{j} + Z'\vec{k}$$
$$\text{and } \vec{G}' = L'\vec{i} + M'\vec{j} + N'\vec{k}.$$

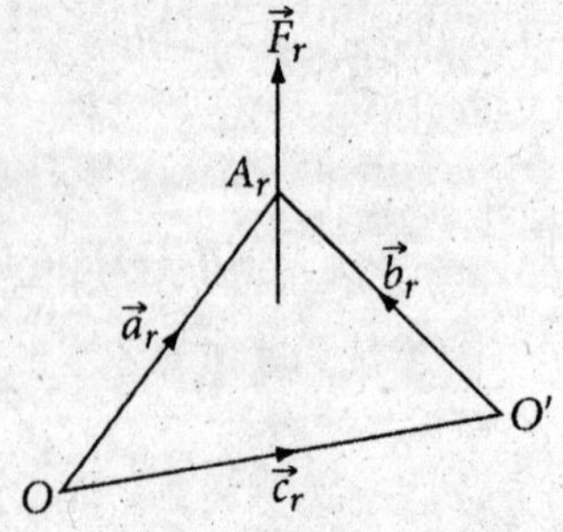

Fig. 6.9

Then, to prove that $X^2+Y^2+Z^2$ and $LX+MY+NZ$ are invariant, i.e., $X^2+Y^2+Z^2 = X'^2+Y'^2+Z'^2$ and $LX+MY+NZ = L'X'+M'Y'+N'Z'$.

We know that

$$\vec{R}=\sum_{r=1}^{n}\vec{F}_r \tag{2}$$

$$\text{and } \vec{G}=\sum_{r=1}^{n}\vec{a}_r\times\vec{F}_r, \tag{3}$$

where $\vec{a}_r$ is the position vector of the point of application A_r of the force $\vec{F}_r$ w.r.t. O as origin of vectors.

Let us take a point O' such that $\overrightarrow{OO'}=\vec{c}$. Let $\overrightarrow{O'A_r}=\vec{b}_r$.

Then,

$$\overrightarrow{OA_r}=\overrightarrow{OO'}+\overrightarrow{O'A_r} \text{ or, } \vec{a}_r=\vec{c}+\vec{b}_r \text{ or, } \vec{a}_r-\vec{c}=\vec{b}_r.$$

Therefore, if the system of forces $\vec{F}_1, \vec{F}_2, \ldots, \vec{F}_n$ reduces to a single force $\vec{R}'$ and a couple G' w.r.t. O', then from (1) and (2), we have

$$\vec{R}'=\sum_{r=1}^{n}\vec{F}_r=\vec{R} \text{ and}$$

$$\vec{G}=\sum_{r=1}^{n}\vec{b}_r\times\vec{F}_r=\sum_{r=1}^{n}(\vec{a}_r-\vec{c})\times\vec{F}_r=\sum_{r=1}^{n}\left(\vec{a}_r\times\vec{F}_r-\vec{c}\times\vec{F}_r\right)$$

$$=\sum_{r=1}^{n}\vec{a}_r\times\vec{F}_r-\sum_{r=1}^{n}\vec{c}\times\vec{F}_r=\vec{G}-\vec{c}\times\sum_{r=1}^{n}\vec{F}_r=\vec{G}-\vec{c}\times\vec{R}.$$

Now, $\vec{R}=\vec{R}'$

$$\Rightarrow|\vec{R}|=|\vec{R}'|\Rightarrow\sqrt{X^2+Y^2+Z^2}=\sqrt{X'^2+Y'^2+Z'^2}$$
$$\Rightarrow X^2+Y^2+Z^2=X'^2+Y'^2+Z'^2.$$

Hence, $X^2+Y^2+Z^2$ is invariant.

Again, $L'X'+M'Y'+N'Z'$.

$$=\left(L'\vec{i}'+M'\vec{j}'+N'\vec{k}'\right)\cdot\left(X'\vec{i}'+Y'\vec{j}'+Z'\vec{k}'\right)=\vec{G}'\cdot\vec{R}'=\vec{G}'\cdot\vec{R}\ \left[\because\ \vec{R}'=\vec{R}\right]$$
$$=\left(\vec{G}-\vec{c}\times\vec{R}\right)\cdot\vec{R}=\vec{G}\cdot\vec{R}-\left(\vec{c}\times\vec{R}\right)\cdot\vec{R}=\vec{G}\cdot\vec{R}=\left(L\vec{i}+M\vec{j}+N\vec{k}\right)\cdot\left(X\vec{i}+Y\vec{j}+Z\vec{k}\right)$$
$$=LX+MY+NZ.$$

Hence, $LX+MY+NZ$ is invariant.

The intensity of the wrench $=|\vec{R}'|=|\vec{R}|=\sqrt{X^2+Y^2+Z^2}=R$ (say).

Also, $\vec{G}'=p\vec{R}'$, because $\vec{G}'$ and R' are parallel. Then, p is called the *pitch of the wrench* $\left(\vec{R}',\vec{G}'\right)$.

Now

$$\vec{G}'=p\vec{R}'\Rightarrow\vec{G}'\cdot\vec{R}'=p\vec{R}'\cdot\vec{R}'\Rightarrow\vec{G}\cdot\vec{R}=p\vec{R}\cdot\vec{R}\ \left[\because\vec{G}\cdot\vec{R}'=\vec{G}\cdot\vec{R}\text{ and }\vec{R}=\vec{R}'\right]$$

$$\Rightarrow p=\frac{\vec{G}\cdot\vec{R}}{\vec{R}\cdot\vec{R}}=\frac{LX+MY+NZ}{X^2+Y^2+Z^2}.$$

Theorem 1. *To find the equations of the central axis of a system of forces acting on a rigid body.*

[RU(H) 2000; LNMU(H) 1995; MU(H) 1998; PU(H) 1996; TMBU(H) 2009; KU(H) 2008]

Classical Method. We know that a system of forces acting on a rigid body can be reduced to a single force R together with a couple G. When the axis of the couple and line of action of R coincides, the axis is called *central axis.* Let the given system of forces be equivalent to X, Y, Z, L, M, N referred to coordinate axis, where

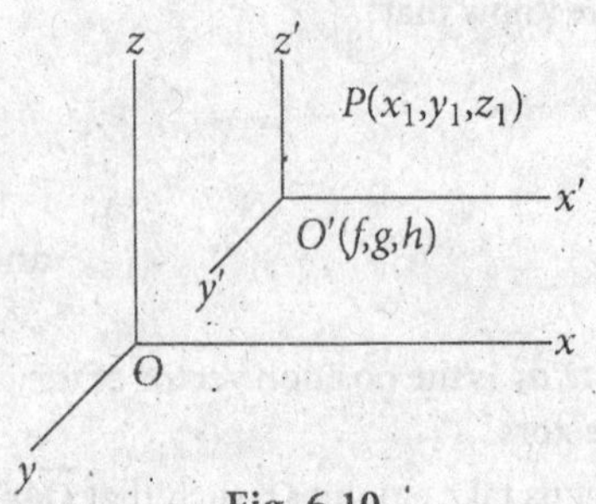

Fig. 6.10

$$X=\Sigma X_1, Y=\Sigma Y_1, Z=\Sigma Z_1$$

and $L=\Sigma(y_1Z_1-z_1Y_1), M=\Sigma(z_1X_1-x_1Z_1)$

$$N=\Sigma(x_1Y_1-y_1X_1),$$

(x_1,y_1,z_1) being the coordinates of a point at which a force (X_1, Y_1, Z_1) acts.

Let O' be a point on the central axis, whose coordinates referred to Ox, Oy, Oz as axes be (f,g,h).

Let P be a point whose coordinates are (x_1,y_1,z_1) w.r.t. Ox, Oy, Oz and hence, coordinates of P referred to $O'x'$, $O'y'$, $O'z'$ are (x_1-f, y_1-g, z_1-h).

If the values of L, M, N be L', M', N', then

$$L'=\Sigma[(y_1-g)Z_1-(Z_1-h)Y_1]=\Sigma[(y_1Z_1-z_1Y_1)-gZ_1+hY_1]$$
$$=\Sigma(y_1Z_1-z_1Y_1)-g\Sigma Z_1+h\Sigma Y_1=L-gZ+hY.$$

Similarly,

$$M'=M-hX+fZ$$
$$N'=N-fY+gX.$$

Also, the components of the resultant force are the same for all points such as O' and are thus, X, Y, and Z.

Since O' lies on central axis, the direction cosines of the axis of the couple corresponding to it are proportional to those of the resultant force.

$$\therefore \frac{L'}{X}=\frac{M'}{Y}=\frac{N'}{Z}$$

or,
$$\frac{L-gZ+hY}{X}=\frac{M-hX+fZ}{Y}=\frac{N-fY+gX}{Z}=\frac{LX+MY+NZ}{X^2+Y^2+Z^2}=\frac{RK}{R^2}=\frac{K}{R}.$$

Hence, the equation of the locus of the point (f,g,h), i.e., the required equation of the central axis is

$$\frac{L-yZ+zY}{X}=\frac{M-zX+xZ}{Y}=\frac{N-xY+yX}{Z}=\frac{K}{R}=\text{the pitch } p \text{ of the wrench.}$$

Vector Method. We take some conveniently chosen point O as the origin and some mutually perpendicular lines OX, OY, OZ as the coordinate axes. Also, we suppose that w.r.t. O as the base point the system reduces to a single force $\vec{R}=(X,Y,Z)$, i.e., $X\vec{i}+Y\vec{j}+Z\vec{k}$ and a couple $\vec{G}=(L,M,N)$.

Let $P(x,y,z)$ be any point on the central axis of the system. If $\overrightarrow{OP}=\vec{r}$, then $\vec{r}=x\vec{i}+y\vec{j}+z\vec{k}$.

Since P is on the central axis of the system, the system must reduce to a wrench, say, $(\vec{R}',\vec{G}')$ at P. We have $\vec{R}'=\vec{R}$ and $\vec{G}'=\vec{G}-\vec{r}\times\vec{R}$.

Now $(\vec{R}', \vec{G}')$ is a wrench.

$$\Rightarrow \vec{G}' = p\vec{R}, \text{ where } p \text{ is the pitch of the wrench}$$
$$\Rightarrow \vec{G} - \vec{r} \times \vec{R} = p\vec{R}$$
$$\Rightarrow \left(L\vec{i} + M\vec{j} + N\vec{j}\right) - \left(x\vec{i} + y\vec{j} + z\vec{k}\right) \times \left(X\vec{i} + Y\vec{j} + Z\vec{k}\right) = p\left(X\vec{i} + Y\vec{j} + Z\vec{k}\right)$$
$$\Rightarrow (L - yZ + zY)\vec{i} + (M - zX + xZ)\vec{j} + (N - xY + yX)\vec{k} = pX\vec{i} + pY\vec{j} + pZ\vec{k}.$$

Equating the coefficients of $\vec{i}, \vec{j}, \vec{k}$, we get $L - yZ + zY = pX,$

$$M - zX + xZ = pY,$$
$$N - xY + yX = pZ$$

$$\text{or, } \frac{L - yZ + zY}{X} = \frac{M - zX + xZ}{Y} = \frac{N - xY + yX}{Z} = p,$$

which is the required equation of the central axis.

Theorem 2. *Show that a given system of forces can be replaced by two forces equivalent to the given system in an infinite number of ways and the tetrahedron formed by the two forces is of constant volume.*

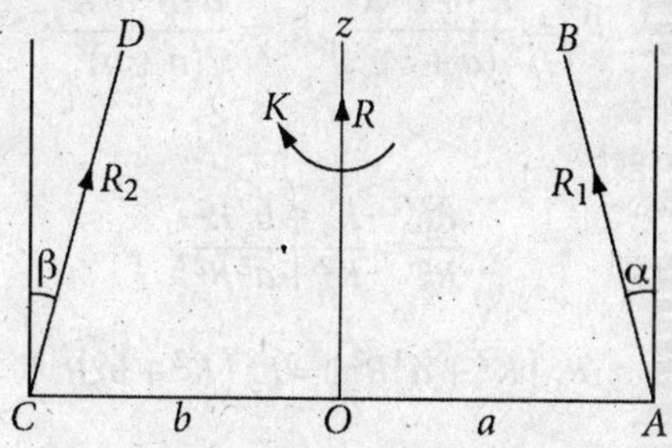

Fig. 6.11

Let the central axis of the given system be taken as Oz.

Let the resultant force be R along Oz and the resultant couple K about Oz.

We draw a line COA perpendicular to Oz and on it take two points A and C on opposite sides of O at distances a and b from O, i.e., $OA = a$ and $OC = b$.

Let the given system be equivalent to (R, K). Let R_1 and R_2 be the two forces equivalent to the given system. Then, directions of R_1 and R_2 each must be perpendicular to Oz, that is, plane of R_1 and that of R_2 each must be perpendicular to Oz.

We assume that R_1 acts at A along AB making an angle α with Oz and R_2 acts at C along CD making the angle β with Oz, the plane of each force, certainly being at right angles to Oz.

Since the two forces are equivalent, therefore, resolving them along and at right angles to Oz, we have

$$R = R_1 \cos\alpha + R_2 \cos\beta \tag{1}$$
$$O = R_1 \sin\alpha - R_2 \sin\beta \tag{2}$$
$$\text{and } K = aR_1 \sin\alpha + bR_2 \sin\beta \tag{3}$$
$$O = aR_1 \cos\alpha - bR_2 \cos\beta. \tag{4}$$

From (1) and (4), we have

$$R_1 \cos\alpha + R_2 \cos\beta - R = 0$$
$$aR_1 \cos\alpha - bR_2 \cos\beta - 0 = 0.$$

$$\therefore\ R_1 = \frac{-bR\cos\beta}{-\cos\alpha\cos\beta(a+b)} = \frac{bR}{\cos\alpha(a+b)}$$

$$\text{or,}\ \ R_1\cos\alpha = \frac{bR}{a+b},\ R_2\cos\beta = \frac{aR}{a+b}. \tag{5}$$

Similarly, solving (2) and (3), we obtain

$$R_1 \sin\alpha = R_2 \sin\beta = \frac{K}{a+b}. \tag{6}$$

From (5) and (6), squaring and adding, we get

$$R_1^2 = \frac{K^2 + b^2R^2}{(a+b)^2},\ R_2^2 = \frac{a^2R^2 + K^2}{(a+b)^2}.$$

Dividing these two, we get

$$\frac{R_1^2}{R_2^2} = \frac{K^2 + b^2R^2}{K^2 + a^2R^2}.$$

$$\therefore\ R_1^2\left(K^2 + a^2R^2\right) = R_2^2\left(K^2 + b^2R^2\right). \tag{7}$$

So R_1 and R_2 are found in terms of a, b, R and K.

Also, dividing (6) by (5), we get

$$\tan\alpha = \frac{K}{bR}\ \text{ and }\ \tan\beta = \frac{K}{aR}. \tag{8}$$

So, α and β are also found in terms of a and b.

But as a and b are arbitrary quantities, therefore, there can be infinite values of R_1 and R_2 and α, β. Hence, there are infinite ways in which the given system can be replaced by two forces.

Again, from (5) and (6), we have

$$R_1\cos\alpha R_2\sin\beta = \frac{bR}{a+b}\frac{K}{a+b}\ \text{ or, } bRK = (a+b)^2 R_1R_2\cos\alpha\sin\beta$$
$$\text{and }\ R_2\cos\beta R_1\sin\alpha = \frac{aR}{a+b}\frac{K}{a+b}\ \text{ or, } aRK = (a+b)^2 R_1R_2\sin\alpha\cos\beta.$$

$$\therefore\ RK(a+b) = (a+b)^2 R_1R_2\sin(\alpha+\beta)\ \text{or, } RK = (a+b)R_1R_2\sin(\alpha+\beta).$$

Let AB and CD represent the forces R_1 and R_2.

Volume of tetrahedron $ABCD$

$$= \frac{1}{3}\text{area of } \Delta ABC \times \text{perpendicular from } D \text{ upon } ABC$$
$$= \frac{1}{3}\cdot\frac{1}{2}AB\cdot AC\cdot CD\sin(\alpha+\beta) = \frac{1}{6}R_1R_2(a+b)\sin(\alpha+\beta)$$
$$= \frac{1}{6}RK \text{ [from (9)]}$$

which is constant, being independent of a and b.

Theorem 3. *To find the resultant wrench of two given wrenches.*

[PU(H) 2001, RU(H) 1999, MU(H) 2001]

Let the one wrench be (R_1, K_1) whose axis is AB and the other wrench is (R_2, K_2) whose axis is CD.

Let θ be the angle between AB and CD. Let $AC = a$ be the shortest distance between these two axes.

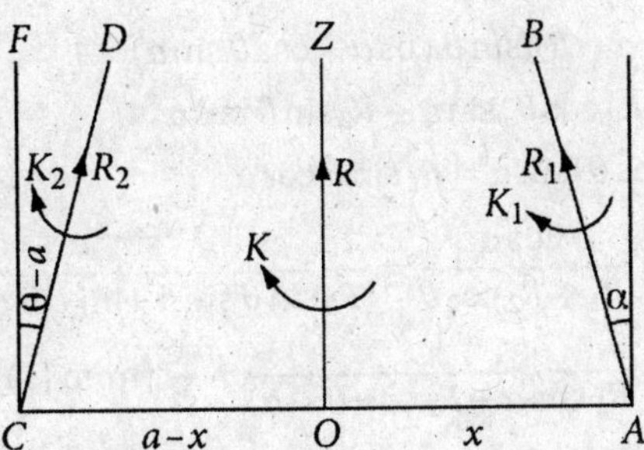

Fig. 6.12

We know that central axis is parallel to the resultant of the forces R_1 and R_2, and is perpendicular to the shortest distance AC. Moments of the two given component wrenches vanish about AC, because the two axes AB and CD intersect AC, hence, moment of the resultant wrench too must vanish about it. This requires that the central axis must cut AC in some point O. Let $AO = x$ so that $CO = a - x$.

Let OZ be inclined at an angle α to AB, then OZ is inclined at an angle $\theta - \alpha$ to CD.

Let (R, K) be the resultant wrench. By resolving forces along and perpendicular to OZ, we have

$$R = R_1\cos\alpha + R_2\cos(\theta-\alpha) \tag{1}$$
$$O = R_1\sin\alpha - R_2\sin(\theta-\alpha) \tag{2}$$

and by taking moments about the above lines, we have

$$K = K_1\cos\alpha + K_2\cos(\theta-\alpha) + xR_1\sin\alpha + (a-x)R_2\sin(\theta-\alpha) \tag{3}$$
$$O = 14\sin\alpha - K_2\sin(\theta-\alpha) - xR_1\cos\alpha + (a-x)R_2\cos(\theta-\alpha). \tag{4}$$

From (3), we get

$$K = K_1 \cos\alpha + K_2 \cos(\theta-\alpha) + aR_2 \sin(\theta-\alpha) + x\{R_1 \sin\alpha - R_2 \sin(\theta-\alpha)\}$$
$$K = K_1 \cos\alpha + K_2 \cos(\theta-\alpha) + aR_2 \sin(\theta-\alpha) \text{ [from (2)]}. \tag{5}$$

From (4), we have

$$O = K_1 \sin\alpha - K_2 \sin(\theta-\alpha) + a\cos(\theta-\alpha) - x\{R_1 \cos\alpha + R_2 \cos(\theta-\alpha)\}$$
$$O = K_1 \sin\alpha - K_2 \sin(\theta-\alpha) + aR_2 \cos(\theta-\alpha) - xR \text{ [from (1)]}. \tag{6}$$

Squaring and adding (1) and (2), we obtain

$$R^2 = R_1^2 + R_2^2 + 2R_1R_2 \cos(\theta-\alpha+\alpha)$$
$$R^2 = R_1^2 + R_2^2 + 2R_1R_2 \cos\theta. \tag{7}$$

This gives R.

From (2), we have

$$O = R_1 \sin\alpha - R_2(\sin\theta\cos\alpha - \cos\theta\sin\alpha)$$
$$O = (R_1 + R_2\cos\theta)\sin\alpha - R_2\sin\theta\cos\alpha$$
$$\text{or, } (R_1 + R_2\cos\theta)\sin\alpha = R_2\sin\theta\cos\alpha$$
$$\text{or, } \frac{\sin\alpha}{R_2\sin\theta} = \frac{\cos\alpha}{R_1 + R_2\cos\theta} = \frac{\sin^2\alpha + \cos^2\alpha}{R_2\sin\theta\sin\alpha + R_1\cos\alpha + R_2\cos\theta\cos\alpha}$$
$$= \frac{1}{R_1\cos\alpha + R_2\cos(\theta-\alpha)} = \frac{1}{R} \text{ [from (1)]}. \tag{8}$$

From (5), we have

$$K = K_1\cos\alpha + K_2(\cos\theta\cos\alpha + \sin\theta\sin\alpha) + aR_2(\sin\theta\cos\alpha - \cos\theta\sin\alpha)$$
$$= (K_1 + K_2\cos\theta + aR_2\sin\theta)\cos\alpha + (K_2\sin\theta - aR_2\cos\theta)\sin\alpha.$$
$$\therefore RK = (K_1 + K_2\cos\theta + aR_2\sin\theta)R\cos\alpha + (K_2\sin\theta - aR_2\cos\theta)R\sin\alpha$$
$$RK = (K_1 + K_2\cos\theta + aR_2\sin\theta)(R_1 + R_2\cos\theta) + (K_2\sin\theta - aR_2\cos\theta)R_2\sin\theta \quad \text{[from (8)]}$$
$$RK = K_1R_1 + K_2R_2\cos\theta + aR_1R_2\sin\theta + R_2K_1\cos\theta$$
$$+ R_2K_2\cos^2\theta + aR_2^2\sin\theta\cos\theta$$
$$+ R_2K_2\sin^2\theta - aR_2^2\sin\theta\cos\theta.$$
$$RK = K_1R_1 + K_2R_2 + \cos\theta(R_2K_1 + K_2R_1) + aR_1R_2\sin\theta. \tag{9}$$

Since θ is given, so when R is known from (7), then (9) gives K.

Equation (6) gives

$$xR = K_1\sin\alpha - K_2(\sin\theta\cos\alpha - \cos\theta\sin\alpha) + aR_2(\cos\theta\cos\alpha + \sin\theta\sin\alpha)$$
$$xR = (K_1 + K_2\cos\theta + aR_2\sin\theta)\sin\alpha - (K_2\sin\theta - aR_2\cos\theta)\cos\alpha$$
$$xR^2 = (K_1 + K_2\cos\theta + aR_2\sin\theta)R\sin\alpha - (K_2\sin\theta - aR_2\cos\theta)R\cos\alpha$$

$$xR^2 = (K_1 + K_2 \cos\theta + aR_2 \sin\theta) R_2 \sin\theta + (aR_2 \cos\theta - K_2 \sin\theta)(R_2 \cos\theta + R_1)$$

[from (8)]

$$= K_1 R_2 \sin\theta + K_2 R_2 \sin\theta \cos\theta + aR_2^2 \sin^2\theta + aR_2^2 \cos^2\theta - R_2 K_2 \sin\theta \cos\theta + aR_1 R_2 \cos\theta - K_2 R_1 \sin\theta$$

$$= \sin\theta (K_1 R_2 - K_2 R_1) + aR_2^2 + aR_1 R_2 \cos\theta$$

$$xR^2 = \sin\theta (K_1 R_1 - K_2 R_1) + aR_2 (R_2 + R_1 \cos\theta). \tag{10}$$

This equation gives x, that is, position of the central axis.

Deductions. To find the resultant wrench of two given forces R_1 and R_2 inclined at an angle θ. [MU(H) 2012]

Proof. Here $K_1 = 0$, $K_2 = 0$.

Hence, from (7), we get

$$R^2 = R_1^2 + R_2^2 + 2R_1 R_2 \cos\theta. \tag{11}$$

From (9),

$$RK = 0 + 0 + 0 + aR_1 R_2 \sin\theta$$

$$RK = aR_1 R_2 \sin\theta. \tag{12}$$

From (10),

$$xR^2 = 0 + aR_2 (R_2 + R_1 \cos\theta)$$

$$xR^2 = aR_2 (R_2 + R_1 \cos\theta). \tag{13}$$

The equation (11) gives the intensity R of the wrench (R, K). The equation (12) gives the value of K and the (13) declares the position of central axis.

Theorem 4. *Definition of Null Lines, Null Plane and Null Point.* [RU(H) 1998]

Null Lines. Null lines of a given system of forces, referred to origin O', are those lines about which the moment of the system vanishes.

Null Plane. The Plane in which all these null lines lie is called the *null plane of the point O'*.

Null Point. The point O' itself is called *null point.*

Example. Let R be the resultant force and G the resultant couple corresponding to the origin O. Let $O'C$ be perpendicular to $O'B$. Since R meets $O'C$ at O' and $O'C$ is perpendicular to the axis of G, therefore, the sum of the moments of the forces of the system about this line is zero.

The line $O'C$ is called the *null line* and the locus of $O'C$, which is the line perpendicular to the axis of G, is called the *null plane* of O'. The point O' is called the *null point.*

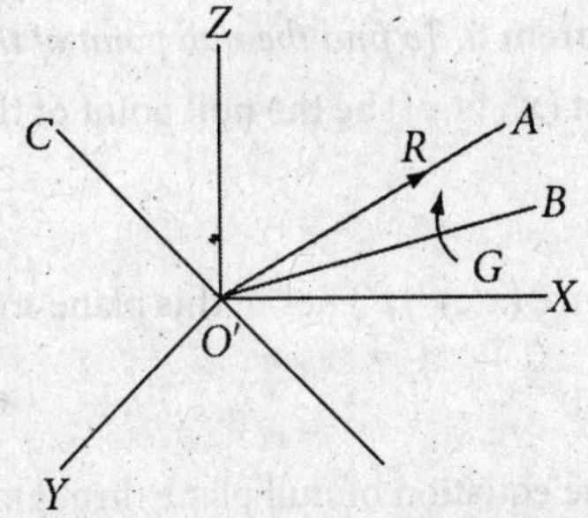

Fig. 6.13

Theorem 5. *To find the equation of null plane of a given point (a, b, c) referred to rectangular axis Ox, Oy, Oz.* [RU(H) 1997]

Let the given system of forces reduce to forces X, Y, Z parallel to axis and component couple L, M, N about the axis respectively.

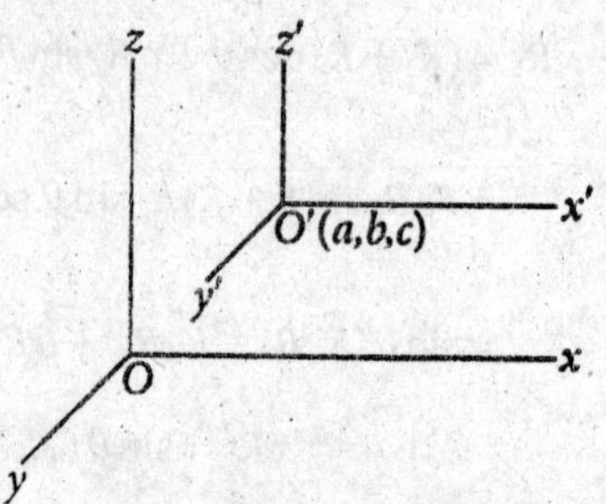

Fig. 6.14

Let moments of couples about the lines through (f, g, h) parallel to the axis be L', M', N'.

$$L' = L-(yZ-zY) = L-bZ+cY$$
$$M' = M-cX+aZ$$
$$N' = N-aY+bX.$$

But L', M', N' are direction ratios of the axis of the resultant couple at (a,b,c), which is normal to the null plane there.

Next we suppose that (x,y,z) are the coordinates of any point in the null plane of (a,b,c), so that the line joining (x,y,z) to (a,b,c) is a null line and its direction ratios are $x-a$, $y-b$, $z-c$.

By definition, null line is perpendicular to the axis of the couple and hence,

$$(x-a)L' + (y-b)M' + (z-c)N' = 0$$
$$(x-a)[L-bZ+cY]+(y-b)(M-cX+aZ)+(z-c)(N-aY+bX)=0$$
$$x(L-bZ+cY)+y(M-cX+aZ)+z(N-aY+bX)$$
$$-aL+abZ-acY-bM+bcX-abZ-cN+acY-bcX=0$$
$$x(L-bZ+cY)+y(M-cX+aZ)+z(N-aY+bX)=aL+bM+cN.$$

This is the locus of (x,y,z) which is any point in the null plane of (a,b,c). Therefore, (1) is the equation of null plane of (a,b,c). (1) is also expressible in determinant form as

$$\begin{vmatrix} x & y & z \\ a & b & c \\ X & Y & Z \end{vmatrix} = L(x-a)+M(y-b)+N(z-c).$$

Theorem 6. *To find the null point of the plane $lx+my+nz=1$.*

Let (x', y', z') be the null point of the plane

$$lx+my+nz=1. \tag{1}$$

Since (x', y', z') lies in this plane so that

$$lx'+my'+nz'=1. \tag{2}$$

The equation of null plane through (x', y', z') is

$$x((L-y'Z+z'Y)+y(M-z'X+x'Z)+z(N-x'Y+y'X)=Lx'+My'+Nz'. \tag{3}$$

Since (1) and (3) represent the same plane and hence, they must be identical so that on comparing, we get

$$\frac{L-y'Z+z'Y}{l}=\frac{M-z'X+x'Z}{m}=\frac{N-x'Y+y'X}{n}=\frac{Lx'+My'+Nz'}{1}. \tag{4}$$

Taking first and second ratios of (4), we have

$$Lm - y'Zm + z'Ym = Ml - z'Xl + x'Zl$$

$$\text{or, } (Lm - lM) - (my' + lx')Z + mz'Y + lz'X = 0. \tag{5}$$

From first and third ratios of (4), we have

$$Ln - y'Zn + z'Yn = lN - x'Yl + y'Xl$$

$$\text{or, } (Ln - lN) - y'nZ + (lx' + nz')Y - ly'X = 0. \tag{6}$$

From second and third ratios of (4), we have

$$nM - nz'X + nx'Z = mN - mx'Y + my'X$$

$$\text{or, } (nM - mN) - (my' + nz')X + nx'Z + mx'Y = 0. \tag{7}$$

Writing (5), (6) and (7) with the help of (2), we get

$$(Lm - lM) + (nz' - 1)Z + mz'Y + lz'X = 0 \tag{8}$$

$$(Ln - lN) - y'nZ + (1 - my')Y - ly'X = 0 \tag{9}$$

$$(nM - mN) + (-1 + lx')X + nx'Z + mx'Y = 0. \tag{10}$$

Again, from (8), (9) and (10), we have

$$-(Lm - lM - Z) = (lX + mY + nZ)z'$$

$$Ln - lN + Y = (lX + mY + nZ)y'$$

$$-(nM - mN - X) = (lX + mY + nZ)x'$$

$$\Rightarrow \frac{x'}{X - nM + mN} = \frac{y'}{Y - lN + Ln} = \frac{z'}{Z - Lm + lM} = \frac{1}{lX + mY + nZ}$$

$$\text{or, } \frac{x'}{X - \begin{vmatrix} M & N \\ m & n \end{vmatrix}} = \frac{y'}{Y - \begin{vmatrix} N & L \\ n & l \end{vmatrix}} = \frac{z'}{Z - \begin{vmatrix} L & M \\ l & m \end{vmatrix}} = \frac{1}{lX + mY + nZ}.$$

This gives the coordinates of the null point of the given plane.

Theorem 7. *To find the condition that the straight line*

$$\frac{x - f}{l} = \frac{y - g}{m} = \frac{z - h}{n},$$

may be a null line for the same system of forces. **[RU(H) 1997, TMBU(H) 2011]**

Let the given system of forces reduce to forces X, Y, Z parallel to the axis and component couples L, M, N about the axis respectively.

Clearly, the line

$$\frac{x - f}{l} = \frac{y - g}{m} = \frac{z - h}{n}$$

passes through the point $O'(f,g,h)$ and has the direction ratio (l,m,n). We draw $O'x'$, $O'y'$, $O'z'$ parallel to Ox, Oy, Oz respectively. Let the components of resultant couple about $O'x'$, $O'y'$, $O'z'$ be L', M', N'. Then,

$$L' = L - gZ + hY,$$
$$M' = M - hX + fZ,$$
$$N' = N - fY + gX.$$

Fig. 6.15

If the given line is a null line, then it should be perpendicular to the axis of the couple.

Hence, $lL' + mM' + nN' = 0$

$$\text{or, } l(L - gZ + hY) + m(M - hX + fZ) + n(N - fY + gX) = 0$$
$$\text{or, } X(mh - ng) + Y(nf - lh) + Z(lg - mf) = Ll + Mm + Nn$$
$$\text{or, } \begin{vmatrix} X & Y & Z \\ l & m & n \\ f & g & h \end{vmatrix} = Ll + Mm + Nn.$$

This is the required condition that the given line may be a null line of the system.

Theorem 8. *To show that a given system of forces may be replaced by two forces, one of which acts along a given line OA.*

Proof. Let the given system of forces reduce to a resultant force R and a resultant couple G referred to O as origin.

The plane of the couple G is plane BOC so that its axis is perpendicular to this plane.

Now through the given line OA and the line of action of the R we draw a plane and let it cut the plane BOC of the couple in the line OD. Hence, R, OA and OD are in the same plane. As such, R can be resolved into two forces, say, P along OA and Q along OD. Therefore, Q along OD together with the couple (i.e., two forces of the couple) give rise to a force Q (in the plane BOC) parallel to OD. But the plane BOC is the null plane of O.

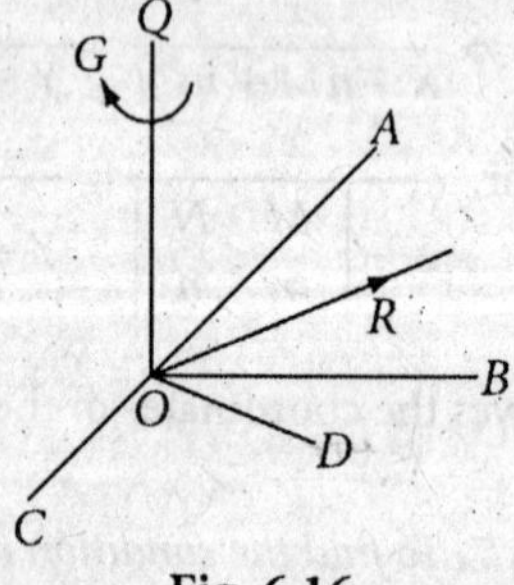

Fig. 6.16

Finally, the system is reduced to a force P along OA and other force Q in the null plane of O but parallel to OB.

Such forces P and Q are called *conjugate forces* and their lines of action are called *conjugate lines.*

Theorem 9. *To find the equation of conjugate line of the line*

$$\frac{x-f}{l} = \frac{y-g}{m} = \frac{z-h}{n}.$$

Let the system of forces be (X, Y, Z, L, M, N) referred to O as the origin.

In order to find the equation of the conjugate line of

$$\frac{x-f}{l}=\frac{y-g}{m}=\frac{z-h}{n}, \tag{1}$$

we have to find out the equation of null planes of any two chosen points on the line (1). Let one point be (f, g, h) and the other point $(l\rho, m\rho, n\rho)$ at infinity, where $\rho \to \infty$. Evidently, these two points lie on (1).

Equation of null plane through (f, g, h) is

$$\begin{vmatrix} x & y & z \\ f & g & h \\ X & Y & Z \end{vmatrix} = L(x-f)+M(y-g)+N(z-h) \tag{2}$$

and the null plane through $(l\rho, m\rho, n\rho)$ is

$$\begin{vmatrix} x & y & z \\ l\rho & m\rho & n\rho \\ X & Y & Z \end{vmatrix} = L(x-l\rho)+M(y-m\rho)+N(z-n\rho).$$

Dividing by ρ,

$$\begin{vmatrix} x & y & z \\ l & m & n \\ X & Y & Z \end{vmatrix} = L\left(\frac{x}{\rho}-l\right)+M\left(\frac{y}{\rho}-m\right)+N\left(\frac{z}{\rho}-n\right).$$

Making $\rho \to \infty$, we get

$$\begin{vmatrix} x & y & z \\ l & m & n \\ X & Y & Z \end{vmatrix} = -(Ll+Mm+Nn). \tag{3}$$

The line of intersection of null planes (2) and (3) is a conjugate line to the given line (1).

Working rule to solve problems.

The values of L_r, M_r, N_r are given by the determinant

$$L_r\vec{i}+M_r\vec{j}+N_r\vec{k} = \begin{vmatrix} \vec{i} & \vec{j} & \vec{k} \\ x_r & y_r & z_r \\ X_r & Y_r & Z_r \end{vmatrix}.$$

By equating the coefficients of $\vec{i}$, $\vec{j}$, $\vec{k}$ on both the sides, we get $L=\Sigma L_r$, $M=\Sigma M_r$, $N=\Sigma N_r$.

Theorem 10. *Conditions of equilibrium of a system of forces acting on a rigid body using principle of virtual work.*

Let a system of forces $P_1, P_2, P_3, \ldots$ act at given points of a rigid body and let a small displacement be given to the system consistent with its geometrical conditions. If δp_1 be the displacement at the points of application of P_1 along its line of action and similarly, $\delta p_2, \delta p_3, \ldots$ the displacements of the points of application of $P_2, P_3, \ldots$, then, if the forces $P_1, P_2, \ldots$ are in equilibrium the algebraic sum of their virtual works is zero, i.e.,

$$P_1 \delta p_1 + P_2 \delta p_2 + P_3 \delta p_3 + \cdots = 0.$$

Conversely, if the sum of the virtual work done is zero, i.e., if $P_1 \delta p_1 + P_2 \delta p_2 + P_3 \delta p_3 + \cdots = 0$, the system of forces is in equilibrium.

First method. We know that if a number of forces act on a particle and displace it, then the algebraic sum of the works done by these forces is equal to the work done by their resultant. Hence, if X_1, Y_1, Z_1 be the components of P_1 parallel to the axes and $\delta x_1, \delta y_1, \delta z_1$ the components in the same direction of δp_1, then

$$P_1 \delta p_1 = X_1 \delta x_1 + Y_1 \delta x_1 + Z_1 \delta x_1.$$

Similarly,
$$P_2 \delta p_2 = X_2 \delta x_2 + Y_2 \delta y_2 + Z_2 \delta z_2$$
$$P_3 \delta p_3 = X_3 \delta x_3 + Y_3 \delta y_3 + Z_3 \delta z_3$$
$$\ldots\ldots\ldots\ldots\ldots\ldots\ldots\ldots\ldots$$

$$\begin{aligned} \therefore P_1 \delta p_1 + P_2 \delta p_2 + P_3 \delta p_3 + \ldots \\ = (X_1 \delta x_1 + X_2 \delta x_2 + X_3 \delta x_3 + \ldots) + (Y_1 \delta y_1 + Y_2 \delta y_2 + Y_3 \delta y_3 + \ldots) \\ + (Z_1 \delta z_1 + Z_2 \delta z_2 + Z_3 \delta z_3 + \ldots). \end{aligned} \tag{1}$$

Now suppose that the displacement is a general one, i.e., of translation and rotation both. On account of translation let the displacements parallel to the coordinate axes be a, b, c and on account of rotation, let the body turn about the axes.

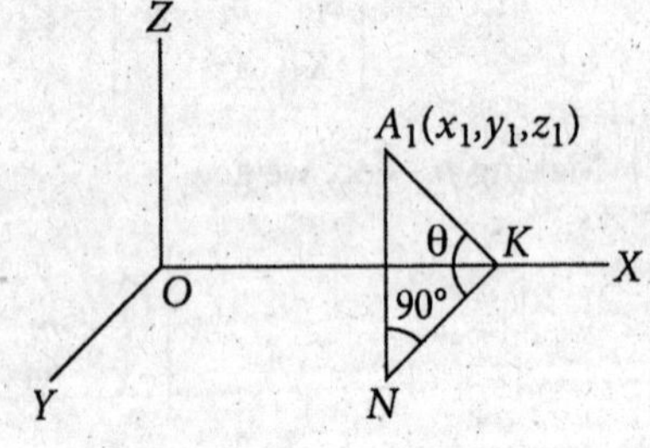

Fig. 6.17

We draw A_1K perpendicular to OX, A_1N perpendicular to the plane XOY, and let $\angle A_1KN = \theta$.

The y-coordinate of A_1 = KN = $A_1K \cos\theta$.

But the rotation about OX will change this into $A_1K\cos(\theta + \theta_1)$.

Hence, the change in the y-coordinate

$$\begin{aligned} &= A_1K\cos(\theta + \theta_1) - A_1K\cos\theta = A_1K[\cos(\theta + \theta_1) - \cos\theta] \\ &= A_1K[\cos\theta\cos\theta_1 - \sin\theta\sin\theta_1 - \cos\theta]. \end{aligned}$$

Since θ_1 is small, so that $\cos\theta_1 = 1, \sin\theta_1 = \theta$.

$\therefore$ change in y-coordinate

$$\begin{aligned} &= A_1K[\cos\theta \cdot 1 - \sin\theta \cdot \theta_1 - \cos\theta] = -\theta_1 A_1K\sin\theta \\ &= -\theta_1 A_1N = -\theta_1 z_1. \end{aligned}$$

Similarly, the change in the z-coordinate of A_1

$$= KA_1[\sin(\theta+\theta_1)-\sin\theta] = KA_1[\sin\theta\cos\theta_1+\cos\theta\sin\theta_1-\sin\theta]$$
$$= KA_1[\sin\theta\cdot 1+\theta_1\cos\theta-\sin\theta] = KA_1\cos\theta\,\theta_1 = y_1\theta_1.$$

In a similar manner it can be shown that a rotation of θ_2 about y-axis will produce change equal to $-\theta_2 x_1$ in z-coordinate and $\theta_2 z_1$ is x-coordinate. A rotation of θ_3 about z-axis will produce changes equal to $-\theta_3 y_1$ in x-coordinate and $\theta_3 x_1$ in y-coordinate.

Hence, due to rotation the changes are $\theta_2 z_1 - \theta_3 y_1$ parallel to OX, $\theta_3 x_1 - \theta_1 z_1$ parallel to OY, $\theta_1 y_1 - \theta_2 x_1$ parallel to OZ.

As stated before if a, b, c be the small displacements parallel to axes due to displacement of translation, then the total displacements parallel to the axes are

$$\delta x_1 = a + (\theta_2 z_1 - \theta_3 y_1) \text{ parallel to } OX$$
$$\delta y_1 = b + (\theta_3 x_1 - \theta_1 z_1) \text{ parallel to } OY$$
$$\delta z_1 = c + (\theta_1 y_1 - \theta_2 x_1) \text{ parallel to } OZ,$$

where $a, b, c, \theta_1, \theta_2$ are same for all points of the system.

$$\therefore\ P_1\delta p_1 = X_1\delta x_1 + Y_1\delta y_1 + Z_1\delta z_1$$
$$= X_1(a+\theta_2 z_1-\theta_3 y_1) + Y_1(b+\theta_3 x_1-\theta_1 z_1) + Z_1(c+\theta_1 y_1-\theta_2 x_1)$$
$$= aX_1 + bY_1 + cZ_1 + \theta_1(y_1 Z_1 - z_1 Y_1) + \theta_2(z_1 X_1 - x_1 Z_1) + \theta_3(x_1 Y_1 - y_1 X_1)$$

and $$\Sigma P_1\delta p_1 = a\Sigma X_1 + b\Sigma Y_1 + c\Sigma Z_1 + \theta_1\Sigma(y_1 Z_1 - z_1 Y_1) + \theta_2\Sigma(z_1 X_1 - x_1 Z_1) + \theta_3\Sigma(x_1 Y_1 - y_1 X_1).$$

($\because$ $a, b, c, \theta_1, \theta_2, \theta_3$ are the same for all the points of the system.)

$$\Rightarrow \Sigma P_1\delta p_1 = aX + bY + cZ + \theta_1 L + \theta_2 M + \theta_3 N, \quad (2)$$

where $L = \Sigma(y_1 Z_1 - z_1 Y_1)$, $M = \Sigma(z_1 X_1 - x_1 Z_1)$, $N = \Sigma(x_1 Y_1 - y_1 X_1)$.

If the system of forces be in equilibrium, then $X = \Sigma X_1 = 0$, $Y = \Sigma Y_1 = 0$, $Z = \Sigma Z_1 = 0$, $L = 0$, $M = 0$, $N = 0$.

Hence, $\Sigma P_1\delta p_1$, i.e., $P_1\delta p_1 + P_2\delta p_2 + P_3\delta p_3 + \cdots = 0$, i.e., sum of the virtual works of the system is zero.

Converse. The sum of virtual works of the system is zero for all displacements translator as well as rotatory, i.e.,

$$aX + bY + cZ + \theta_1 L + \theta_2 M + \theta_3 N = 0, \quad (3)$$

then to prove that the forces are in equilibrium.

Let us choose a small displacement parallel to axis of x alone, i.e., $b = 0, c = \theta, \theta_1 = 0, \theta_2 = 0, \theta_3 = 0$, but $a \neq 0$.

So from (3),

$$aX = 0 \Rightarrow X = 0.$$

Similarly, considering small displacements parallel to axis of y alone and axis of z alone respectively, we have $Y = 0, Z = 0$.

Again, suppose that the displacement is rotatory and that too about axis of x alone so that $a=0, b=0, c=0, \theta_2=0, \theta_3=0$.

Hence, from (3), $\theta_1 L=0$.

But θ_1 is not zero, $\therefore$ $L=0$.

Similarly, we can show that $M=0, N=0$.

Hence, if the equation of virtual work holds, all the conditions of equilibrium are satisfied and as such the system of forces is in equilibrium.

Vector Method. When the rigid body undergoes a small displacement, which is the sum of constant translation $\vec{u}=a\vec{i}+b\vec{j}+c\vec{k}$ (say) and a small rotation $\vec{e}=\theta_1\vec{i}+\theta_2\vec{j}+\theta_3\vec{k}$ (say) about O, then if $\vec{p}$ is displaced to $\vec{p}+d\vec{p}$, we have $d\vec{p}=\vec{u}+\vec{e}\times\vec{p}$, etc.

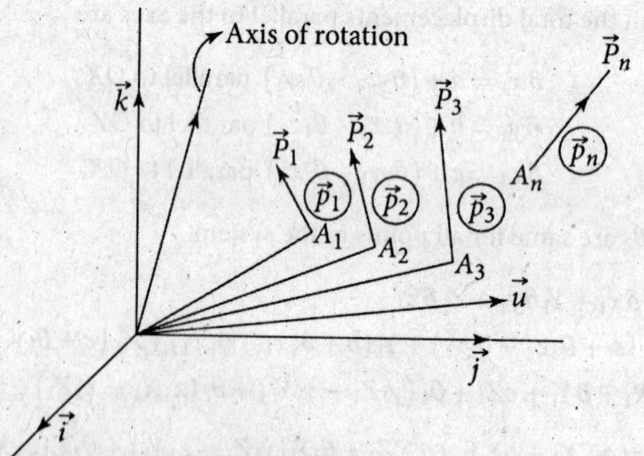

Fig. 6.18

Now, work done by the system of forces

$$=\vec{u}\cdot\left(\vec{P}_1+\vec{P}_2+\cdots+\vec{P}_n\right)+\vec{e}\cdot\left(\vec{p}_1\times\vec{P}_1+\vec{p}_2\cdot\vec{P}_2+\cdots+\vec{p}_n\times\vec{P}_n\right)$$

$$=\vec{u}\cdot\vec{P}+\vec{e}\cdot\vec{G}\text{ (say),} \tag{1}$$

where $\vec{P}=\vec{P}_1+\vec{P}_2+\ldots\vec{P}_n$ and

$$\vec{G}=\sum_{r=1}^{n}p_r\times P_r=\sum_{r=1}^{n}\left(x_r\vec{i}+y_r\vec{j}+z_r\vec{k}\right)\times\left(X_1\vec{i}+Y_1\vec{j}+Z_1\vec{k}\right).$$

Again, work done

$$=\left(a\vec{i}+b\vec{j}+c\vec{k}\right)\cdot\left(X\vec{i}+Y\vec{j}+Z\vec{k}\right)+\left(\theta_1\vec{i}+\theta_2\vec{j}+\theta_3\vec{k}\right)\cdot\left(L\vec{i}+M\vec{j}+N\vec{k}\right)$$

$$\left[\because\ \vec{P}=\sum_{r=1}^{n}\vec{P}_0=(X,Y,Z)\text{ and }\vec{G}=(L,M,N)\right]$$

$$=aX+bY+cZ+\theta_1L+\theta_2M+\theta_3N.$$

Now, proceed as in first method.

WORKED-OUT EXAMPLES

1. Forces X, Y and Z act along the straight lines $y=b$, $z=-c$, $z=c$, $x=-a$ and $x=a$, $y=-b$ respectively, show that they will have a single resultant of $\frac{a}{X}+\frac{b}{Y}+\frac{c}{Z}=0$.

[LNMU(H) 1993, BVABU(H) 1997]

And that the equations to its line of action are any two of three:

$$\frac{y}{Y}-\frac{z}{Z}-\frac{a}{X}=0, \qquad \frac{z}{Z}-\frac{x}{X}-\frac{b}{Y}=0, \qquad \frac{x}{X}-\frac{y}{Y}-\frac{c}{Z}=0.$$

Solution: The given lines can be written as

$$\frac{x}{1}=\frac{y-b}{O}=\frac{z+c}{O}, \qquad \frac{x+a}{O}=\frac{y}{1}=\frac{z-c}{O}, \qquad \frac{x-a}{O}=\frac{y+b}{O}=\frac{z}{1}.$$

Forces acting along them are X, Y, Z respectively. Their components parallel to the axis are (X,O,O), (O,Y,O), (O,O,Z), i.e., (X_1,Y_1,Z_1), (X_2,Y_2,Z_2), (X_3,Y_3,Z_3).

$$X=\Sigma X_1=X, \qquad Y=\Sigma Y_1=Y, \qquad Z=\Sigma Z_1=Z.$$

Also,

$$L_1\vec{i}+M_1\vec{j}+N_1\vec{k}=\begin{vmatrix}\vec{i} & \vec{j} & \vec{k}\\ x_1 & y_1 & z_1\\ X_1 & Y_1 & Z_1\end{vmatrix}=\begin{vmatrix}\vec{i} & \vec{j} & \vec{k}\\ O & b & -c\\ X & O & O\end{vmatrix}$$

$$=\vec{i}(O)-\vec{j}(O+cX)+\vec{k}(O-bX).$$

$$\therefore\ L_1=O, \quad M_1=-cX, \quad N_1=-bX.$$

$$L_2\vec{i}+M_2\vec{j}+N_2\vec{k}=\begin{bmatrix}\vec{i} & \vec{j} & \vec{k}\\ x_2 & y_2 & z_2\\ X_2 & Y_2 & Z_2\end{bmatrix}=\begin{vmatrix}\vec{i} & \vec{j} & \vec{k}\\ -a & O & c\\ O & Y & O\end{vmatrix}$$

$$=\vec{i}(O-cY)=\vec{j}(O)+\vec{k}(-aY-O)$$

$$\Rightarrow L_2=-cY_2, \quad M_2=O, \quad N_2=-aY.$$

$$L_3\vec{i}+M_3\vec{j}+N_3\vec{k}=\begin{vmatrix}\vec{i} & \vec{j} & \vec{k}\\ x_3 & y_3 & z_3\\ X_3 & Y_3 & Z_3\end{vmatrix}=\begin{vmatrix}\vec{i} & \vec{j} & \vec{k}\\ a & -b & O\\ O & O & Z\end{vmatrix}$$

$$=\vec{i}(-bZ-O)-\vec{j}(aZ-O)+\vec{k}(O)$$

$$\Rightarrow L_3=-bZ, \quad M_3=-aZ, \quad N_3=O.$$

$$L=\Sigma L_1=O-bZ-cY=-(bZ+cY)$$

$$M=\Sigma M_1=-(cX+aZ)$$

$$N=\Sigma N_1=-(bX+aY).$$

The systems will reduce to a single force, if

$$LX+MY+N=0$$

$$\text{or, }-[(bZ+cY)X+(cX+aZ)Y+(bX+aY)Z]=0$$

$$\text{or, }2[aYZ+bZX+CXY]=0$$

$$\text{or, }\frac{a}{X}+\frac{b}{Y}+\frac{c}{Z}=0. \tag{1}$$

This is the required condition.

To determine the line of action of single force.

The equation of central axis are

$$\frac{L-yZ+zY}{X}=\frac{M-zX+xZ}{Y}=\frac{N-xY+yX}{N}$$
$$=p=\frac{LX+MY+NZ}{X^2+Y^2+Z^2}=0,\ [\text{by } (1)].$$

The above gives

$$L-yZ-zY=0,\ M-zX+xZ=0,\ N-xY+yX=0,$$

$$\text{or, } -(bZ+cY)-yZ+zY=0$$
$$-(cX-aZ)-zX+xZ=0$$
$$-(bX+aY)-xY+yX=0.$$

Dividing these equations by YZ, ZX and XY respectively, we get

$$-\left(\frac{b}{Y}+\frac{c}{Z}\right)-\frac{y}{Y}+\frac{z}{Z}=0$$
$$-\left(\frac{c}{Z}+\frac{a}{X}\right)-\frac{z}{Z}+\frac{x}{X}=0$$
$$-\left(\frac{b}{Y}+\frac{a}{X}\right)-\frac{x}{X}+\frac{y}{Y}=0.$$

Using (1), the last three equations become

$$\frac{a}{X}-\frac{y}{Y}+\frac{z}{Z}=0, \qquad \frac{y}{Y}-\frac{z}{Z}-\frac{a}{X}=0,$$
$$\frac{b}{Y}-\frac{z}{Z}+\frac{x}{X}=0, \qquad \frac{z}{Z}-\frac{x}{X}-\frac{b}{Y}=0,$$
$$\frac{c}{Z}-\frac{x}{X}+\frac{y}{Y}=0, \qquad \frac{x}{X}-\frac{y}{Y}-\frac{c}{Z}=0.$$

If we take any two of the above, then we get the third by means of (1). Hence, any two of the above equations represent the line of action of single force.

2. A force P acts along the axis of x and another force nP along a generator of the cylinder $x^2+y^2=a^2$; show that the central axis lies on the cylinder $n^2(nx-z)^2+(1+n^2)^2y^2=n^4a^2$. [IAS 1992, RU(H) 2000, LNMU(H) 1995]

Solution: Let $(a\cos\theta, a\sin\theta, 0)$ be a point on the cylinder and a force nP acts at it in a direction parallel to the axis of z whose $d.c.$'s are $(0,0,1)$. The second force P acts at $(0,0,0)$ along axis of x whose $d.c.$'s are $(1,0,0)$.

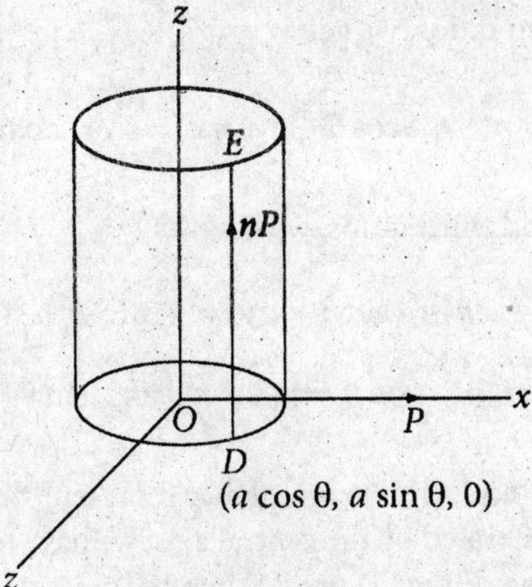

Fig. 6.19

Thus, the forces nP and P act along the lines

$$\frac{x-a\cos\theta}{O}=\frac{y-a\sin\theta}{O}=\frac{z-O}{1} \text{ and } \frac{x-O}{1}=\frac{y-O}{O}=\frac{z-O}{O}.$$

Let the system be (X, Y, Z, L, M, N).
The components of nP and P along the axis are (O, O, P), (P, O, O), i.e., (X_1, Y_1, Z_1) and (X_2, Y_2, Z_2)

$$X=\Sigma X_1=P, \quad Y=\Sigma Y_1=O, \quad Z=\Sigma Z_1=nP.$$

Also,

$$L_1\vec{i}+M_1\vec{j}+N_1\vec{k}=\begin{vmatrix}\vec{i} & \vec{j} & \vec{k}\\ x_1 & y_1 & z_1\\ X_1 & Y_1 & Z_1\end{vmatrix}=\begin{vmatrix}\vec{i} & \vec{j} & \vec{k}\\ a\cos\theta & a\sin\theta & O\\ O & O & nP\end{vmatrix}$$

$$=\vec{i}(a\sin\theta\, nP)-\vec{j}(a\cos\theta\, nP)+\vec{k}O.$$

$$\Rightarrow L_1=a\sin\theta\, nP, \; M_1=-nPa\cos\theta N_1=0.$$

$$L_2\vec{i}+M_2\vec{j}+N_2\vec{k}=\begin{vmatrix}\vec{i} & \vec{j} & \vec{k}\\ x_2 & y_2 & z_2\\ X_2 & Y_2 & Z_2\end{vmatrix}=\begin{vmatrix}\vec{i} & \vec{j} & \vec{k}\\ O & O & O\\ P & O & O\end{vmatrix}=\vec{i}.O+\vec{j}.O+\vec{k}.O.$$

$$\Rightarrow L_2=0, \; M_2=0, \; N_2=0.$$

$$L=\Sigma L_1=a\sin\theta\, nP, \quad M=\Sigma M_1=-nPa\cos\theta, \quad N=\Sigma N_1=0.$$

Equations of central axis are

$$\frac{L-(yZ-zY)}{X}=\frac{M-(zX-xZ)}{Y}=\frac{N-(xY-yX)}{Z}$$

or, $$\frac{na\sin\theta P-(nPy-O)}{P}=\frac{-nPa\cos\theta-(zP-xnP)}{O}=\frac{O-(O-yP)}{nP},$$

or, $$\frac{na\sin\theta-ny}{1}=\frac{-na\cos\theta-z+nx}{0}=\frac{y}{n}. \qquad (1)$$

Taking the last two ratios, we get

$$n(-na\cos\theta - z + nx) = 0 \text{ or, } \cos\theta = \frac{nx - z}{na}. \tag{2}$$

Taking the first and third ratios, we have

$$n(na\sin\theta - ny) = y \text{ or, } n^2a\sin\theta - n^2y = y$$

$$\text{or, } n^2a\sin\theta = y(1+n^2) \text{ or, } \sin\theta = \frac{y(1+n^2)}{n^2a}. \tag{3}$$

The equations (2) and (3) taken together represent the central axis. In order to find out the surface generated by the central axis, we have to eliminate θ from (2) and (3). Squaring and then adding (2) and (3), we get

$$1 = \frac{(nx-z)^2}{n^2a^2} + \frac{y^2(1+n^2)^2}{n^4a^2} \text{ or, } n^4a^2 = n^2(nx-z)^2 + y^2(1+n^2)^2.$$

This is the required surface.

3. Three forces each equal to P act on a body one at the point $(a,0,0)$ parallel to Oy, the second at the point $(0,b,0)$ parallel to Oz and the third at the point $(0,0,c)$ parallel to Ox, the axis being rectangular. Find the resultant wrench in magnitude and position.

[CU(H) 2004, NBU(H) 2003]

Solution: The lines of action of the three equal forces, each being P, are

$$\frac{x-a}{O} = \frac{y}{1} = \frac{z}{O}, \quad \frac{x}{O} = \frac{y-b}{O} = \frac{z}{1} \text{ and } \frac{x}{1} = \frac{y}{O} = \frac{z-c}{O}.$$

The components parallel to axes are

$$(0,P,0),(0,0,P),(P,0,0), \text{ i.e., } (X_1,Y_1,Z_1)(X_2,Y_2,Z_2)(X_3,Y_3,Z_3)$$

$$X = \Sigma X_1 = P, \quad Y = \Sigma Y_1 = P, \quad Z = \Sigma Z_1 = P.$$

Also,

$$L_1\vec{i} + M_1\vec{j} + N_1\vec{k} = \begin{vmatrix} \vec{i} & \vec{j} & \vec{k} \\ x_1 & y_1 & z_1 \\ X_1 & Y_1 & Z_1 \end{vmatrix} = \begin{vmatrix} \vec{i} & \vec{j} & \vec{k} \\ a & 0 & 0 \\ 0 & P & 0 \end{vmatrix} = \vec{i}(0) - \vec{j}(0) + \vec{k}(aP)$$

$$\Rightarrow L_1 = 0, \quad M_1 = 0, \quad N_1 = aP.$$

Similarly,

$$L_2 = bP, M_2 = 0, N_2 = 0$$
$$L_3 = 0, M_3 = cP, N_3 = 0$$
$$L = \Sigma L_1 = bP, M = \Sigma M_1 = cP,$$
$$N = \Sigma N_1 = aP.$$

Also, $R^2 = X^2 + Y^2 + Z^2 = 3P^2$ so that $R = P\sqrt{3}$ and $LX + MY + NZ = bP^2 + cP^2 + aP^2 = P^2(a+b+c)$.

The resultant wrench is given by

$$\frac{K}{R} = \frac{LX+MY+NZ}{X^2+Y^2+Z^2} = \frac{LX+MY+NZ}{R^2}.$$

$$\therefore\ K = \frac{LX+MY+NZ}{R} = \frac{P^2(a+b+c)}{P\sqrt{3}} \Rightarrow K = \frac{(a+b+c)P}{\sqrt{3}}.$$

Equations of central axis are

$$\frac{L-(yZ-zY)}{X} = \frac{M-(zX-xZ)}{Y} = \frac{N-(xY-yX)}{Z} = p = \frac{K}{R}$$

$$\text{or, } \frac{bP-(yP-zP)}{P} = \frac{cP-(zP-xP)}{P} = \frac{aP-(xP-yP)}{P} = \frac{(a+b+c)P}{\sqrt{3}P\sqrt{3}}$$

$$\text{or, } b-y+z = c-z+x = a-x+y = \frac{a+b+c}{3}. \tag{1}$$

From the first and last ratios of (1), we get

$$3(b-y+z) = a+b+c$$

$$z = \frac{a+b+c}{3} - b + y.$$

Again, from second and last ratios of (1), we have

$$z = c + x - \frac{a+b+c}{3}.$$

$$\therefore\ \frac{a+b+c}{3} - b + y = c + x - \frac{a+b+c}{3}$$

$$\text{or, } a+b+c-3b+3y = 3c+3x-a-b-c$$

$$\text{or, } 3y + a + c - 2b = 3x - a - b + 2c$$

$$\text{or, } 3x + b + 2c = 3y + 2a + c.$$

Adding $a+b+c$ to both sides, we get

$$3x + a + 2b + 3c = 3y + 3a + b + 2c$$

$$\text{or, } x + \frac{a+2b+3c}{3} = y + \frac{3a+b+2c}{3} = z + \frac{c+2a+3b}{3} \text{ (by symmetry).} \tag{2}$$

By magnitude of the resultant wrench, we mean the values of K and R. The position of wrench means the equation of central axis. We have shown that

$$K = \left(\frac{a+b+c}{\sqrt{3}}\right)P,\ R = P\sqrt{3}$$

and (2) represents the equation of central axis.

4. OA, OB, OC are the three coterminous edges of a cube and AA', BB', CC' and OO' are diagonals. Along BC', CA', AB' and OO' act forces equal to X, Y, Z and R respectively. Show that they are equivalent to single resultant if $(YZ + ZX + XY)\sqrt{3} + R(X+Y+Z) = 0$.

[CU(H) 2007, KU(H) 2005]

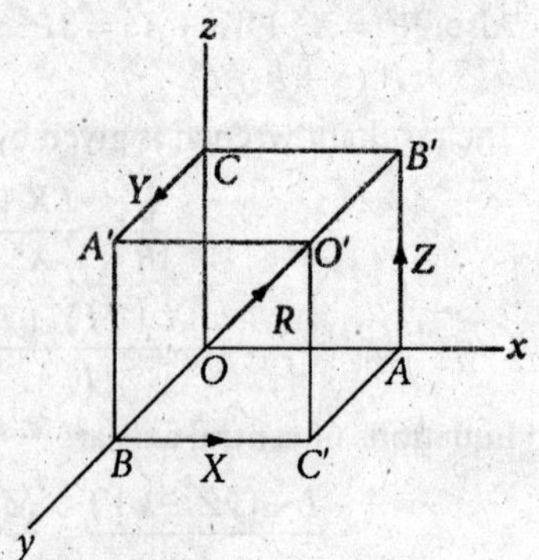

Fig. 6.20

Solution: Taking O as origin and the axis along OA, OB and OC. If a be the edge of a cube, then coordinates of A, B, C, O' are $(a,0,0)$, $(0,a,0)$, $(0,0,a)$, (a,a,a).

The direction cosines of OO' are

$$\frac{a}{a\sqrt{3}}, \frac{a}{a\sqrt{3}}, \frac{a}{a\sqrt{3}}, \text{ i.e., } \frac{1}{\sqrt{3}}, \frac{1}{\sqrt{3}}, \frac{1}{\sqrt{3}}.$$

The lines of action of X, Y and Z are parallel to the axis. Thus, the lines of action are

$$\frac{x}{1}=\frac{y-a}{O}=\frac{z}{O} \text{ for } X; \quad \frac{x}{O}=\frac{y}{1}=\frac{z-a}{O} \text{ for } Y;$$

$$\frac{x-a}{O}=\frac{y}{O}=\frac{z}{1} \text{ for } Z; \quad \frac{x}{\frac{1}{\sqrt{3}}}=\frac{y}{\frac{1}{\sqrt{3}}}=\frac{z}{\frac{1}{\sqrt{3}}} \text{ for } R.$$

Their components parallel to the axis are

$$(X,0,0),(0,Y,0),(0,0,Z),\left(\frac{R}{\sqrt{3}},\frac{R}{\sqrt{3}},\frac{R}{\sqrt{3}}\right),$$

$$\text{i.e., } (X_1,Y_1,Z_1),(X_2,Y_2,Z_2),(X_3,Y_3,Z_3),(X_4,Y_4,Z_4).$$

Let the system be equivalent to (X',Y',Z',L',M',N')

$$X'=\Sigma X_1=X+\frac{R}{\sqrt{3}}, \quad Y'=\Sigma Y_1=Y+\frac{R}{\sqrt{3}}, \quad Z'=\Sigma Z_1=Z+\frac{R}{\sqrt{3}}.$$

Also,

$$L_1\vec{i}+M_1\vec{j}+N_1\vec{k}=\begin{vmatrix} \vec{i} & \vec{j} & \vec{k} \\ x_1 & y_1 & z_1 \\ X_1 & Y_1 & Z_1 \end{vmatrix}=\begin{vmatrix} \vec{i} & \vec{j} & \vec{k} \\ 0 & a & 0 \\ X & 0 & 0 \end{vmatrix}=\vec{i}(0)-\vec{j}(0)+K(-aX)$$

$$\Rightarrow L_1=0,\ M_1=0,\ N_1=-aX.$$

Similarly,

$$L_2=-aY, \quad M_2=N_2=0$$
$$L_3=N_3=0, \quad M_3=-aZ$$

$$L_4\vec{i}+M_4\vec{j}+N_4\vec{k}=\begin{vmatrix} \vec{i} & \vec{j} & \vec{k} \\ 0 & 0 & 0 \\ \frac{R}{\sqrt{3}} & \frac{R}{\sqrt{3}} & \frac{R}{\sqrt{3}} \end{vmatrix}=0 \Rightarrow L_4=M_4=N_4=0.$$

Now

$$L'=\Sigma L_1=L_1+L_2+L_3+L_4=-aY$$
$$M'=\Sigma M_1=-aZ, N'=-aX.$$

Let the system reduce to a single force so that

$$L'X' + M'Y' + N'Z' = 0$$

$$\text{or, } -aY\left(X + \frac{R}{\sqrt{3}}\right) - aZ\left(Y + \frac{R}{\sqrt{3}}\right) - aX\left(Z + \frac{R}{\sqrt{3}}\right) = 0$$

$$\text{or, } \frac{R}{\sqrt{3}}(X+Y+Z) + (XY+YZ+ZX) = 0$$

$$\text{or, } (XY+YZ+ZX)\sqrt{3} + R(X+Y+Z) = 0.$$

5. If P and Q be two non-intersecting forces whose directions are perpendicular, show that the ratio of distances of the central axis from their lines of action is Q^2 to P^2.

Solution: Let the resultant wrench be (R, K). Also, let $OA = a$, $OB = b$.

Angle between P and Q is $\theta + \phi = \frac{\pi}{2}$.

Resolving forces along the axis of z and y, we get

$$P\cos\theta + Q\cos\phi = R$$

$$P\sin\theta - Q\sin\phi = 0$$

$$\text{or, } P\cos\theta + Q\sin\theta = R \qquad (1)$$

$$P\sin\theta - Q\cos\theta = 0. \qquad (2)$$

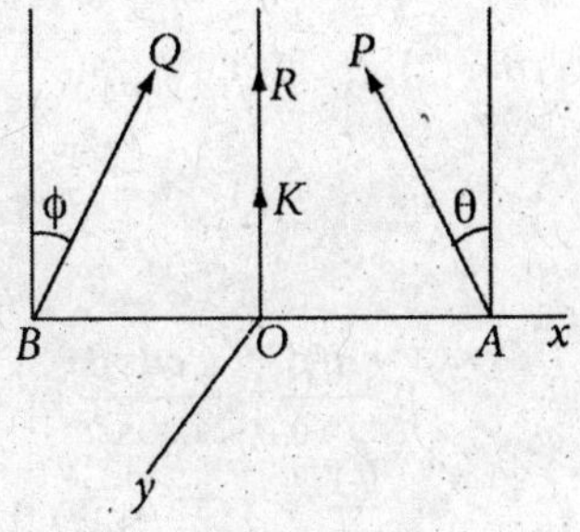

Fig. 6.21

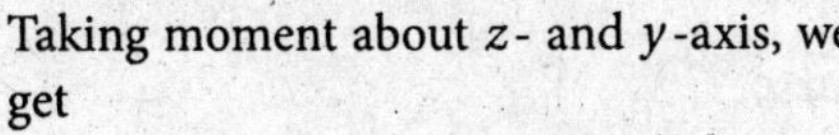

Taking moment about z- and y-axis, we get

$$P\sin\theta\, a + Q\sin\phi\, b = K$$

$$-P\cos\theta\, a + Q\cos\phi\, b = 0$$

$$aP\sin\theta + bQ\cos\theta = K \qquad (3)$$

$$-aP\cos\theta + bQ\sin\theta = 0. \qquad (4)$$

From (4), we get $aP\cos\theta = bQ\sin\theta$.

$$\frac{a}{b} = \frac{Q\sin\theta}{P\cos\theta} = \frac{Q}{P}\tan\theta. \qquad (5)$$

From (2),

$$P\sin\theta = Q\cos\theta; \quad \frac{\sin\theta}{\cos\theta} = \frac{Q}{P}.$$

So from (5),

$$\frac{a}{b} = \frac{Q}{P}\frac{Q}{P} = \frac{Q^2}{P^2}.$$

6. Two forces P and Q are such that their central axis is given in position and the line of action of P is given. Show that the locus of the line of action of Q is a conchoid.

Solution: Let the lines of action of P and Q make angle α, β with Z-axis. Let the line of action of P be fixed so that α is constant but β is not constant.

Take any point $(-x, -y, z)$ on BQ.

Resolving forces along z- and y-axis,

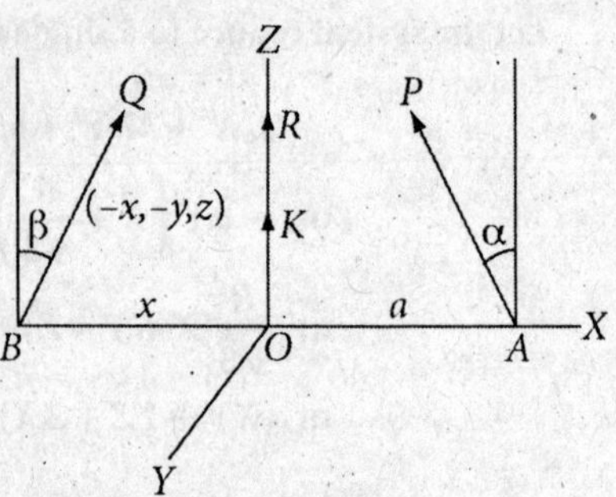

Fig. 6.22

$$P\cos\alpha + Q\cos\beta = R \qquad (1)$$
$$P\sin\alpha - Q\sin\beta = 0. \qquad (2)$$

Taking moment about z- and y-axis,

$$P\sin\alpha.a + Q\sin\beta x = K \qquad (3)$$
$$-P\cos\alpha.a + Q\cos\beta x = 0. \qquad (4)$$

From (4),

$$\frac{P}{Q} = \frac{x\cos\beta}{a\cos\alpha}.$$

From (2),

$$\frac{P}{Q} = \frac{\sin\beta}{\sin\alpha}.$$

$$\therefore \frac{\sin\beta}{\sin\alpha} = \frac{x\cos\beta}{a\cos\alpha} \text{ or, } a\sin\beta\cos\alpha = x\cos\beta\sin\alpha, \text{ or, } a\tan\beta = x\tan\alpha$$

$$\text{or, } \frac{ay}{z} = x\tan\alpha, \text{ or, } ay = zx\tan\alpha,$$

where $\tan\beta = \frac{y}{z}$.

Thus, the locus of line of action of Q is obviously a conchoid.

7. Find the null point of the plane $x + y + z = 0$ for the force-system (X, Y, Z, L, M, N).

[RU(H) 1998]

Solution: Let (f, g, h) be the null point of the plane

$$x + y + z = 0. \qquad (1)$$

As (f, g, h) lies on (1), and hence,

$$f + g + h = 0. \qquad (2)$$

But the null plane of (f, g, h) is

$$x(L - gZ + hY) + y(M - hX + fZ) + z(N - fY + gX) = Lf + Mg + Nh.$$

Comparing this with the plane (1), we get

$$\frac{L - gZ + hY}{1} = \frac{M - hX + fZ}{1} = \frac{N - fY + gX}{1} = \frac{Lf + Mg + Nh}{O}.$$

Taking first, second; second, third and third, first,

$$L-gZ+hY=M-hX+fZ$$
$$M-hX+fZ=N-fY+gX$$
$$N-fY+gX=L-gZ+hY$$
$$\text{or, } L-M=-hY-hX+(f+g)Z,$$
$$\text{or, } L-M=-hY-hX-hZ \text{ [from (2)]}$$
$$M-N=(g+h)X-fZ-fY=-fX-fZ-fY \quad \text{[from (2)]}$$
$$N-L=(f+h)Y-gX-gZ=-gY-gX-gZ \quad \text{[from (2)]}$$
$$\text{or, } h=-(L-M)K,\ f=-(M-N)K,\ g=-(N-L)K,$$

where $K=\frac{1}{X+Y+Z}$.

Thus,

$$\left(\frac{M-L}{X+Y+Z}, \frac{N-M}{X+Y+Z}, \frac{L-N}{X+Y+Z}\right)$$

are the coordinates of null point.

8. Show that the wrench (X,Y,Z,L,M,N) is equivalent to two forces, one along the line $x=y=z$ and the other along the line given by

$$Lx+My+Nz=0$$
$$x(Y-Z)+y(Z-X)+z(X-Y)=L+M+N$$

and find the magnitude of the two forces.

Solution: Let a force P act along the line

$$\frac{x}{1}=\frac{y}{1}=\frac{z}{1}, \text{ i.e., } \frac{x-0}{\frac{1}{\sqrt{3}}}=\frac{y-0}{\frac{1}{\sqrt{3}}}=\frac{z-0}{\frac{1}{\sqrt{3}}}.$$

Then, P acts at $(0,0,0)$ and its components parallel to the axis are $X_1=KP$, $Y_1=KP$, $Z_1=KP$, where $K=\frac{1}{\sqrt{3}}$.

Now

$$L_1\vec{i}+M_1\vec{j}+N_1\vec{k}=\begin{vmatrix}\vec{i} & \vec{j} & \vec{k}\\ x_1 & y_1 & z_1\\ X_1 & Y_1 & Z_1\end{vmatrix}=\begin{vmatrix}\vec{i} & \vec{j} & \vec{k}\\ 0 & 0 & 0\\ KP & KP & KP\end{vmatrix}=0.$$
$$\Rightarrow L_1=0=M_1=N_1.$$

Since the system reduces to (X,Y,Z,L,M,N). Then,

$$X_2=X-KP,\quad Y_2=Y-KP,\quad Z_2=Z-KP.$$

Let this force act at $(f,g,0)$.

Now

$$L_2\vec{i}+M_2\vec{j}+N_2\vec{k}=\begin{vmatrix}\vec{i} & \vec{j} & \vec{k}\\ x_2 & y_2 & z_2\\ X_2 & Y_2 & Z_2\end{vmatrix}=\begin{vmatrix}\vec{i} & \vec{j} & \vec{k}\\ f & g & 0\\ X-KP & Y-KP & Z-KP\end{vmatrix}$$
$$=\vec{i}[g(Z-KP)-0]-\vec{j}[f(Z-KP)-0]+\vec{k}[f(Y-KP)-g(X-KP)]$$

$$\Rightarrow L_2 = g(Z-KP), \quad M_2 = -f(Z-KP), \quad N_2 = f(Y-KP) - g(X-KP).$$

Now

$$L = L_1 + L_2 = 0 + L_2 = L_2, \quad M = M_2, \quad N = N_2.$$

$$\therefore L = g(Z-KP), \quad M = -f(Z-KP), \quad N = f(Y-KP) - g(X-KP),$$

$$\text{or, } \frac{L}{g} = -\frac{M}{f} = Z - KP$$

$$N = -\frac{(Y-KP)M}{(Z-KP)} - \frac{L(X-KP)}{Z-KP}$$

$$\text{or, } N(Z-KP) = -M(Y-KP) - L(X-KP)$$

$$\text{or, } N(Z-KP) + M(Y-KP) + L(X-KP) = 0$$

$$\text{or, } LX + MY + NZ - KP(L+M+N) = 0$$

$$\text{or, } P = \frac{LX+MY+NZ}{K(L+M+N)} = \frac{(LX+MY+NZ)\sqrt{3}}{L+M+N}. \tag{6.1}$$

This gives the magnitude of the first force acting along $x = y = z$.

Now

$$X - KP = X - \frac{(LX+MY+NZ)\sqrt{3}}{L+M+N} \cdot \frac{1}{\sqrt{3}} = \frac{LX+MX+NX-LX-MY-NZ}{L+M+N}$$

$$= \frac{X(M+N)-(MY+NZ)}{L+M+N}.$$

Similarly,

$$Y - KP = \frac{(L+N)Y-(LX+NZ)}{L+M+N}$$

$$Z - KP = \frac{Z(L+M)-(LX+MY)}{L+M+N}.$$

The magnitude of the second force

$$= \sqrt{(X-KP)^2 + (Y-KP)^2 + (Z-KP)^2}$$

$$= \frac{1}{L+M+N}\sqrt{\Sigma[X(M+N)-(MY+NZ)]^2}.$$

To determine the line of action of the second force.

Equation of null plane at (f, g, h) is

$$\begin{vmatrix} x & y & z \\ f & g & h \\ X & Y & Z \end{vmatrix} = L(x-f) + M(y-g) + N(z-h).$$

Hence, equation of null plane at $(0, 0, 0)$ is

$$\begin{vmatrix} x & y & z \\ 0 & 0 & 0 \\ X-KP & Y-KP & Z-KP \end{vmatrix} = L(x-0) + M(y-0) + N(x-0),$$

$$\text{or, } 0 = Lx + My + Nz. \tag{2}$$

Point at infinity on the line (1) is (ρ,ρ,ρ) where $\rho \to \infty$.

The equation of null plane at this point is

$$\begin{vmatrix} x & y & z \\ \rho & \rho & \rho \\ X-KP & Y-KP & Z-KP \end{vmatrix} = L(x-\rho)+M(y-\rho)+z(z-\rho). \qquad (3)$$

Dividing by ρ and then making $\rho \to \infty$.

$$\begin{vmatrix} x & y & z \\ 1 & 1 & 1 \\ X-KP & Y-KP & Z-KP \end{vmatrix} = L(0-1)+M(0-1)+N(0-1),$$

$$\text{or, } x[Z-KP-Y+KP]-y[Z-KP-X+KP]$$
$$+z[Y-KP-X+KP]=-(L+M+N)$$
$$\text{or, } x(Y-Z)+y(Z-X)+z(X-Y)=L+M+N. \qquad (4)$$

Equation (3) and (4) together represent the line of action of the other force.

9. Three forces act along the straight lines $x=0, y-z=a, y=0, z-x=a; z=0, x-y=a$. Show that they cannot reduce to a couple. Prove also that if the system reduces to a single force its line of action lies on the surface $x^2+y^2+z^2-2yz-2zx-2xy=a^2$.

[TMBU(H) 1994, BVABU(H) 1996, NB(H) 2006, CU(H) 1991]

Solution: The given line can be written as

$$\frac{x}{0}=\frac{y}{1}=\frac{z+a}{1}; \; \frac{x+a}{1}=\frac{y}{0}=\frac{z}{1}; \; \frac{x}{1}=\frac{y+a}{1}=\frac{z}{0},$$

i.e., along the lines

$$\frac{x}{0}=\frac{y}{\frac{1}{\sqrt{2}}}=\frac{z+a}{\frac{1}{\sqrt{2}}}; \; \frac{x+a}{\frac{1}{\sqrt{2}}}=\frac{y}{0}=\frac{z}{\frac{1}{\sqrt{2}}}; \; \frac{x}{\frac{1}{\sqrt{2}}}=\frac{y+a}{\frac{1}{\sqrt{2}}}=\frac{z}{0}.$$

Let P_1, P_2 and P_3 be the forces acting along these lines.

Their components parallel to the axes are

$$\left(0, \frac{P_1}{\sqrt{2}}, \frac{P_1}{\sqrt{2}}\right), \left(\frac{P_2}{\sqrt{2}}, 0, \frac{P_2}{\sqrt{2}}\right), \left(\frac{P_3}{\sqrt{2}}, \frac{P_3}{\sqrt{2}}, 0\right),$$

i.e.,
$$(X_1, Y_1, Z_1), (X_2, Y_2, Z_2), (X_3, Y_3, Z_3).$$

$$\therefore X=\Sigma X_1=\frac{P_2+P_3}{\sqrt{2}}, Y=\Sigma Y_1=\frac{P_1+P_3}{\sqrt{2}}, Z=\Sigma Z_1=\frac{P_1+P_2}{\sqrt{2}}.$$

Now,
$$L_1\vec{i}+M_1\vec{j}+N_1\vec{k}=\begin{vmatrix} \vec{i} & \vec{j} & \vec{k} \\ 0 & 0 & -a \\ 0 & \frac{P_1}{\sqrt{2}} & \frac{P_1}{\sqrt{2}} \end{vmatrix}=\vec{i}\left(\frac{aP_1}{\sqrt{2}}\right)-\vec{j}\cdot 0+\vec{k}\cdot 0.$$

$$\Rightarrow L_1=\frac{aP_1}{\sqrt{2}}, \; M_1=0, \; N_1=0.$$

$$L_2\vec{i}+M_2\vec{j}+N_2\vec{k}=\begin{vmatrix} \vec{i} & \vec{j} & \vec{k} \\ -a & 0 & 0 \\ \frac{P_2}{\sqrt{2}} & 0 & \frac{P_2}{\sqrt{2}} \end{vmatrix}=\vec{i}(0)-\vec{j}\left(-\frac{aP_2}{\sqrt{2}}\right)+\vec{k}(0)$$

$$\Rightarrow L_2=0,\ M_2=\frac{aP_2}{\sqrt{2}},\ N_2=0.$$

$$L_3\vec{i}+M_3\vec{j}+N_3\vec{k}=\begin{vmatrix} \vec{i} & \vec{j} & \vec{k} \\ 0 & -a & 0 \\ \frac{P_3}{\sqrt{2}} & \frac{P_3}{\sqrt{2}} & 0 \end{vmatrix}=\vec{i}(0)-\vec{j}(0)+\vec{k}\left(0+\frac{aP_3}{\sqrt{2}}\right)$$

$$\Rightarrow L_3=0, M_3=0, N_3=\frac{aP_3}{\sqrt{2}}$$

$$\therefore\ L=\Sigma L_1=\frac{aP_1}{\sqrt{2}}, M=\Sigma M_1=\frac{aP_2}{\sqrt{2}}, N=\Sigma N_1=\frac{aP_3}{\sqrt{2}}.$$

Now,
$$R=\sqrt{X^2+Y^2+Z^2}=\sqrt{\left(\frac{P_2+P_3}{\sqrt{2}}\right)^2+\left(\frac{P_3+P_1}{\sqrt{2}}\right)^2+\left(\frac{P_1+P_2}{\sqrt{2}}\right)^2}$$
$$=\tfrac{1}{2}\sqrt{(P_1+P_2)^2+(P_2+P_3)^2+(P_3+P_1)^2}.$$

Since $R\neq 0$, hence, the system cannot be reduced to a single couple.
Further, if the system reduces to a single force, then the couple $k=0$, the equation of the central axis are

$$\frac{L-yZ+zY}{X}=\frac{M-zX+xZ}{Y}=\frac{N-nY+yX}{Z}=\frac{0}{R}$$
$$\Rightarrow L-yZ+zY=0, M-zX+xZ=0, N-xY+yX=0$$
$$\Rightarrow \frac{aP_1}{\sqrt{2}}-y\frac{(P_1+P_2)}{\sqrt{2}}+z\left(\frac{P_1+P_3}{\sqrt{2}}\right)=0$$
$$\Rightarrow aP_1-y(P_1+P_2)+z(P_1+P_3)=0$$
$$\Rightarrow P_1(a-y+z)-yP_1+zP_3=0. \tag{1}$$

Similarly,
$$P_1x+P_2(a+x-z)-P_3z=0 \tag{2}$$
$$-P_1x+P_2y+P_3(a-x+y)=0. \tag{3}$$

Eliminating P_1, P_2, P_3 from (1), (2), (3), the required surface is given by

$$\begin{vmatrix} a-y+z & -y & z \\ x & a+x-z & -z \\ -x & y & a-x+y \end{vmatrix}=0.$$

Now $P_1\rightarrow R_1+R_2$ and $R_2\rightarrow R_2+R_3$ gives

$$\begin{vmatrix} a-y+z+x & a+x-y-z & 0 \\ 0 & a+x+y-z & a-x+y-z \\ -x & y & a-x+y \end{vmatrix}=0.$$

Now $C_2' \to C_2-(C_1+C_3)$ gives

$$\begin{vmatrix} a+x-y+z & -2z & 0 \\ 0 & 2x & a-x+y-z \\ -x & 2x-a & a-x+y \end{vmatrix}=0.$$

By $R_3 \to R_3-R_2$, we get

$$\begin{vmatrix} a+x-y-z & -2z & 0 \\ 0 & 2x & a-x+y-z \\ -x & -a & z \end{vmatrix}=0$$

$$\Rightarrow (a+x-y+z)[2xz+a(a-x+y-z)]+2z[0+x(a-x+y-z)]=0$$
$$\Rightarrow (a+x+z-y)\left(2xz+a^2-ax+ay-az\right)+2zx(a-x+y-z)=0$$
$$\Rightarrow 2axz+a^3-a^2x+a^2y-a^2z+2x^2z+xa^2-x^2a+axy-azx$$
$$+2xz^2+a^2z-axz+ayz-az^2-2xyz-ya^2+axy$$
$$-ay^2+ayz+2azx-2x^2z+2xyz-2xz^2=0$$
$$\Rightarrow 2axy+2azx+a^3+2ayz-ax^2-az^2-ay^2=0$$
$$\Rightarrow 2xy+2zx+2yz+a^2-x^2-y^2-z^2=0$$
$$\Rightarrow x^2+y^2-z^2-2xy-2yz-2zx=a^2.$$

10. Two equal forces act along generators of the same system of the hyperboloid $\frac{x^2+y^2}{a^2}-\frac{z^2}{b^2}=1$ and cut the plane $z=0$ at the ends of perpendicular diameters of the circle $x^2+y^2=a^2$, show that the pitch of the equivalent wrench is $\frac{a^2b}{a^2+2b^2}$.

[CU(H) 2008, VU(H) 2007]

Solution: Let the coördinates of the ends of the perpendicular diameters of the circle $x^2+y^2=a^2, z=0$ be $A(a\cos\theta, a\sin\theta, 0)$ and $B(a\cos\left(\frac{\pi}{2}+\theta\right), a\sin\left(\frac{\pi}{2}+\theta\right), 0)$ be the coordinates of B are $(-a\sin\theta, a\cos\theta, 0)$.

Let two equal forces P, P act along the generators of the same system of the hyperboloid $\frac{x^2+y^2}{a^2}-\frac{z^2}{b^2}=1$ through A and B, whose equations are $\frac{x-a\cos\theta}{a\sin\theta}=\frac{y-a\sin\theta}{-a\cos\theta}=\frac{z}{b}$ and

$$\frac{x+a\sin\theta}{a\cos\theta}=\frac{y-a\cos\theta}{a\sin\theta}=\frac{z}{b}.$$

Direction cosine of the two generators are

$$\frac{a\sin\theta}{\sqrt{a^2+b^2}}, \frac{-a\cos\theta}{\sqrt{a^2+b^2}}, \frac{b}{\sqrt{a^2+b^2}}$$

and

$$\frac{a\cos\theta}{\sqrt{a^2+b^2}}, \frac{a\sin\theta}{\sqrt{a^2+b^2}}, \frac{b}{\sqrt{a^2+b^2}}.$$

Components of P parallel to the axes are

$$\left(\frac{Pa\sin\theta}{\sqrt{a^2+b^2}}, \frac{-Pa\cos\theta}{\sqrt{a^2+b^2}}, \frac{Pb}{\sqrt{a^2+b^2}}\right)$$

and

$$\left(\frac{Pa\cos\theta}{\sqrt{a^2+b^2}}, \frac{Pa\sin\theta}{\sqrt{a^2+b^2}}, \frac{Pb}{\sqrt{a^2+b^2}}\right),$$

i.e., (X_1, Y_1, Z_1) and (X_2, Y_2, Z_2).

$$\therefore\ X = \Sigma X_1 = \frac{Pa(\sin\theta+\cos\theta)}{\sqrt{a^2+b^2}}$$

$$Y = \Sigma Y_1 = \frac{Pa(\sin\theta-\cos\theta)}{\sqrt{a^2+b^2}}$$

$$Z = \Sigma Z_1 = \frac{2Pb}{\sqrt{a^2+b^2}}.$$

Also

$$L_1\vec{i}+M_1\vec{j}+N_1\vec{k} = \begin{vmatrix} \vec{i} & \vec{j} & \vec{k} \\ a\cos\theta & a\sin\theta & 0 \\ X_1 & Y_1 & Z_1 \end{vmatrix}$$

$$= \vec{i}(a\sin\theta Z_1 - 0) - \vec{j}(a\cos\theta Z_1 - 0) + \vec{k}(a\cos\theta Y_1 - a\sin\theta X_1)$$

$$= \frac{\vec{i}a\sin\theta Pb}{\sqrt{a^2+b^2}} - \frac{\vec{j}a\cos\theta Pb}{\sqrt{a^2+b^2}} + \vec{k}\left[\frac{a\cos\theta(-Pa\cos\theta)}{\sqrt{a^2+b^2}} - \frac{a\sin\theta Pa\sin\theta}{\sqrt{a^2+b^2}}\right]$$

$$= \frac{\vec{i}a\sin\theta Pb}{\sqrt{a^2+b^2}} - \frac{\vec{j}a\cos\theta Pb}{\sqrt{a^2-b^2}} + \frac{\vec{k}Pa^2}{\sqrt{a^2+b^2}}\left(-\cos^2\theta - \sin^2\theta\right)$$

$$= \frac{\vec{i}a\sin\theta Pb}{\sqrt{a^2+b^2}} - \frac{\vec{j}a\cos\theta Pb}{\sqrt{a^2+b^2}} - \frac{\vec{k}Pa^2}{\sqrt{a^2+b^2}}$$

$$\Rightarrow L_1 = \frac{Pab\sin\theta}{\sqrt{a^2+b^2}},\ M_1 = -\frac{Pab\cos\theta}{\sqrt{a^2+b^2}},\ N = \frac{-Pa^2}{\sqrt{a^2+b^2}}.$$

Again,

$$L_2\vec{i}+M_2\vec{j}+N_2\vec{k} = \begin{vmatrix} \vec{i} & \vec{j} & \vec{k} \\ -a\sin\theta & a\cos\theta & 0 \\ X_2 & Y_2 & Z_2 \end{vmatrix}$$

$$= \vec{i}a\cos\theta Z_2 + \vec{j}a\sin\theta Z_2 + \vec{k}(-a\sin\theta Y_2 - a\cos\theta X_2)$$

$$= \frac{ia\cos\theta Pb}{\sqrt{a^2+b^2}} + \frac{\vec{j}a\sin\theta Pb}{\sqrt{a^2+b^2}} + \vec{k}\left[\frac{-a\sin\theta Pa\sin\theta}{\sqrt{a^2+b^2}} - \frac{a\cos\theta Pa\cos\theta}{\sqrt{a^2+b^2}}\right]$$

$$\Rightarrow L_2 = \frac{Pab\cos\theta}{\sqrt{a^2+b^2}},\ M_2 = \frac{Pab\sin\theta}{\sqrt{a^2+b^2}},\ N = \frac{-Pa^2}{\sqrt{a^2+b^2}}.$$

Now,

$$L = \Sigma L_1 = \frac{Pab(\sin\theta+\cos\theta)}{\sqrt{a^2+b^2}} = bX.$$

$$M = \Sigma M_1 = \frac{Pab(\sin\theta-\cos\theta)}{\sqrt{a^2+b^2}} = bY.$$

$$N=\Sigma N_1=\frac{-2Pa^2}{\sqrt{a^2+b^2}}=\frac{a^2}{b}Z.$$

$\therefore$ the pitch p of the equivalent wrench is given by

$$p=\frac{LX+MY+NZ}{X^2+Y^2+Z^2}=\frac{bX^2+bY^2-\frac{a^2}{b}Z^2}{X^2+Y^2+Z^2}=\frac{b\left(X^2+Y^2\right)-\frac{a^2}{b}\cdot\frac{4P^2b^2}{a^2+b^2}}{\left(X^2+Y^2\right)+\frac{4P^2b^2}{a^2+b^2}}$$

$$=\frac{\frac{b2P^2a^2}{a^2+b^2}-\frac{4P^2a^2b}{a^2+b^2}}{\frac{2P^2a^2}{a^2+b^2}+\frac{4P^2b^2}{a^2+b^2}}=\frac{-2P^2a^2b}{2P^2\left(a^2+2b^2\right)}=\frac{a^2b}{a^2+2b^2}\text{ (in magnitude).}$$

11. Equal forces act along two perpendicular diagonals of opposite faces of a cube of side a; show that they are equivalent to a single force R acting along a line through the centre of the cube and a couple $\frac{a}{2}R$ with the same line for axis. [CU(H) 2004]

Solution: Let the edges OA, OB, OC of the cube be taken as the three coordinate axes and a be the length of each edge of the cube. Let the two equal forces P and P act along the two perpendicular, diagonals OC' and $A'B'$ of the opposite faces $OAC'B$ and $CB'O'A'$ of the cube respectively.

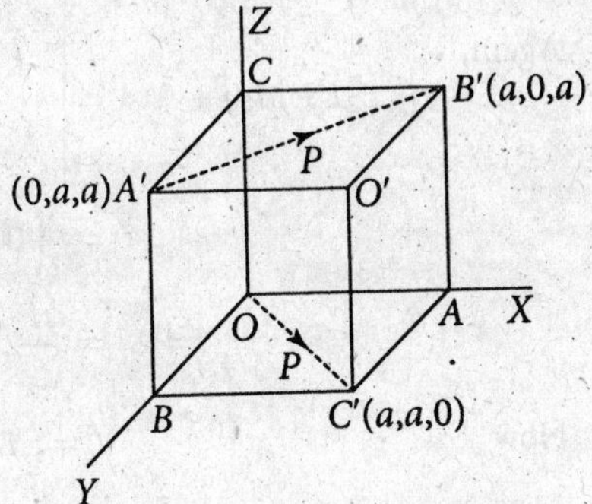

Fig. 6.23

The coordinates of the ends of these diagonals are as follows: $O(0,0,0)$, $C'(a,a,0)$, $A'(0,a,a)$ and $B'(a,0,a)$.

Direction ratios of the diagonal OC' are $a-0$, $a-0$, $0-0$, i.e., $a,a,0$.

$\therefore$ its direction cosines are

$$\frac{a}{\sqrt{a^2+a^2+0}},\ \frac{a}{\sqrt{a^2+a^2+0}},\ \frac{0}{\sqrt{a^2+a^2+a^2}},\text{ i.e., }\frac{1}{\sqrt{2}},\frac{1}{\sqrt{2}},0.$$

Also direction ratios of the diagonal $A'B'$ are $a-0, 0-a, a-a$, i.e., $a,-a,0$.

$\therefore$ its direction cosines are

$$\frac{a}{\sqrt{a^2+a^2+0}},\ \frac{-a}{\sqrt{a^2+a^2+0}},\ \frac{0}{\sqrt{a^2+a^2+0}},\text{ i.e., }\frac{1}{\sqrt{2}},-\frac{1}{\sqrt{2}},0.$$

Therefore, the forces acting on the body are as follows:

(1) the force P acting at the point $O(0,0,0)$ along OC' whose d.c.'s are $\frac{1}{\sqrt{2}},\frac{1}{\sqrt{2}},0$ and

(2) the force P acting at the point $A'(0,a,a)$ along $A'B'$ whose d.c.'s are $\frac{1}{\sqrt{2}},-\frac{1}{\sqrt{2}},0$.

The component of these forces parallel to the coordinate axes are

$$X_1=P\cdot\frac{1}{\sqrt{2}}=\frac{P}{\sqrt{2}},\ Y_1=P\cdot\frac{1}{\sqrt{2}}=\frac{P}{\sqrt{2}},\ Z_1=P\cdot 0=0.$$

$$X_2 = P \cdot \frac{1}{\sqrt{2}} = \frac{P}{\sqrt{2}},\ Y_2 = P \cdot \left(-\frac{1}{\sqrt{2}}\right) = -\frac{P}{\sqrt{2}},\ Z_2 = P \cdot 0 = 0.$$

$$\therefore\ X = \Sigma X_1 = \frac{P}{\sqrt{2}} + \frac{P}{\sqrt{2}} = P\sqrt{2}$$

$$Y = \Sigma Y_1 = \frac{P}{\sqrt{2}} - \frac{P}{\sqrt{2}} = 0$$

$$Z = \Sigma Z_1 = 0 \neq 0 = 0.$$

Also

$$L_1\vec{i} + M_1\vec{j} + N_1\vec{k} = \begin{bmatrix} \vec{i} & \vec{j} & \vec{k} \\ 0 & 0 & 0 \\ X_1 & Y_1 & Z_1 \end{bmatrix} = \vec{i}\cdot 0 + \vec{j}\cdot 0 + \vec{k}\cdot 0$$

$$\Rightarrow L_1 = 0,\ M_1 = 0,\ N_1 = 0.$$

Again,

$$L_2\vec{i} + M_2\vec{j} + N_2\vec{k} = \begin{vmatrix} \vec{i} & \vec{j} & \vec{k} \\ 0 & a & a \\ X_2 & Y_2 & Z_2 \end{vmatrix} = \begin{vmatrix} \vec{i} & \vec{j} & \vec{k} \\ 0 & a & a \\ \frac{P}{\sqrt{2}} & -\frac{P}{\sqrt{2}} & 0 \end{vmatrix}$$

$$= \vec{i}\left(0 + \frac{aP}{\sqrt{2}}\right) - \vec{j}\left(0 - \frac{aP}{\sqrt{2}}\right) + \vec{k}\left(0 - \frac{aP}{\sqrt{2}}\right)$$

$$\Rightarrow L_2 = \frac{aP}{\sqrt{2}},\ M_2 = \frac{aP}{\sqrt{2}},\ N_2 = -\frac{aP}{\sqrt{2}}.$$

Now

$$L = \Sigma L_1 = 0 + \frac{aP}{\sqrt{2}} = \frac{aP}{\sqrt{2}}$$

$$M = \Sigma M_1 = 0 + \frac{aP}{\sqrt{2}} = \frac{aP}{\sqrt{2}}$$

$$N = \Sigma N_1 = 0 - \frac{aP}{\sqrt{2}} = -\frac{aP}{\sqrt{2}}.$$

If the system of given forces is equivalent to a wrench (R, K), i.e., a single force R and a couple K having the same line of action as that of R, then

$$R = \sqrt{X^2 + Y^2 + Z^2} = \sqrt{\left(P\sqrt{2}\right)^2 + 0 + 0} = P\sqrt{2}$$

and from

$$KR = LX + MY + NZ = \frac{aP}{\sqrt{2}} \cdot P\sqrt{2} + \frac{aP}{\sqrt{2}} \cdot 0 - \frac{aP}{\sqrt{2}} \cdot 0 = aP^2 \Rightarrow KP\sqrt{2} = aP^2$$

$$\Rightarrow K = \frac{aP}{\sqrt{2}} = \frac{a\sqrt{2}P}{2} = \frac{aR}{2}.$$

Also the equation of central axis are

$$\frac{L - yZ + zY}{X} = \frac{M - zX + xZ}{Y} = \frac{N - xY - yX}{Z}$$

$$\Rightarrow \frac{\frac{aP}{\sqrt{2}} - y \cdot 0 + z \cdot 0}{P\sqrt{2}} = \frac{\frac{aP}{\sqrt{2}} - z \cdot P\sqrt{2} + x \cdot 0}{0} = \frac{-\frac{aP}{\sqrt{2}} - x \cdot 0 + y \cdot P\sqrt{2}}{0}$$

$$\Rightarrow \frac{a}{2} = \frac{\frac{a}{\sqrt{2}} - z\sqrt{2}}{0} = \frac{-\frac{a}{\sqrt{2}} + y\sqrt{2}}{0}$$
$$\Rightarrow z = \frac{a}{2} \text{ and } y = \frac{a}{2}.$$

Clearly, this line passes through the centre of the cube whose coordinates are $\left(\frac{a}{2}, \frac{a}{2}, \frac{a}{2}\right)$. Hence, the system is equivalent to a wrench (R, K), where $k = \frac{1}{2}aR$ acting along a line through the centre of the cube.

12. Three forces each equal to Q act on a rigid body, one at the point $(1,0,0)$ parallel to Oy; the second at the point $(0,1,0)$ parallel to z-axis, the third at the point $(0,0,1)$ parallel to x-axis, the axes being rectangular; find the equation of the central axis. [CU(H) 2001]

Solution: The line of action of the three equal forces each Q are

$$\frac{x-1}{0} = \frac{y}{1} = \frac{z}{0}, \frac{x}{0} = \frac{y-1}{0} = \frac{z}{1} \text{ and } \frac{x}{1} = \frac{y}{0} = \frac{z-1}{0}.$$

The components parallel to axes are

$$(0,Q,0), (0,0,Q), (Q,0,0), \text{ i.e., } (X_1, Y_1, Z_1), (X_2, Y_2, Z_2), (X_3, Y_3, Z_3).$$

$$\therefore X = \Sigma X_1 = Q, \ Y = \Sigma Y_1 = Q, \ Z = \Sigma Z_1 = Q.$$

Also,
$$L_1\vec{i} + M_1\vec{j} + N_1\vec{k} = \begin{vmatrix} \vec{i} & \vec{j} & \vec{k} \\ 1 & 0 & 0 \\ 0 & Q & 0 \end{vmatrix} = \vec{i}(0) - \vec{j}(0) + \vec{k}(0)$$
$$\Rightarrow L_1 = 0, M_1 = 0, N_1 = Q.$$

Similarly,
$$L_2 = Q, M_2 = 0, N_2 = 0$$
$$L_3 = 0, M_3 = Q, N_3 = 0.$$

$$\therefore L = \Sigma L_1 = Q, M = \Sigma M_1 = Q, N = \Sigma N_1 = Q.$$

So equation of central axis are

$$\frac{L - yZ + zY}{X} = \frac{M - zX + xZ}{Y} = \frac{N - xY + yX}{Z}$$
$$\Rightarrow \frac{Q - yQ + zQ}{Q} = \frac{Q - zQ + xQ}{Q} = \frac{Q - xQ + yQ}{Q}$$
$$\Rightarrow 1 - y + z = 1 - z + x = 1 - x + y$$
$$\Rightarrow z - y = x - z = y - x = \frac{0}{3} = 0$$
$$\Rightarrow x = y = z.$$

13. Equal forces act along the straight lines

$$\frac{x - \alpha_1}{l_i} = \frac{y - \beta_i}{m_i} = \frac{z - \gamma_i}{n_i} \ (i = 1,2,3,\ldots,n).$$

Find the equation of the central axis of the system. [CU(H) 2003]

Solution: Let the equal forces P act along the straight lines

$$\frac{x-\alpha_i}{l_i}=\frac{y-\beta_i}{m_i}=\frac{z-\gamma_i}{n_i},$$

i.e., along the time

$$\frac{x-\alpha_i}{l_i k_i}=\frac{y-\beta_i}{m_i k_i}=\frac{z-\gamma_i}{n_i k_i},$$

where $k_i=\dfrac{1}{\sqrt{l_i^2+m_i^2+n_i^2}}$.

The component of these forces along the given lines

$$(Pl_i k_i, Pm_i k_i, Pn_i k_i)=(X_i, Y_i, Z_i).$$

$$\therefore\ X=\Sigma X_i=\sum_{l=1}^{n} Pl_i k_i,\quad Y=\Sigma Y_i=\sum_{l=1}^{n} Pm_i k_i,\quad Z=\Sigma Z_i=\sum_{l=1}^{n} Pn_i k_i.$$

Also,

$$L_i\vec{i}+M_i\vec{j}+N_i\vec{k}=\begin{vmatrix}\vec{i} & \vec{j} & \vec{k}\\ \alpha_i & \beta_i & \gamma_i\\ X_i & Y_i & Z_i\end{vmatrix}$$

$$=\vec{i}(\beta_i Z_i-Y_i\gamma_i)-\vec{j}(\alpha_i Z_i-\gamma_i X_i)+\vec{k}(\alpha_i Y_i-\beta_i X_i)$$

$$\Rightarrow L_i=\beta_i Z_i-Y_i\gamma_i,\ M_i=\gamma_i X_i-\alpha_i Z_i,\ N_i=\alpha_i Y_i-\beta_i X_i.$$

$$\therefore\ L=\Sigma L_i=\Sigma(\beta_i Z_i-\gamma_i Y_i),=P\sum_{i=1}^{n}(\beta_i n_i k_i-\gamma_i m_i k_i)$$

$$M=\Sigma M_i=\sum_{i=1}^{n}(\gamma_i X_i-\alpha_i Z_i)=P\sum_{i=1}^{n}(\gamma_i l_i k_i-\alpha_i n_i k_i)$$

$$N=\Sigma N_i=\sum(\alpha_i Y_i-\beta_i X_i)=P\sum_{i=1}^{n}(\alpha_i m_i k_i-\beta_i l_i k_i).$$

So equation of central axis are

$$\frac{L-yZ+zY}{X}=\frac{M-zX+xZ}{Y}=\frac{N-xY+yX}{Z}$$

$$\Rightarrow\frac{P\sum_{i=1}^{n}(\beta_i n_i k_i-\gamma_i m_i k_i)-yP\sum_{i=1}^{n} n_i k_i+zP\sum_{i=1}^{n} m_i k_i}{P\sum_{i=1}^{n} l_i k_i}$$

$$=\frac{P\sum_{i=1}^{n}(\gamma_i l_i k_i-\alpha_i n_i k_i)-zP\sum_{i=1}^{n} l_i k_i+xP\sum_{i=1}^{n} n_i k_i}{P\sum_{i=1}^{n} m_i k_i}$$

$$=\frac{P\sum_{i=1}^{n}(\alpha_i m_i k_i-\beta_i l_i k_i)-xP\sum_{i=1}^{n}m_i k_i+yP\sum_{i=1}^{n}l_i k_i}{P\sum_{i=1}^{n}n_i k_i}$$

$$=\frac{\sum_{i=1}^{n}(\beta_i n_i-\gamma_i m_i)\,k_i-y\sum_{i=1}^{n}n_i k_i+z\sum_{i=1}^{n}m_i k_i}{\sum_{i=1}^{n}l_i k_i}$$

$$=\frac{\sum_{i=1}^{n}(\gamma_i l_i-\alpha_i n_i)\,k_i-z\sum_{i=1}^{n}l_i k_i+x\sum_{i=1}^{n}n_i k_i}{\sum_{i=1}^{n}m_i k_i}$$

$$=\frac{\sum_{i=1}^{n}(\alpha_i m_i-\beta_i l_i)\,k_i-x\sum_{i=1}^{n}m_i k_i+y\sum_{i=1}^{n}l_i k_i}{\sum_{i=1}^{n}n_i k_i},$$

where $k_i=\dfrac{1}{\sqrt{l_i^2+m_i^2+n_i^2}}$.

This is the required equation of central axis.

14. Two forces P and Q act along the straight lines whose equations are $y=x\tan\alpha, z=c$ and $y=-x\tan\alpha, z=-c$ respectively, show that their central axis lies on a straight line

$$y=x\frac{P-Q}{P+Q}\tan\alpha,\quad \frac{z}{c}=\frac{P^2-Q^2}{P^2+PQ\cos 2\alpha+Q^2}.$$

For all values of P and Q, prove that this line is a generator of the surface $(x^2+y^2)\,z\sin 2\alpha=2cxy$. **[CU(H) 2008, KU(H) 2004, NBU(H) 06, VU(H) 2008]**

Solution: The equation of the given lines are

$$\frac{x}{\cos\alpha}=\frac{y}{\sin\alpha}=\frac{z-c}{0}\ \text{and}\ \frac{x}{\cos\alpha}=\frac{y}{-\sin\alpha}=\frac{z+c}{0}.$$

Forces acting along them are P and Q respectively. Their components parallel to the axes are $(P\cos\alpha, P\sin\alpha, 0)$ and $(Q\cos\alpha, -\theta\sin\alpha, 0)$, i.e., (X_1, Y_1, Z_1) and (X_2, Y_2, Z_2).

$$\therefore\ X=\Sigma X_1=P\cos\alpha+Q\cos\alpha=(P+Q)\cos\alpha$$
$$Y=\Sigma Y_1=P\sin\alpha-Q\sin\alpha=(P-Q)\sin\alpha$$
$$Z=\Sigma Z_1=0+0=0.$$

Also,

$$L_1\vec{i}+M_1\vec{j}+N_1\vec{k}=\begin{vmatrix}\vec{i} & \vec{j} & \vec{k}\\ 0 & 0 & c\\ P\cos\alpha & P\sin\alpha & 0\end{vmatrix}$$

$$=\vec{i}(0-P\sin\alpha c)-\vec{j}(0-P\cos\alpha c)+\vec{k}(0)$$

$$=\vec{i}(-P\sin\alpha c)+\vec{j}(P\cos\alpha c)+\vec{k}(0)$$

$$\Rightarrow L_1=-P\sin\alpha c, M_1=P\cos\alpha c, N_1=0$$

$$\text{and } L_2\vec{i}+M_2\vec{j}+N_2\vec{k}=\begin{vmatrix}\vec{i} & \vec{j} & \vec{k}\\ 0 & 0 & -c\\ Q\cos\alpha & -Q\sin\alpha & 0\end{vmatrix}$$

$$=\vec{i}(0-Q\sin\alpha c)-\vec{j}(0+Q\cos\alpha c)+\vec{k}(0)$$

$$\Rightarrow L_2=-Q\sin\alpha c,\ M_2=-Q\cos\alpha c,\ N_2=0.$$

$$\therefore\ L=\Sigma L_1=-P\sin\alpha c-Q\sin\alpha c=-P(P+Q)\sin\alpha c$$

$$M=\Sigma M_1=P\cos\alpha c-Q\cos\alpha c=(P-Q)\cos\alpha c$$

$$N=\Sigma N_1=0+0=0.$$

$\therefore$ the equation of the central axis are

$$\frac{L-yZ+zY}{X}=\frac{M-zX+nZ}{Y}=\frac{N-xY+yX}{Z}$$

$$\Rightarrow\frac{-(P+Q)c\sin\alpha-y\cdot 0+z\cdot(P-Q)\sin\alpha}{(P+Q)\cos\alpha}$$

$$=\frac{(P-Q)\cos\alpha c-z(P+Q)\cos\alpha+x\cdot 0}{(P-Q)\sin\alpha}=\frac{0-x\cdot(P-Q)\sin\alpha+y\cdot(P+Q)\cos\alpha}{0}.$$

$$\Rightarrow\frac{-(P+Q)c\sin\alpha+z(P-Q)\sin\alpha}{(P+Q)\cos\alpha}=\frac{(P-Q)\cos\alpha c-z(P+Q)\cos\alpha}{(P-Q)\sin\alpha}$$

$$=\frac{-x(P-Q)\sin\alpha+y(P+Q)\cos\alpha}{0}. \tag{1}$$

Last two members of (1) give

$$-x(P-Q)\sin\alpha+y(P+Q)\cos\alpha=0\Rightarrow y(P+Q)\cos\alpha=x(P-Q)\sin\alpha$$

$$\Rightarrow y=x\frac{P-Q}{P+Q}\tan\alpha. \tag{2}$$

And the first two members of (1) give

$$z(P-Q)^2\sin^2\alpha-c\sin^2\alpha\left(P^2-Q^2\right)=\left(P^2-Q^2\right)c\cos^2\alpha-z(P+Q)^2\cos^2\alpha$$

$$\Rightarrow z\left[(P-Q)^2\sin^2\alpha+(P+Q)^2\cos^2\alpha\right]=c\left(P^2-Q^2\right)\left(\cos^2\alpha+\sin^2\alpha\right)$$

$$\Rightarrow\frac{z}{c}=\frac{(P^2-Q^2)}{(R^2-2PQ+Q^2)\sin^2\alpha+(P^2+2PQ+Q^2)\cos^2\alpha}$$

$$\Rightarrow\frac{z}{c}=\frac{P^2-Q^2}{P^2+Q^2+2PQ\left(\cos^2\alpha-\sin^2\alpha\right)}$$

$$\Rightarrow\frac{z}{c}=\frac{P^2-Q^2}{P^2+Q^2+2PQ\cos 2\alpha}. \tag{3}$$

∴ the equations of the central axis are given by (2) and (3).

Again, first two members of (1) give

$$-c\tan\alpha+z\left(\frac{P-Q}{P+Q}\right)\tan\alpha=c\cot\alpha-z\left(\frac{P+Q}{P-Q}\right)\cot\alpha. \tag{4}$$

Eliminating $\frac{P-Q}{P+Q}$ from (2) and (4), we get

$$-c\tan\alpha+z\frac{y}{x}=c\cot\alpha-z\frac{x\tan\alpha}{y}\cot\alpha \Rightarrow -c\tan\alpha+\frac{zy}{x}=c\cot\alpha-\frac{zx}{y}$$

$$\Rightarrow z\left(\frac{y}{x}+\frac{x}{y}\right)=c(\cot\alpha+\tan\alpha)\Rightarrow\frac{z\left(x^2+y^2\right)}{xy}=c\left(\frac{\cos\alpha}{\sin\alpha}+\frac{\sin\alpha}{\cos\alpha}\right)$$

$$\Rightarrow\frac{z\left(x^2+y^2\right)}{xy}=\frac{2c}{2\sin\alpha\cos\alpha}\Rightarrow z\left(x^2+y^2\right)\sin 2\alpha=2cxy \text{ (Proved).}$$

15. Two forces act, one along the line $y=0, z=0$ and the other along the line $x=0, z=c$. As the forces vary show that the surface generated by the axis of their equivalent wrench is $\left(x^2+y^2\right)z=cy^2$. [CU(H) 2007]

Solution: Given lines can be written as

$$\frac{x}{1}=\frac{y}{0}=\frac{z}{0} \text{ and } \frac{x}{0}=\frac{y}{1}=\frac{z-c}{0}.$$

Let the two forces along these lines be P and Q. Their components parallel to the axis are $(P,0,0)$ and $(0,Q,0)$, i.e., (X_1,Y_1,Z_1) and (X_2,Y_2,Z_2).

$$X=\Sigma X_1=P+0=P$$
$$Y=\Sigma Y_1=0+Q=Q$$
$$Z=\Sigma Z_1=0+0=0.$$

Also,

$$L_1\vec{i}+M_1\vec{j}+N_1\vec{k}=\begin{vmatrix}\vec{i}&\vec{j}&\vec{k}\\0&0&0\\P&0&0\end{vmatrix}=\vec{i}(0)+\vec{j}(0)+\vec{k}(0)$$

$$\Rightarrow L_1=0, M_1=0, N_1=0$$

and

$$L_2\vec{i}+M_2\vec{j}+N_2\vec{k}=\begin{vmatrix}\vec{i}&\vec{j}&\vec{k}\\0&0&c\\0&Q&0\end{vmatrix}=\vec{i}(0-cQ)-\vec{j}(0)+\vec{k}(0)$$

$$\Rightarrow L_2=-cQ, M_2=0, N_2=0.$$

$$\therefore\ L=\Sigma L_1=0-cQ=-cQ$$
$$M=\Sigma M_1=0+0=0$$
$$N=\Sigma N_1=0+0=0.$$

Hence, the equations to the central axis are

$$\frac{L-yZ+zY}{X}=\frac{M-zX+xZ}{Y}=\frac{N-xY+Yx}{Z}$$

$$\Rightarrow\frac{-cQ-y\cdot 0+zQ}{P}=\frac{0-zP+x\cdot 0}{Q}=\frac{0-xQ+yP}{0}$$

$$\Rightarrow\frac{(z-c)Q}{P}=\frac{-ZP}{Q}=\frac{yP-xQ}{0}. \tag{1}$$

From the third member of (1),

$$yP-xQ=0\Rightarrow\frac{P}{Q}=\frac{x}{y}. \tag{2}$$

Also from the first two members,

$$\frac{Q}{P}(z-c)=-\frac{P}{Q}z. \tag{3}$$

(2) and (3) are the equations of the central axis.

Hence, the surface generated by the central axis is obtained by eliminating P and Q from (2) and (3).

∴ surface is

$$(z-c)\frac{y}{x}=-\frac{x}{y}z\Rightarrow(z-c)y^2=-x^2z\Rightarrow z\left(y^2+z^2\right)=cy^2.$$

16. A force parallel to the axis of z acts at the point $(a,0,0)$ and an equal force perpendicular to the axis of z acts at the point $(-a,0,0)$. Show that the central axis of the system lies on the surface $z^2(x^2+y^2)=(x^2+y^2-ax)^2$.

[CU(H) 1998, VU(H) 2007, KU(H) 2008, TMBU(H) 2006]

Solution: Let a force P act at the point $(a,0,0)$ parallel to z-axis, i.e., along the line

$$\frac{x-a}{0}=\frac{y-0}{0}=\frac{z-0}{1}.$$

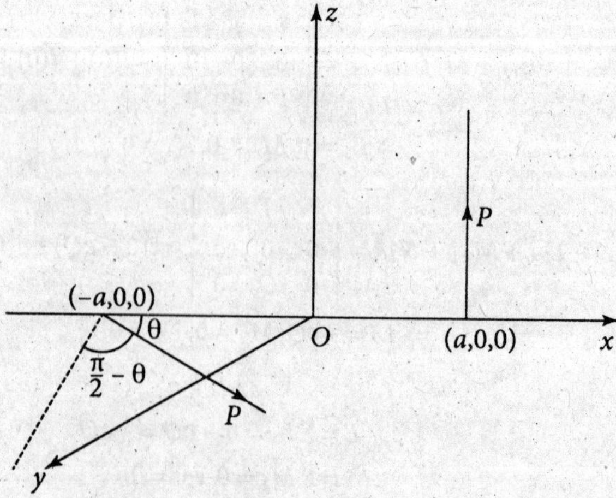

Fig. 6.24

Also, let an equal force P act at the point $(-a,0,0)$ perpendicular to the z-axis. The line of action of this force obviously lies in the xy-plane. If it makes an angle θ with the axis of x, then its d.c.'s are $\cos\theta, \cos\left(\frac{\pi}{2}-\theta\right), \cos 90°$, i.e., $\cos\theta, \sin\theta, 0$.
$\therefore$ the second force P acts along the line whose equations are

$$\frac{x+a}{\cos\theta}=\frac{y-0}{\sin\theta}=\frac{z-0}{0}.$$

Components of P parallel to the axes are $(0,0,P)$ and $(P\cos\theta, P\sin\theta, 0)$, i.e., (X_1, Y_1, Z_1) and (X_2, Y_2, Z_2).

$$\therefore\ X=\Sigma X_1=0+P\cos\theta=P\cos\theta$$
$$Y=\Sigma Y_1=0+P\sin\theta=P\sin\theta$$
$$Z=\Sigma Z_1=P+0=P.$$

Also,
$$L_1\vec{i}+M_1\vec{j}+N_1\vec{k}=\begin{vmatrix}\vec{i} & \vec{j} & \vec{k}\\ a & 0 & 0\\ 0 & 0 & P\end{vmatrix}=\vec{i}(0)-\vec{j}(aP)+\vec{k}(0)$$
$$\Rightarrow L_1=0, M_1=-aP, N_1=0$$

and
$$L_2\vec{i}+M_2\vec{j}+N_2\vec{k}=\begin{vmatrix}\vec{i} & \vec{j} & \vec{k}\\ -a & 0 & 0\\ P\cos\theta & P\sin\theta & 0\end{vmatrix}=\vec{i}(0)-\vec{j}(0)+\vec{k}(-aP\sin\theta)$$
$$\Rightarrow L_2=0, M_2=0, N_2=-aP\sin\theta.$$

Now,
$$L=\Sigma L_1=0+0=0$$
$$M=\Sigma M_1=-aP+0=-aP$$
$$N=\Sigma N_1=0-aP\sin\theta.$$

$\therefore$ the equation of the central axis are

$$\frac{L-yZ+zY}{X}=\frac{M-zX+xZ}{Y}=\frac{N-xY+yX}{Z}$$
$$\Rightarrow\frac{0-yP+zP\sin\theta}{P\cos\theta}=\frac{-aP-zP\cos\theta+xP}{P\sin\theta}=\frac{-aP\sin\theta-xP\sin\theta+yP\cos\theta}{P}$$
$$\Rightarrow\frac{-y+z\sin\theta}{\cos\theta}=\frac{-a-z\cos\theta+x}{\sin\theta}=\frac{-a\sin\theta-x\sin\theta+y\cos\theta}{1}. \quad (1)$$

From first and second ratios of (1), we have

$$z\sin^2\theta-y\sin\theta=-a\cos\theta-z\cos^2\theta+x\cos\theta$$
$$\Rightarrow z=x\cos\theta+y\sin\theta-a\cos\theta. \quad (2)$$

Also from first and third ratios of (1), we have

$$-y+z\sin\theta=-a\sin\theta\cos\theta-x\sin\theta\cos\theta+y\cos^2\theta$$
$$\Rightarrow z\sin\theta=y\left(1+\cos^2\theta\right)-(x+a)\sin\theta\cos\theta. \quad (3)$$

Multiplying (2) by $\sin\theta$ and then subtracting from (3),

$$0=2x\sin\theta\cos\theta+y\left(\sin^2\theta-1-\cos^2\theta\right)$$
$$\Rightarrow 2x\sin\theta\cos\theta=y\left(1-\sin^2\theta+\cos^2\theta\right)$$
$$\Rightarrow 2x\sin\theta\cos\theta=y2\cos^2\theta$$
$$\Rightarrow \tan\theta=\frac{y}{x}. \tag{4}$$

(2) and (4) are equations of the central axis.

The required surface is obtained by eliminating from (2) and (4).

From (2),

$$z=(x-a)\cos\theta+y\sin\theta \Rightarrow z=\frac{(x-a)}{\sqrt{1+\frac{y^2}{x^2}}}+y\cdot\frac{1}{\sqrt{1+\frac{x^2}{y^2}}}$$
$$\Rightarrow z=\frac{(x-a)x}{\sqrt{x^2+y^2}}+\frac{y^2}{\sqrt{x^2+y^2}}$$
$$\Rightarrow z^2\left(x^2+y^2\right)=\left(x^2+y^2-ax\right)^2.$$

17. If a force (X,Y,Z) acts along the generator of the hyperbolic paraboloid $\frac{x^2}{a^2}-\frac{y^2}{b^2}=\frac{2z}{c}$ and be equivalent to an equal force (X,Y,Z) at the origin together with a couple (L,M,N), show that $aL\pm bM=0$, $bX+aY=0$ and $cN\pm abZ=0$. [VU(H) 2006]

Solution: The generators of the given hyperboloid are given by

$$\frac{x}{a}-\frac{y}{b}=\frac{2z}{c\mu},\quad \frac{x}{a}-\frac{y}{b}=\mu \tag{1}$$

and

$$\frac{x}{a}-\frac{y}{b}=\frac{2z}{c\lambda},\quad \frac{x}{a}+\frac{y}{b}=\lambda. \tag{2}$$

Now, we shall reduce the generator (1) to the symmetrical form. Let (l,m,n) be the direction cosine of the line (1).

$$\therefore\ \frac{l}{a}+\frac{m}{b}-\frac{2n}{c\mu} \text{ and } \frac{l}{a}-\frac{m}{b}=0.$$

Solving this, we get

$$\frac{l}{a}=\frac{m}{b}=\frac{n}{c\mu}.$$

Let the line (1) intersect xy-plane so that $z=0$ in this plane.

$$\therefore\ \frac{x}{a}+\frac{y}{b}=0,\quad \frac{x}{a}-\frac{y}{b}=\mu,$$

by which we get

$$x=\frac{a\mu}{2},\quad y=-\frac{b\mu}{2}.$$

Hence, the generator (1) in symmetrical form is given by

$$\frac{x-\frac{a\mu}{2}}{a/\sqrt{a^2+b^2+c^2\mu^2}}=\frac{y+\frac{b\mu}{2}}{b/\sqrt{a^2+b^2+c^2\mu^2}}=\frac{z-0}{c\mu/\sqrt{a^2+b^2+c^2\mu^2}}.$$

Now if R be the force whose components are (X, Y, Z), then

$$X=\frac{Ra}{\sqrt{a^2+b^2+c^2\mu^2}},\quad Y=\frac{Rb}{\sqrt{a^2+b^2+c^2\mu^2}},\quad Z=\frac{Rc\mu}{\sqrt{a^2+b^2+c^2\mu^2}}.$$

Also,

$$L\vec{i}+M\vec{j}+N\vec{k}=\begin{vmatrix}\vec{i} & \vec{j} & \vec{k}\\ \frac{a\mu}{2} & -\frac{b\mu}{2} & 0\\ X & Y & Z\end{vmatrix}=\vec{i}\left(-\frac{b\mu}{2}Z\right)-\vec{j}\left(\frac{a\mu}{2}Z\right)+\vec{k}\left(\frac{a\mu}{2}Y+\frac{b\mu}{2}X\right).$$

$$\Rightarrow L=-\frac{b\mu}{2}Z,\quad M=-\frac{a\mu}{2}Z,\quad N=\frac{\mu}{2}(aY+bX).$$

Now

$$aL-bM=-\frac{ab\mu}{2}Z+\frac{ba\mu}{2}Z=0.$$

$$bX-aY=\frac{bRa}{\sqrt{a^2+b^2+c^2\mu^2}}-\frac{aRb}{\sqrt{a^2+b^2+c^2\mu^2}}=0.$$

$$cN-abZ=\frac{c\mu}{2}(aY+bX)-abZ$$

$$=\frac{c\mu}{2}\left[\frac{aRb}{\sqrt{a^2+b^2+c^2\mu^2}}+\frac{bRa}{\sqrt{a^2+b^2+c^2\mu^2}}\right]-\frac{abRc\mu}{\sqrt{a^2+b^2+c^2\mu^2}}=0.$$

The results $aL+bM=0, bX+aY=0, cN+abZ=0$ follow from the generator (2) of the other kind.

18. OA, OB, OC are edges of a cube of side a and OO', AA', BB', CC' are its diagonals; along OB', $O'A$, BC and $C'A'$ act forces equal to P, $2P$, $3P$ and $4P$ respectively; show that they are equivalent to a force $P\sqrt{35}$ at O along a line whose direction cosines are proportion to $-3, -5, 6$ together with a couple $\frac{Pa}{2}\sqrt{114}$ about a line whose direction cosines are proportional to $7, -2, 2$. [IAS 2001]

Solution: Let the edges OA, OB, OC of a cube of side a be taken as the three coordinate axes as shown in the figure.

The coordinates of the vertices of the cube are as follows: $O(0,0,0)$, $A(a,0,0)$, $B(0,a,0)$, $C(0,0,a)$, $A'(0,a,a)$, $B'(a,0,a)$, $C'(a,a,0)$, $O'(a,a,a)$.

Direction ratios of the line OB' are $a-0, 0-0, a-0$, i.e., $a, 0, a$.

$\therefore$ its direction cosines are

$$\frac{a}{\sqrt{a^2+0^2+a^2}},\frac{0}{\sqrt{a^2+0^2+a^2}},\frac{a}{\sqrt{a^2+0^2+a^2}},\text{ i.e., }\frac{1}{\sqrt{2}},0,\frac{1}{\sqrt{2}}.$$

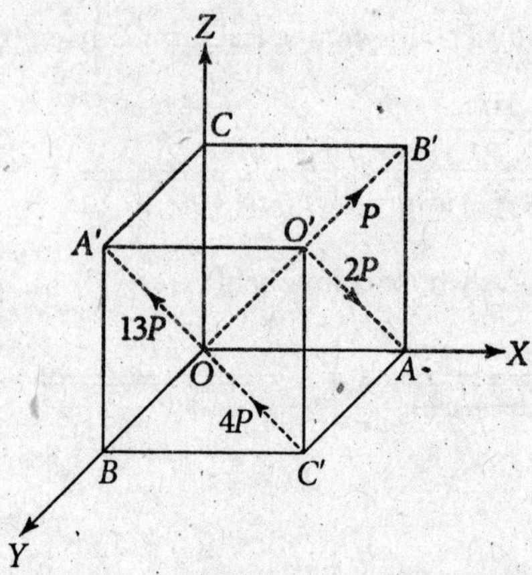

Fig. 6.25

Similarly, the direction cosines of $O'A$, BC and $C'A'$ are

$$0,-\frac{1}{\sqrt{2}},-\frac{1}{\sqrt{2}};0,-\frac{1}{\sqrt{2}},\frac{1}{\sqrt{2}} \text{ and } -\frac{1}{\sqrt{2}},0,\frac{1}{\sqrt{2}} \text{ respectively.}$$

So, equations of the line OB', $O'A$, BC and $C'A'$ are

$$\frac{x-a}{1/\sqrt{2}}=\frac{y-0}{0}=\frac{z-a}{1/\sqrt{2}},\frac{x-a}{0}=\frac{y-a}{-1/\sqrt{2}}=\frac{z-a}{-1/\sqrt{2}}.$$

$$\frac{x-0}{0}=\frac{y-a}{-1/\sqrt{2}}=\frac{z-0}{1/\sqrt{2}} \text{ and } \frac{x-a}{-1/\sqrt{2}}=\frac{y-a}{0}=\frac{z-0}{1/\sqrt{2}}.$$

Forces acting along them are P, $2P$, $3P$ and $4P$ respectively. Their components parallel to the axis are

$$\left(\frac{P}{\sqrt{2}},0,\frac{P}{\sqrt{2}}\right),\left(0,-\frac{2P}{\sqrt{2}},-\frac{2P}{\sqrt{2}}\right),\left(0,-\frac{3P}{\sqrt{2}},\frac{3P}{\sqrt{2}}\right) \text{ and } \left(-\frac{4P}{\sqrt{2}},0,\frac{4P}{\sqrt{2}}\right),$$

i.e., (X_1, Y_1, Z_1), (X_2, Y_2, Z_2), (X_3, Y_3, Z_3) and (X_4, Y_4, Z_4).

$$\therefore\ X=\Sigma X_1=\frac{P}{\sqrt{2}}+0+0-\frac{4P}{\sqrt{2}}=-\frac{3P}{\sqrt{2}}$$

$$Y=\Sigma Y_1=0-\frac{2P}{\sqrt{2}}-\frac{3P}{\sqrt{2}}+0=-\frac{5P}{\sqrt{2}}$$

$$Z=\Sigma Z_1=\frac{P}{\sqrt{2}}-\frac{2P}{\sqrt{2}}+\frac{3P}{\sqrt{2}}+\frac{4P}{\sqrt{2}}=\frac{6P}{\sqrt{2}}.$$

Now

$$L_1\vec{i}+M_1\vec{j}+N_1\vec{k}=\begin{vmatrix}\vec{i} & \vec{j} & \vec{k}\\ a & 0 & a\\ X_1 & Y_1 & Z_1\end{vmatrix}$$

$$=\vec{i}(0-aY_1)-\vec{j}(aZ_1-aX_1)+\vec{k}(aY_1)$$

$$\Rightarrow L_1=-aY_1, M_1=a(X_1-Z_1), N_1=aY_1$$

$$\Rightarrow L_1 = 0, M_1 = a\left(\frac{P}{\sqrt{2}} - \frac{P}{\sqrt{2}}\right) = 0, N_1 = 0.$$

$$L_2\vec{i} + M_2\vec{j} + N_2\vec{k} = \begin{vmatrix} \vec{i} & \vec{j} & \vec{k} \\ a & a & a \\ X_2 & Y_2 & Z_2 \end{vmatrix}$$
$$= \vec{i}a(Z_2 - Y_2) - \vec{j}a(Z_2 - X_2) + \vec{k}a(Y_2 - X_2)$$

$$\Rightarrow L_2 = a(Z_2 - Y_2), M_2 = a(X_2 - Z_2), N_2 = a(Y_2 - X_2)$$

$$\Rightarrow L_2 = a\left(-\frac{2P}{\sqrt{2}} + \frac{2P}{\sqrt{2}}\right) = 0, M_2 = a\left(0 + \frac{2P}{\sqrt{2}}\right) = \frac{a2P}{\sqrt{2}}$$

$$N_2 = a\left(-\frac{2P}{\sqrt{2}} - 0\right) = -\frac{2aP}{\sqrt{2}}.$$

$$L_3\vec{i} + M_3\vec{j} + N_3\vec{k} = \begin{vmatrix} \vec{i} & \vec{j} & \vec{k} \\ 0 & a & 0 \\ X_3 & Y_3 & Z_3 \end{vmatrix} = \vec{i}(aZ_3 - 0) - \vec{j}(0) + \vec{k}(0 - aX_3).$$

$$\Rightarrow L_3 = aZ_3 = \frac{3aP}{\sqrt{2}}, M_3 = 0, N_3 = -aX_3 = 0$$

and

$$L_4\vec{i} + M_4\vec{j} + N_4\vec{k} = \begin{vmatrix} \vec{i} & \vec{j} & \vec{k} \\ a & a & 0 \\ X_4 & Y_4 & Z_4 \end{vmatrix} = \vec{i}(aZ_4 - 0) - \vec{j}(aZ_4 - 0) + \vec{k}(aY_4 - aX_4)$$

$$\Rightarrow L_4 = aZ_4 = \frac{a4P}{\sqrt{2}}, M_4 = -aZ_4 = -\frac{a4P}{\sqrt{2}}, N_4 = a(Y_4 - X_4) = a\left(0 + \frac{4P}{\sqrt{2}}\right) = \frac{a4P}{\sqrt{2}}.$$

Now

$$L = \Sigma L_1 = 0 + 0 + \frac{3aP}{\sqrt{2}} + \frac{4aP}{\sqrt{2}} = \frac{7aP}{\sqrt{2}}$$

$$M = \Sigma M_1 = 0 + \frac{2aP}{\sqrt{2}} + 0 - \frac{4aP}{\sqrt{2}} = -\frac{2aP}{\sqrt{2}}$$

$$N = \Sigma N_1 = 0 + \left(-\frac{2aP}{\sqrt{2}}\right) + 0 + \frac{4aP}{\sqrt{2}} = \frac{2aP}{\sqrt{2}}.$$

The magnitude of the single force acting at O is

$$R = \sqrt{X^2 + Y^2 + Z^2} = \sqrt{\left(\frac{9}{2} + \frac{25}{2} + \frac{36}{2}\right) P^2} = \sqrt{35}P.$$

Direction cosines of the line of action of this single force R are in the ratio $X : Y : Z$, i.e.,

$$-\frac{3P}{\sqrt{2}} : -\frac{5P}{\sqrt{2}} : \frac{6P}{\sqrt{2}},$$

i.e., they are proportional to $-3, -5, 6$.

Also magnitude of the resultant couple is

$$G = \sqrt{L^2 + M^2 + N^2} = \sqrt{\left(\frac{49}{2} + \frac{4}{2} + \frac{4}{2}\right) a^2P^2} = \frac{aP}{2}\sqrt{114}$$

and the direction cosines of the line about which this couple of moment G acts are in the ratio

$$L:M:N, \text{ i.e., } \frac{7Pa}{\sqrt{2}}, -\frac{2Pa}{\sqrt{2}}\frac{2Pa}{\sqrt{2}}, \text{ i.e., } 7:-2:2,$$

i.e., they are proportional to $7,-2,2$.

19. On a rigid body, the forces $10\left(\vec{i}+2\vec{j}+2\vec{k}\right), 10\left(\vec{i}+2\vec{j}+2\vec{k}\right)N, 5\left(-2\vec{i}-\vec{j}+2\vec{k}\right)N$ and $6\left(2\vec{i}+2\vec{j}-\vec{k}\right)N$ are acting at points with position vectors $\vec{i}-\vec{j}, 2\vec{i}+5\vec{k}$ and $4\vec{i}-\vec{k}$ respectively. Reduce this system to a single force $\vec{R}$ acting at the point $\left(4\vec{i}+2\vec{j}\right)$ together with a couple $\vec{G}$ whose axis passes through this point. Does the point $\left(4\vec{i}+2\vec{j}\right)$ lie on the central axes? **[IAS 2009]**

Solution: Let $\vec{F_1}=10\left(\vec{i}+2\vec{j}+2\vec{k}\right)$, $\vec{F_2}=5\left(-2\vec{i}-\vec{j}+2\vec{k}\right)$ and $\vec{F_3}=6\left(2\vec{i}+2\vec{j}-\vec{k}\right)$.

The resultant force $R=\vec{F_1}+\vec{F_2}+\vec{F_3}$

$$=10\vec{i}+20\vec{j}+20\vec{k}-10\vec{i}-5\vec{j}+10\vec{k}+12\vec{i}+12\vec{j}-6\vec{k}$$
$$=12\vec{i}+27\vec{j}+24\vec{k}.$$

The resultant couple

$$\vec{b}=\vec{r_1}\times\vec{F_1}+\vec{r_2}\times\vec{F_2}+\vec{r_3}\times\vec{F_3}, \tag{1}$$

where $\vec{r_1}=\vec{i}-\vec{j}, \vec{r_2}=2\vec{i}+5\vec{k}, \vec{r_3}=4\vec{i}-\vec{k}$.

Now

$$\vec{r_1}\times\vec{F_1}=\begin{vmatrix}\vec{i} & \vec{j} & \vec{k}\\ 1 & -1 & 0\\ 10 & 20 & 20\end{vmatrix}=\vec{i}(-20-0)-\vec{j}(20-0)+\vec{k}(20+10)$$
$$=-20\vec{i}-20\vec{j}+30\vec{k}$$

$$\vec{r_2}\times\vec{F_2}=\begin{vmatrix}\vec{i} & \vec{j} & \vec{k}\\ 2 & 0 & 5\\ -10 & -5 & 10\end{vmatrix}=\vec{i}(0+25)-\vec{j}(20+50)+\vec{k}(-10-0)$$
$$=25\vec{i}-70\vec{j}-10\vec{k}$$

$$\text{and } \vec{r_3}\times\vec{F_3}=\begin{vmatrix}\vec{i} & \vec{j} & \vec{k}\\ 4 & 0 & -1\\ 12 & 12 & -6\end{vmatrix}=\vec{i}(0+12)-\vec{j}(-24+12)+\vec{k}(48-0)$$
$$=12\vec{i}+12\vec{j}+48\vec{k}.$$

So from (1),

$$\vec{G}=-20\vec{i}-20\vec{j}+30\vec{k}+25\vec{i}-70\vec{j}-10\vec{k}+12\vec{i}+12\vec{j}+48\vec{k}$$
$$=17\vec{i}-78\vec{j}+68\vec{k}.$$

Since a force and a couple can be reduced to a single force, the above system is equivalent to a single force $\vec{F}=\vec{R}=12\vec{i}+27\vec{j}+24\vec{k}$ acting through the point $\vec{r}=4\vec{i}+2\vec{j}$ and its moment about O is

$$\vec{r}\times\vec{F}=\begin{vmatrix}\vec{i} & \vec{j} & \vec{k}\\ 4 & 2 & 0\\ 12 & 27 & 24\end{vmatrix}=\vec{i}48-\vec{j}96+\vec{k}(108-24)$$

$$=48\vec{i}-96\vec{j}+84\vec{k}.$$

Here $L=17, M=-78, N=68$ and $X=12, Y=27, Z=24$.

Now equation of central axis is

$$\frac{L-yZ+zY}{X}=\frac{M-zX+xZ}{Y}=\frac{N-xY+yX}{Z}$$

$$\Rightarrow\frac{17-y24+z27}{12}=\frac{-78-12z+24x}{27}=\frac{68-x27+y12}{24}$$

$$\Rightarrow\frac{17-24y+27z}{4}=\frac{-78-12z+24x}{9}=\frac{68-27x+12y}{8}.$$

If the point $\left(4\vec{i}+2\vec{j}\right)$, i.e., $(4,2,0)$ lies on it, then

$$\frac{17-48}{4}=\frac{-78+96}{9}=\frac{68-108+24}{8}\Rightarrow-\frac{31}{4}=\frac{18}{9}=-\frac{16}{8}\Rightarrow-\frac{31}{4}=2=-2,$$

which is not true.

Hence, the point $\left(4\vec{i}+2\vec{j}\right)$ does not lie on central axis.

EXERCISE 6

1. Equal forces act along the coordinate axis and along the straight line

$$\frac{x-\alpha}{l}=\frac{y-\beta}{m}=\frac{z-\gamma}{n}.$$

Find the equation of the central axis of the system.

2. Forces X, Y, Z act along the three lines given by the equations $y=0,\ z=c,\ z=0,\ x=a,\ x=0,\ y=b$.

Prove that the pitch of the equivalent wrench is

$$\frac{aYZ+bZX+cXY}{X^2+Y^2+Z^2}.$$

If the wrench reduces to a single force, show that the line of action of force lies on the hyperboloid $(x-a)(y-b)(z-c)=xyz$. [KU(H) 2001, VU(H) 2001, BU(H) 1995]

3. Forces (or four forces) act along generators of the system of a hyperboloid. Their magnitudes are such that if they are transferred parallel to themselves to act at one point, then they would be in equilibrium. Show that they are in equilibrium when acting along generators.

4. Two forces act, one along the line $y=0,\ z=0$ and the other along the line $x=0$, $z=c$. As the forces vary, show that the surface generated by the central axis (the axis of their equivalent wrench) is $(x^2+y^2)z=cy^2$. [CU(H) 2000]

5. A single force is equivalent to component forces X, Y and Z along the axes of coordinates and to couples L, M, N about three axes. Prove that the magnitude of the single force $(X^2+Y^2+Z^2)$ and the equations to its line of action are
$$\frac{yZ-zY}{L}=\frac{zX-xZ}{M}=\frac{xY-yX}{N}=1.$$ [VU(H) 2000]

6. The equal forces act along each of the straight lines
$$\frac{x\pm a\cos\theta}{a\sin\theta}=\frac{y-b\sin\theta}{\pm b\cos\theta}=\frac{z}{c}.$$
Show that their central axis must, for all values of θ, lie on the surface
$$y\left(\frac{x}{z}+\frac{z}{x}\right)=b\left(\frac{a}{c}+\frac{c}{a}\right).$$ [RU(H) 1996, CU(H) 2003, NBU(H) 2005]

7. Forces P, Q, R act along the three non-intersecting edges of a cube; find the central axis. [LNMU(H) 1994, MU(H) 2001, MU(H) 2010]

8. Show that the minimum distance between two forces, which are equivalent to a given system (R,K) and which are inclined at a given angle 2α is $\frac{2K}{R}\cot\alpha$ and the forces are then each equal to $\frac{R}{2}\sec\alpha$.

9. Forces act along the edges of a regular tetrahedron, viz., P and Q along BC and DA, Q along CA and DB and R along AB and DC. Show that the pitch of the equivalent wrench is $\frac{1}{2\sqrt{2}}$ of the edge of tetrahedron.

10. Show that among the null lines of any system of forces four are generators of any hyperboloid, two belonging to one system of generators and two to the other system.

11. A system of forces given by (X,Y,Z,L,M,N) is replaced by two forces, one acting along the axis of x and another force. Show that the magnitudes of the forces are $\frac{LX+MY+NZ}{L}$ and $\left[(MY+NZ)^2+L^2\left(Y^2+Z^2\right)\right]^{\frac{1}{2}}$ and also find the equation of the line of action of the other force. [CU(H) 1969]

12. Prove that, if equal forces act along the edges CA, AB, DA, DB and DC of a regular tetrahedron $ABCD$, the central axis is the perpendicular from D to the plane $ABCD$ and the pitch of the equivalent wrench is $\frac{a}{2\sqrt{2}}$, where a is the edge of the tetrahedron.

13. Show that the following system of forces reduces to a single force:
Force $5\sqrt{2}$ acting through $\left(0,0,\frac{1}{5}\right)$ in the direction $(3:4:-5)$. Force $3\sqrt{11}$ acting through $\left(\frac{1}{3},0,0\right)$ in the direction $(5:5:-7)$. Force $\sqrt{11}$, acting through $\left(0,\frac{1}{4},0\right)$ in the direction $(-3:1:1)$ the system being referred to rectangular axis.

14. Forces act through the angular points of a tetrahedron perpendicular to the opposite faces and proportional to them. Show that they are in equilibrium, if they act either all inwards or outwards. [IAS 1982]

15. Six forces, each equal to P, act along the edges of a cube taken in order which do not meet a given diagonal. Show that their resultant is a couple of moment $2\sqrt{3}Pa$, where a is the edge of the cube. [CU(H) 2002, KU(H) 2004]

16. Four forces, each of magnitude P, act on a rigid body, three of forces act along the rectangular Cartesian coordinate axes of x,y and z, while the fourth force acts along the line $\frac{x-1}{2}=\frac{y-2}{3}=\frac{z-3}{4}$. Find the equation of the central axis. [VU(H) 2003]

17. Three forces each equal to P act on a body one at the point $(a,0,0)$ parallel to Oy, second at a point $(0,b,0)$ parallel to Oz and the third at the point $(0,0,c)$ parallel to Ox, the axes being rectangular. Show that the central axis passes through the point $\left(-\frac{a+2b+3c}{3},-\frac{b+2c+3a}{3},-\frac{c+2a+3b}{3}\right)$ and has the direction ratio $(1,1,1)$. [CU(H) 04]
[*Hints.* See worked-out Example 3.]

18. Two forces $2P$ and P act along the lines, whose equations are $y=x\tan\alpha, z=c$ and $y=-x\tan\alpha, z=-c$ respectively. Find the equation of the central axis.

[CU(H) 2007, VU(H) 1999]

[*Hints.* See worked-out Example 14.]

Answers

1. $\frac{(\beta n-\gamma m)-y(1+n)+z(1+m)}{1+l}=\frac{(\gamma l-\alpha n)-z(1+l)+x(1+n)}{1+m}=\frac{(\alpha m-\beta l)-x((1+m)+y(1+l)}{1+n}$.

4. $\frac{(n\beta-m\gamma)-y(1+n)+z(1+m)}{1+l}=\frac{(l\gamma+n\alpha)-z(1+l)+x(1+n)}{1+m}=\frac{(\alpha m-\beta l)-x(1+m)+y(1+l)}{1+n}$.

7. $\frac{a(R-Q)-yR+zQ}{P}=\frac{-aR-zP+xR}{Q}=\frac{-xQ+yP}{R}$, where $a=$ edge of the cube.

11. Equation of line of action of the other forces are $Lx+My+Nz=0$ and $yZ-zY=L$.

16. $-\frac{1-(4+\sqrt{29})y+z(3+\sqrt{29})}{2+\sqrt{29}}=\frac{2-z(2+\sqrt{29})+x(4+\sqrt{29})}{3+\sqrt{29}}=-\frac{-1-x(3+\sqrt{29})+y(2+\sqrt{29})}{4+\sqrt{29}}$.

18. $y=\frac{1}{3}x\tan\alpha, z=\frac{3c}{\sin^2\alpha+9\cos^2\alpha}$

Chapter 7

Centre of Gravity

7.1 Introduction

A rigid body is made up of particles which are attracted to the centre of the earth with a force proportional to the mass of the particle. If the body be small as compared to the earth, the lines joining the centre of the earth to the different particles will be almost parallel (the radius of the earth is about 6,400 km). Hence, the forces acting on the particles present a good illustration of a system of parallel forces. The resultant of all these parallel forces (which is nothing but the sum of the weights of the different particles) is called the *weight of the body.* The resultant, i.e., the weight, will pass through some fixed point of the body. This fixed point is called the centre of gravity (CG) of the body.

7.2 Definition and General Formulation

Definition. The CG of a body regarded as a system of rigidly connected particles is a point fixed in the body through which the weight of the body always passes, in whatever position the body may be placed, provided its size and shape remain unaltered.

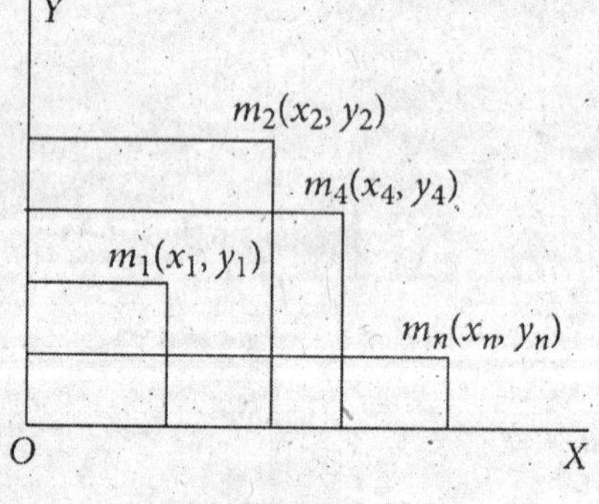

Fig. 7.1

General Formulation. Let there be a system of particles of masses $m_1, m_2, m_3, \ldots, m_n$ in a plane whose coordinates are (x_1, y_1), (x_2, y_2), …, (x_n, y_n) respectively w.r.t. OX and OY as the coordinate axes, O being the origin.

Let $(\bar{x}, \bar{y})$ be the coordinates of their centroid.

Taking moments of all masses about the y-axis, we get

$$(m_1+m_2+\cdots+m_n)\bar{x}=m_1x_1+m_2x_2+\cdots+m_nx_n,$$

$$\text{or, } \left(\sum_{r=1}^{n} m_r\right)\bar{x}=\sum_{r=1}^{m} m_r x_r, \therefore \bar{x}=\frac{\sum_{r=1}^{n} m_r x_r}{\sum_{r=1}^{n} m_r}.$$

Similarly, by taking moments about x-axis, we get

$$\bar{y}=\frac{\Sigma m_r y_r}{\Sigma m_r}.$$

Now in most configurations of practical interest, the mass is not concentrated in discrete mass particles, but is continuously distributed throughout a volume, over a surface or along a curve.

We consider the body to be formed of an infinite number of infinitely small elements of masses.

In this case, the symbol Σ will be replaced by the integral sign.

If δm be an element of mass of a body at a point whose coordinates are (x,y) in two dimensions. Thus, if the coordinates of the centroid G are $(\bar{x},\bar{y})$, then

$$\bar{x}=\frac{\int x\,dm}{\int dm},\ \bar{y}=\frac{\int y\,dm}{\int dm}.$$

Similarly, in three dimensions, let δm be an element of mass of a body at a point whose coordinates are (x,y,z). Then the centroid $(\bar{x},\bar{y},\bar{z})$ will be given by

$$\bar{x}=\frac{\int x\,dm}{\int dm},\ \bar{y}=\frac{\int y\,dm}{\int dm},\bar{z}=\frac{\int z\,dm}{\int dm}.$$

It is to be noted that the limit of integration should be such as to include the whole body.

7.3 Specific Cases

We shall take the body to be homogeneous so that its density ρ is uniform at all points of the body (i.e., it is same at all points).

The above-mentioned formula for the determination of CG of a body takes up the following forms depending upon its construction:

1. CG *of an arc of a curve:*

$$\bar{x} = \frac{\int x\rho\, ds}{\int \rho\, ds}, \; \bar{y} = \frac{\int y\rho\, ds}{\int \rho\, ds}.$$

2. CG *of a plane area:*

$$\bar{x} = \frac{\int x\, dA}{\int dA}, \; \bar{y} = \frac{\int y\, dA}{\int dA}.$$

3. CG *of the area bounded by the curve* $y = f(x)$, *axis of* x *and ordinates at* $x = a$ *and* $x = b$ *is given by*

$$\bar{x} = \frac{\int_a^b xy\, dx}{\int_a^b y\, dx}, \; \bar{y} = \frac{\int_a^b \frac{y}{2}(y\, dx)}{\int_a^b y\, dx}.$$

Corollary 1. *If the area be symmetrical about the x-axis, then* $\bar{y} = 0$.

4. CG *of the area bounded by the curves* $y = f(x)$, $y = \phi(x)$ *and ordinates at* $x = a$ *and* $x = b$:

$$\bar{x} = \frac{\int x(y_1 - y_2)\, dx}{\int (y_1 - y_2)\, dx}, \; \bar{y} = \frac{\frac{1}{2}\int (y_1^2 - y_2^2)\, dx}{\int (y_1 - y_2)\, dx}.$$

5. CG *of a solid of revolution:*

(a) If the axis of revolution is X-axis, then

$$\bar{x} = \frac{\int xy^2\, dx}{\int y^2\, dx}, \; \bar{y} = 0.$$

(b) If the axis of revolution is Y-axis, then

$$\bar{x} = 0, \; \bar{y} = \frac{\int yx^2\, dy}{\int x^2\, dy}.$$

6. CG *of a surface of revolution*: Let the surface be generated by the revolution of the arc of the curve $y = f(x)$ about X-axis included between the ordinates of $x = a$ and $x = b$. Then CG of this surface lies on X-axis.

$$\bar{x} = \frac{\int_a^b \rho xy \, ds}{\int_a^b \rho y \, ds}, \; \bar{y} = 0$$

Similarly, by the CG of a surface generated by the revolution of the curve $x = f(y)$ from $y = c$ to $y = d$ is

$$\bar{x} = 0, \; \bar{y} = \frac{\int_c^d \rho xy \, ds}{\int_c^d \rho x \, ds}.$$

7. CG *of the area bounded between the curve* $r = f(\theta)$ *and radii vectors corresponding to* $\theta = \alpha$ *and* $\theta = \beta$:

$$\bar{x} = \frac{\frac{2}{3}\int_\alpha^\beta r^3 \cos\theta \, d\theta}{\int_\alpha^\beta r^2 \, d\theta}, \; \bar{y} = \frac{\frac{2}{3}\int_\alpha^\beta r^3 \sin\theta \, d\theta}{\int_\alpha^\beta r^2 \, d\theta}.$$

If the area is symmetrical about the initial line $\theta = 0$, then the CG will lie on the initial line and hence, $\bar{y} = 0$.

8. CG *of any area*: The element of area in Cartesian coordinates being $(\delta x, \delta y)$ the element of mass m is $\delta m = \rho \delta x \, \delta y$.

Therefore, the coordinates of the CG of the area are given by

$$\bar{x} = \frac{\iint \rho x \, dx \, dy}{\iint \rho \, dx \, dy}, \; \bar{y} = \frac{\iint \rho y \, dx \, dy}{\iint \rho \, dx \, dy},$$

where the limits of integrations are taken so as to cover the whole area.

Similarly, in polar coordinates, the coordinates of the CG of area are given by

$$\bar{x} = \frac{\iint \rho r \cos\theta \, r \, dr \, d\theta}{\iint \rho r \, d\theta \, dr}, \; \bar{y} = \frac{\iint \rho r \sin\theta \, r \, dr \, d\theta}{\iint \rho r \, d\theta \, dr}.$$

9. *The elements of volume in Cartesian coordinates being* $(\delta x, \delta y, \delta z)$ *the* CG *of volume in three dimensions is given by*

$$\bar{x} - \frac{\iiint \rho x \, dx \, dy \, dz}{\iiint \rho \, dx \, dy \, dz}, \quad \bar{y} = \frac{\iiint \rho y \, dx \, dy \, dz}{\iiint \rho \, dx \, dy \, dz}, \quad \bar{z} = \frac{\iiint \rho z \, dx \, dy \, dz}{\iiint \rho \, dx \, dy \, dz}.$$

In polar coordinates, the element of volume is $\delta r\, r\, \delta\theta\, r \sin\theta\, \delta\phi$, whereas the Cartesian coordinates of a point are

$$(r \sin\theta \cos\phi, r \sin\theta \sin\phi, r \cos\theta)$$

Therefore, the coordinates of the CG of the volume are given by

$$\bar{x} = \frac{\iiint \rho(r \sin\theta \cos\phi) r^2 \sin\theta \, dr \, d\theta d\phi}{\iiint \rho r^2 \sin\theta \, dr \, d\theta d\phi},$$

where the limits of integration are taken so as to cover the entire volume.

7.4 CG of sum or difference of two bodies, i.e., CG of a compound body

1. If the body consists of two parts of weights W_1 and W_2, let G_1 and G_2 be their CG's. We take a point O on G_1G_2 such that $OG_1 = x_1$ and $OG_2 = x$. Let G be the CG of the compound body and let $OG = \bar{x}$.

   Now taking moment about O, we get $(w_1 + w_2)\bar{x} = w_1x_1 + w_2x_2$;

   $$\therefore \bar{x} = \frac{w_1x_1 + w_2x_2}{w_1 + w_2}.$$

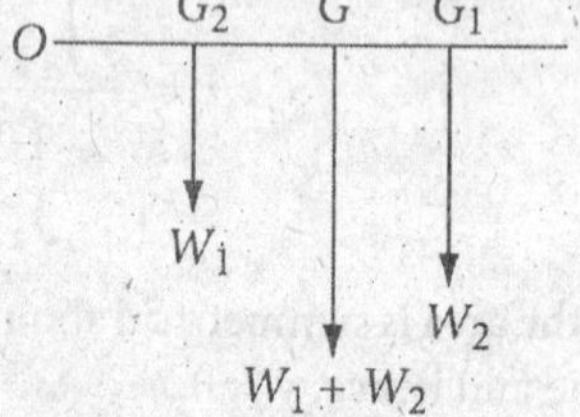

Fig. 7.2

2. Let one of the portions of weight w_1 of the body be removed and we are required to find the CG of the remainder.

   Let W = the weight of the whole body.

   Let G and G_1 be the CG's of the whole body and the portion of the body.

   The weight of the remaining portion is evidently $W - W_1$ and let G_2 be the CG of this portion of the body. It is clear that the points G_2, G and G_1 must lie in the same straight line and we must have

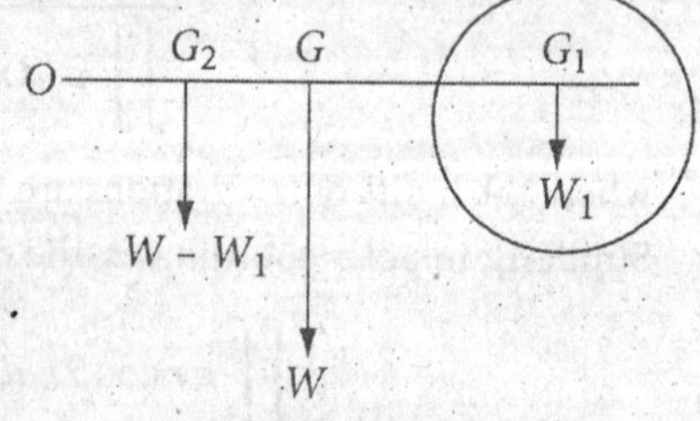

Fig. 7.3

$$W_1G_1G = (W - W_1)G_2G; \; G_2G = \frac{W_1}{W - W_1}G_1G.$$

7.5 The Theorem of Pappus

Statement. If any plane area revolves through any angle about an axis in its down plane, then

1. the volume generated by the area is equal to the product of the area and the length of the path described by the centroid of the area, and
2. the surface generated by the area is equal to the product of the perimeter of the curve and the length of the path described by the centroid of the perimeter.

Proof. Let A be the area of the curve and S the perimeter of the curve, $\bar{y}$ the distance of the centroid of the area and $\bar{y}_1$, that of the perimeter of the curve from the axis of rotation which is taken to be the axis of x.

1. Let P be any point in the area of the curve, whose ordinate is y. If the rotation be through an angle θ, the length of the arc described by P is $y \cdot \theta$.
   Therefore, the volume described by an element δA of area at P is $y\theta\,\delta A$. Hence, the whole volume described by the area

   $$= \int y\theta\, dA = \theta \int y\, dA = \theta \bar{y} A = A\bar{y}\theta$$

   = the area of the curve multiplied by the length of the arc described by the centroid of the area.

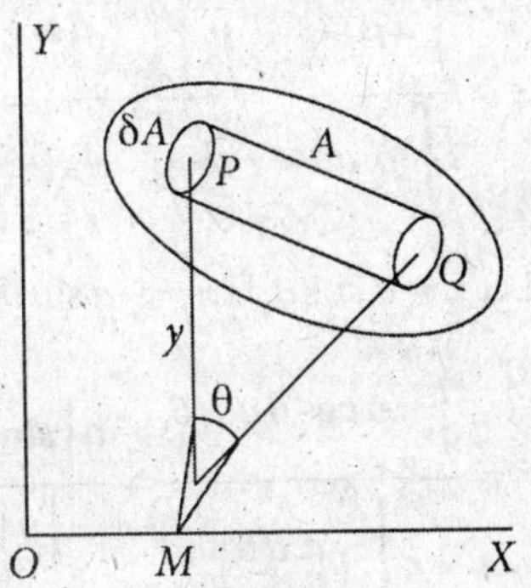

Fig. 7.4

2. Let P_1 be a point on the perimeter of the curve, whose ordinate is y, during the rotation the length of the curve described by P_1 is $y_1 . \theta$.
   Therefore, the surface described by an element δs of the perimeter of P_1 is $y_1\theta\,\delta s$. Hence, the whole surface traced out by the perimeter

   $$= \int y_1\theta\, ds = \theta \int y_1\, ds = \theta \bar{y}_1 s = s\bar{y}_1\theta$$

   = the perimeter of the curve multiplied by the length of the arc described by the centroid of the perimeter.

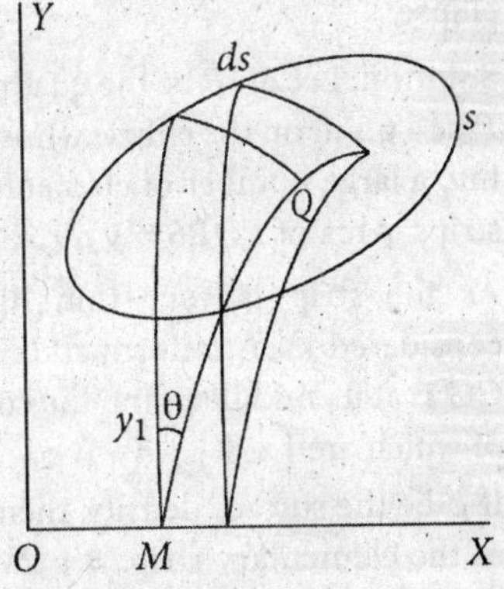

Fig. 7.5

WORKED-OUT EXAMPLES

1. Find the CG of a circular arc subtending an angle 2α at the centre.

   Solution: Let ACB be a circular arc of a circle of radius a subtending an angle 2α at the centre O. Let C be the middle point of the arc AB.

   Taking O as origin and OCX which bisects the arc AB as the X-axis and a line perpendicular to it through O as the Y-axis.

   Let $(\bar{x}, \bar{y})$ be the coordinates of CG. Since the curve is symmetrical about x-axis the CG lies on X-axis. Thus, $\bar{y} = 0$.

   Let us take neighbouring points P and Q on AB such that $\angle XOP = \theta$ and $\angle XOQ = \theta + \delta\theta$. Let arc $CP = s$.

   arc $CQ = s + \delta s$

   $\therefore\ PQ = \delta s.$

$\theta = \text{arc}\,\frac{CP}{a} = \frac{s}{a};\ s = a\theta\ \therefore\ ds = a\,d\theta.$

$$\bar{x} = \frac{\int x\rho\,ds}{\int \rho\,ds} = \frac{\rho\int_{-\alpha}^{\alpha} x\,ds}{\rho\int_{-\alpha}^{\alpha} ds} = \frac{\int_{-\alpha}^{\alpha} xa\,d\theta}{\int_{-\alpha}^{\alpha} a\,d\theta}.$$

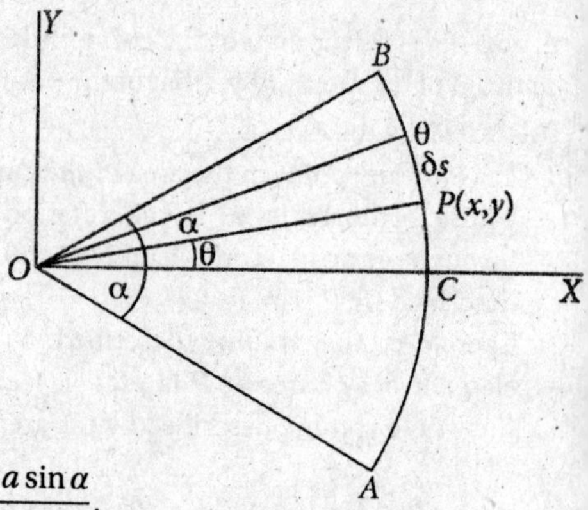

Let $x = a\cos\theta;\ dx = -a\sin\theta\,d\theta.$

$$\bar{x} = \frac{\int_{-\alpha}^{\alpha} a\cos\theta\, a\,d\theta}{\int_{-\alpha}^{\alpha} a\,d\theta} = \frac{a\left[\sin\theta\right]_{-\alpha}^{\alpha}}{\left[\theta\right]_{-\alpha}^{\alpha}};\ \bar{x} = \frac{2a\sin\alpha}{2\alpha}.$$

Fig. 7.6

So coordinates of the CG of the circular arc is $\left(\frac{a\sin\alpha}{\alpha}, 0\right)$.

2. Find the CG of a plane lamina of uniform density in the form of a quadrant of an ellipse. [CU(H) 2007]

Solution: Let AOB be the quadrant of the ellipse $\frac{x^2}{a^2}+\frac{y^2}{b^2}=1$ whose CG is required. Let P be a point on the ellipse whose coordinates are (x, y). We divide the whole quadrant into a large number of elementary strips parallel to Y-axis. Let $PQRS$ be one of such strips. Area of $PQRS = y\,dx$.

As the strip is very thin, it may be considered as a uniform rod. Hence, its CG is at its middle point, the coordinates of which are $\left(x, \frac{y}{2}\right)$.

If ρ be the surface density, then the mass of the elementary strip $= y\,dx\rho$, and if $\bar{x}, \bar{y}$ be the coordinates of CG required, then

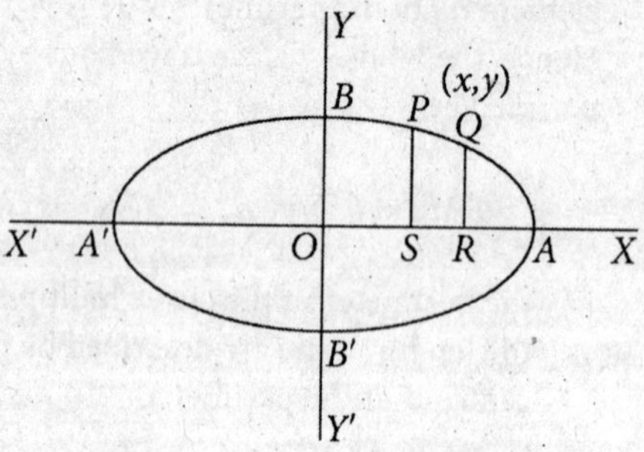

Fig. 7.7

$$\bar{x} = \frac{\int x\,dm}{\int dm} = \frac{\int_0^a xy\,dx\rho}{\int_0^a y\,dx\rho} = \frac{\int_0^a xy\,dx}{\int_0^a y\,dx}.$$

Let $x = a\cos\theta$,
$y = b\sin\theta$,
$dx = -a\sin\theta\,d\theta$.

Also, when $x = a$, $\theta = 0$; when $x = 0$, $\theta = \pi/2$

$$\bar{x} = \frac{\int_{\pi/2}^{0} a\cos\theta\, b\sin\theta(-a\sin\theta\,d\theta)}{\int_{\pi/2}^{0} b\sin\theta(-a\sin\theta\,d\theta)};\ \bar{x} = \frac{a\int_0^{\pi/2}\sin^2\theta\cos\theta\,d\theta}{\int_0^{\pi/2}\sin^2\theta\,d\theta}.$$

For numerator, let $u=\sin\theta, du=\cos\theta\, d\theta$.
When $\theta=\pi/2$, $u=1$, $\theta=0$, $u=0$.

$$Nr=\int_0^1 u^2 du=\left[\frac{u^3}{3}\right]_0^1=\frac{1}{3};\ \bar{x}=\frac{a\cdot\frac{1}{3}}{\frac{2-1}{2}\frac{\pi}{2}}=\frac{a}{3}\frac{4}{\pi}.$$

Also

$$\bar{y}=\frac{\int \frac{y}{2}\,dm}{\int dm}=\frac{\frac{1}{2}\int_0^a y\rho\, dx\, y}{\int_0^a \rho y\, dx}=\frac{\frac{1}{2}\int_0^a y^2\, dx}{\int_0^a y\, dx}=\frac{\frac{1}{2}\int_{\pi/2}^0 b^2\sin^2\theta(-a\sin\theta)\,d\theta}{\int_{\pi/2}^0 b\sin\theta(-a\sin\theta)\, d\theta}$$

$$=\frac{\frac{1}{2}ab^2\int_0^{\pi/2}\sin^3\theta\, d\theta}{ab\int_0^{\pi/2}\sin^2\theta\, d\theta}=\frac{\frac{b}{2}\cdot\frac{3-1}{3}}{\frac{2-1}{2}\frac{\pi}{2}}=\frac{b}{3}\cdot\frac{4}{\pi}.$$

Thus, the coordinates of CG of the quadrant of the ellipse

$$\frac{x^2}{a^2}+\frac{y^2}{b^2}=1 \text{ is } \left(\frac{4a}{3\pi}, \frac{4b}{3\pi}\right).$$

3. Find the position of the centroid of the area of the cardioid $r=a(1+\cos\theta)$.
Solution: Let the coordinates of the CG be $(\bar{x},\bar{y})$.
Since the curve is symmetrical about the initial line, therefore, $\bar{y}=0$.

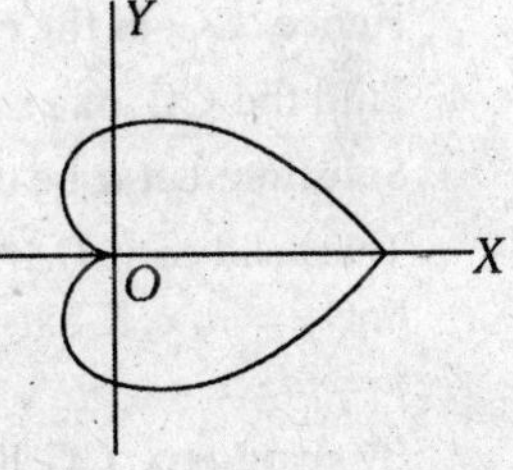

Fig. 7.8

$$\bar{x}=\frac{\frac{2}{3}\int_0^\pi r^3\cos\theta\, d\theta}{\int_0^\pi r^2\, d\theta}=\frac{\frac{2}{3}\int_0^\pi a^3(1+\cos\theta)^3\cos\theta\, d\theta}{\int_0^\pi a^2(1+\cos\theta)^2\, d\theta}$$

$$=\frac{\frac{2}{3}a\int_0^\pi (1+\cos\theta)^3\cos\theta\, d\theta}{\int_0^\pi (1+\cos\theta)^2\, d\theta}.$$

Now Nr of

$$\bar{x}=\frac{2a}{3}\int_0^{\pi}(1+3\cos\theta+3\cos^2\theta+\cos^3\theta)\cos\theta\,d\theta$$

$$=\frac{2a}{3}\int_0^{\pi}(\cos\theta+3\cos^2\theta+3\cos^3\theta+\cos^4\theta)\,d\theta$$

$$=\frac{2a}{3}\cdot 2\int_0^{\pi/2}(3\cos^2\theta+\cos^4\theta)\,d\theta \quad \left[\because \int_0^{\pi}\cos^n\theta=0 \text{ or, } 2\int_0^{\pi/2}\cos^n\theta\,d\theta \text{ according as } n \text{ is even or odd}\right]$$

$$=\frac{4a}{3}\left[3\frac{2-1}{2}\cdot\frac{\pi}{2}+\frac{4-1}{4}\cdot\frac{4-3}{4-2}\cdot\frac{\pi}{2}\right]=\frac{4a}{3}\left[\frac{3\pi}{4}+\frac{3}{4}\cdot\frac{1}{2}\cdot\frac{\pi}{2}\right]$$

$$=3\pi.\frac{4a}{3}\left[\frac{1}{4}+\frac{1}{16}\right]=4\pi\frac{5a}{16}=\frac{5\pi a}{4}.$$

Also Dr of

$$\bar{x}=\int_0^{\pi}(1+2\cos\theta+\cos^2\theta)\,d\theta=\int_0^{\pi}d\theta+2\int_0^{\pi}\cos\theta\,d\theta+2\int_0^{\pi/2}\cos^2\theta\,d\theta$$

$$=\pi+0+2.\frac{2-1}{2}\cdot\frac{\pi}{2}=\frac{3\pi}{2}.$$

$$\bar{x}=\frac{\frac{5\pi a}{4}}{3\frac{\pi}{2}}=\frac{5\pi a}{4}\cdot\frac{2}{3\pi}=\frac{5a}{6}.$$

Hence, CG of the cardioid is at $\left(\frac{5a}{6},0\right)$.

4. Find the CG of a solid right circular cone of height h.

Solution: Let α be the semi-vertical angle of the cone.

This solid cone OAA' is generated by revolving the area OAB enclosed by the straight line $y=mx$, X-axis and the ordinate $x=h$ about the X-axis and O is the origin and $y=mx$.

By symmetry CG lies on X-axis. If $(\bar{x},\bar{y})$ be the coordinates of CG, then $\bar{y}=0$.

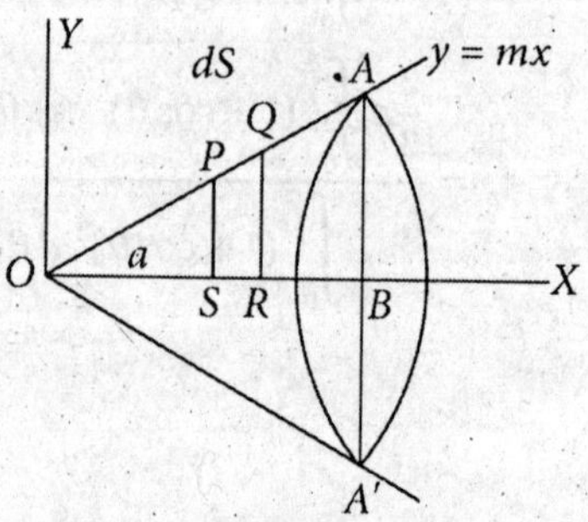

Fig. 7.9

$$\bar{x}=\frac{\int_0^h xy^2\,dx}{\int_0^h y^2\,dx}=\frac{\int_0^h xm^2x^2\,dx}{\int_0^h m^2x^2\,dx}=\frac{\int_0^h x^3\,dx}{\int_0^h x^2\,dx}$$

$$=\frac{\left[\frac{x^4}{4}\right]_0^h}{\left[\frac{x^3}{3}\right]_0^h}=\frac{h^4}{4}\cdot\frac{3}{h^3}=\frac{3}{4}h.$$

5. Find the CG of the arc of the curve $x^{2/3}+y^{2/3}=a^{2/3}$ lying in the first quadrant.

Solution: The figure of the curve is given below.

If $(\bar{x},\bar{y})$ be the coordinates of CG of the given curve, then

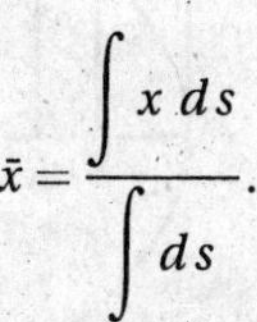

$$\bar{x}=\frac{\int x\,ds}{\int ds}.$$

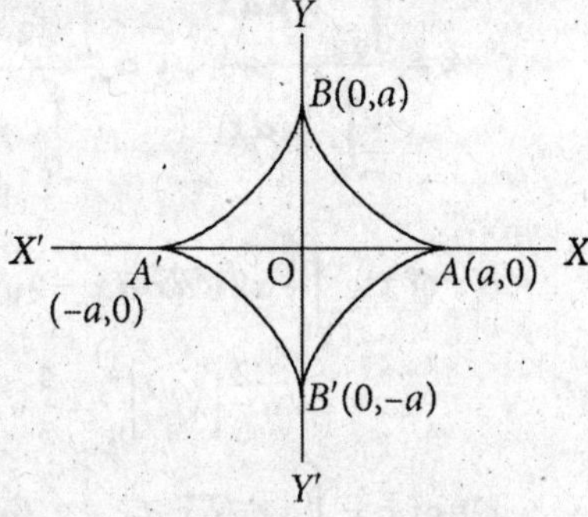

Fig. 7.10

Let $x=a\cos^3\theta,\ y=a\sin^3\theta$.

$$\frac{dx}{d\theta}=-3a\cos^2\theta\sin\theta,$$

$$\frac{dy}{d\theta}=-3a\sin^2\theta\cos\theta.$$

Also

$$\left(\frac{ds}{d\theta}\right)=\sqrt{\left(\frac{dx}{d\theta}\right)^2+\left(\frac{dy}{d\theta}\right)^2}=\sqrt{9a^2\sin^2\theta\cos^4\theta+9a^2\cos^2\theta\sin^4\theta}$$

$$=\sqrt{9a^2\sin^2\theta\cos^2\theta(\cos^2\theta+\sin^2\theta)}=3a\sin\theta\cos\theta.$$

$$\bar{x}=\frac{\int_0^a a\cos^3\theta\, 3a\sin\theta\cos\theta\,d\theta}{\int_0^a 3a\sin\theta\cos\theta\,d\theta}\quad\left[\text{when } x=a,\ \theta=0,\ x=0,\ \theta=\frac{\pi}{2}\right]$$

$$=\frac{a\int_{\pi/2}^{0}\cos^4\theta\sin\theta\,d\theta}{\int_{\pi/2}^{0}\sin\theta\cos\theta\,d\theta}=\frac{a\int_0^{\pi/2}\cos^4\theta\sin\theta\,d\theta}{\int_0^{\pi/2}\sin\theta\cos\theta\,d\theta}.$$

Let $u=\cos\theta,\ du=-\sin\theta\,d\theta$.

When $\theta=\frac{\pi}{2},\ u=0;\ \theta=0,\ u=1$.

$$\therefore\ \bar{x}=\frac{a\int_1^0 u^4(-du)}{-\int_1^0 u\,du}=\frac{a\left[\frac{u^5}{5}\right]_0^1}{\left[\frac{u^2}{2}\right]_0^1}=a\frac{1}{5}\cdot 2=\frac{2a}{5}.$$

Since curve is symmetrical about the line $y=x$, so coordinates of CG are $\left(\frac{2a}{5},\frac{2a}{5}\right)$.

6. Find the CG of the area bounded by the parabola $y^2=4ax$, the axis of x and latus rectum.

Solution: Let us consider a strip *PQRS* parallel to y-axis.

If $(\bar{x}, \bar{y})$ be the coordinates of CG, then

$$\bar{x} = \frac{\int_0^a x(y\,dx)}{\int_0^a y\,dx}, \quad \bar{y} = \frac{\int_0^a \frac{y}{2}(y\,dx)}{\int_0^a y\,dx}.$$

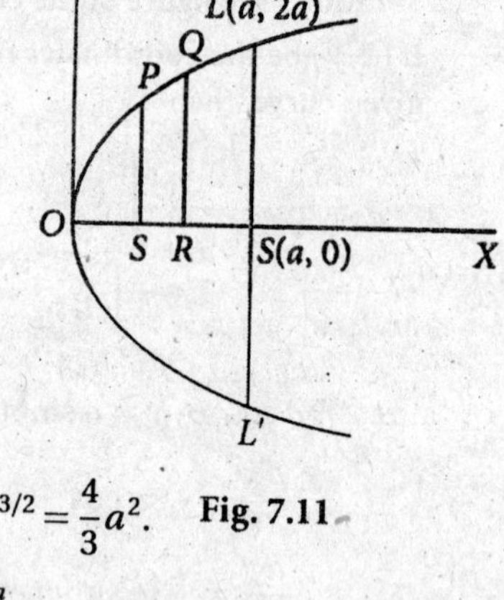

$$Nr \text{ of } \bar{x} = \int_0^a x 2\sqrt{ax}\,dx = 2\sqrt{a}\int_0^a x^{3/2}\,dx$$

$$= 2\sqrt{a}\frac{2}{5}\left[x^{5/2}\right]_0^a = \frac{4}{5}\sqrt{a}a^{5/2} = \frac{4}{5}a^3.$$

$$Dr \text{ of } \bar{x} = \int_0^a 2\sqrt{ax}\,dx = 2\sqrt{a}\frac{2}{3}\left[x^{3/2}\right]_0^a = \frac{4}{3}\sqrt{a}a^{3/2} = \frac{4}{3}a^2.$$

Fig. 7.11

$$Nr \text{ or } \bar{y} = \frac{1}{2}\int_0^a y^2\,dx = \frac{1}{2}\int_0^a 4ax\,dx = 2a\left[\frac{x^2}{2}\right]_0^a = a^3.$$

$$\bar{x} = \frac{\frac{4}{5}a^3}{\frac{4}{3}a^2} = \frac{3}{5}a; \quad \bar{y} = \frac{a^3}{\frac{4}{3}a^2} = \frac{3}{4}a.$$

Hence, the required CG is $\left(\frac{3}{5}a, \frac{3}{4}a\right)$.

7. Show how to cut out of a uniform cylinder a cone, whose base coincides with that of a cylinder, so that the CG of the remaining solid may coincide with the vertex of the cone.

[IAS 1998]

Solution: Let O be the centre of common base and r its radius. Let h_1 be the height of the cylinder and h_2, that of the cone. Let G_1, G_2 be the CG's of cylinder and cone. Then $OG_1 = \frac{1}{2}h_1, OG_2 = \frac{1}{4}h_2$.

Let m be the mass per unit volume.

Mass of the cylinder $= \pi r^2 h_1 m$.

Mass of the cone $= \frac{1}{3}\pi r^2 h_2 m$.

If G be the vertex of the cone, then G is the CG of the remainder.

$OG = h_2$

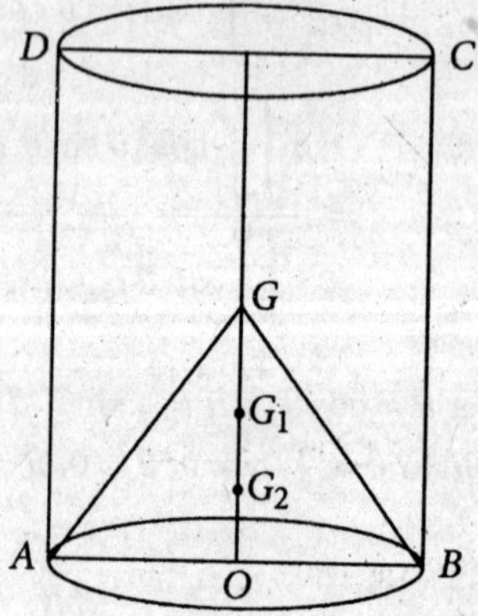

Fig. 7.12

$$\therefore\ h_2 = \frac{\pi r^2 h_1 m \cdot \frac{1}{2}h_1 - \frac{1}{3}\pi r^2 h_2 m \cdot \frac{1}{4}h_2}{\pi r^2 h_1 m - \frac{1}{3}\pi r^2 h_2 m}$$

$$= \frac{\frac{1}{2}h_1^2 - \frac{1}{12}h_2^2}{h_1 - \frac{1}{3}h_2} = \frac{6h_1^2 - h_2^2}{4(3h_1 - h_2)}$$

$$\Rightarrow 12h_1h_2 - 4h_2^2 = 6h_1^2 - h_2^2 \Rightarrow 6h_1^2 - 12h_1h_2 + 3h_2^2 = 0$$

$$\Rightarrow 2\left(\frac{h_1}{h_2}\right)^2 - 4\left(\frac{h_1}{h_2}\right) + 1 = 0 \Rightarrow \frac{h_1}{h_2} = \frac{4 \pm \sqrt{16-8}}{4} = \frac{4 \pm 2\sqrt{2}}{4} = \frac{2 \pm \sqrt{2}}{2}.$$

Rejecting negative value which gives $\frac{h_1}{h_2} < 1$, i.e., $h_1 < h_2$, which is not possible.

$$\therefore \quad \frac{h_1}{h_2} = \frac{2+\sqrt{2}}{2}.$$

8. The density of a hemisphere varies as the nth power of the distance from the centre; show that the CG divides the radius perpendicular to its plane surface in the ratio $n+3:n+5$. [KU(H) 2008]

Solution: Let the hemisphere be made up of hemispherical shells of very small thickness. Let one of them of radius x and thickness δx.

Then the mass of this shell

$$= \rho \cdot \tfrac{1}{2} 4\pi x^2 \delta x \rho = 2\pi x^2 \delta x \rho.$$

Also $\rho \propto x^n$. So let $\rho = kx^n$.

Hence, mass of the shell $= 2\pi x^2 \delta x k x^n = 2\pi k x^{n+2} \delta x$.

The CG of the shell is at a distance $\frac{x}{2}$ from the centre.

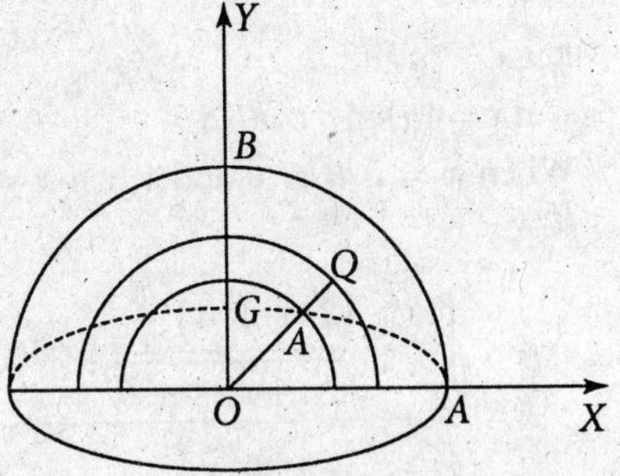

Fig. 7.13

If $(\bar{x}, \bar{y})$ be the required CG of the hemisphere, then $\bar{x} = 0$ (by symmetry about the axis of rotation, i.e., the line OY).

$$\therefore \bar{y} = \frac{\displaystyle\int_0^a \frac{x}{2} \cdot 2\pi k x^{n+2} dx}{\displaystyle\int_0^a 2\pi k x^{n+2} dx} = \frac{\frac{1}{2} \cdot \frac{a^{n+4}}{n+4}}{\frac{a^{n+3}}{n+3}} = \frac{a}{2} \cdot \frac{n+3}{n+4} \Rightarrow OG = \frac{a}{2} \cdot \frac{n+3}{n+4}$$

and

$$GB = OB - OG = a - \frac{a}{2} \cdot \frac{n+3}{n+4} = \frac{a(2n+8-n-3)}{2(n+4)} = \frac{a(n+5)}{2(n+4)}.$$

$$\therefore \quad \frac{OG}{GB} = \frac{n+3}{n+5}.$$

9. Find the CG of the area included between the curve $y^2(2a-x)-x^3$ and its asymptote. [CU(H) 2008]

Solution: The given curve is $y^2(2a-x) = x^3$.

It is symmetrical about the x-axis. It passes through the origin. Equating the coefficient of the highest power of y to zero, we get $2a - x = 0$ or, $x = 2a$ as an asymptote parallel to the y-axis.

We are required to find the CG of the area included between the curve and its asymptote. We consider an elementary strip $PP'Q'Q$ of this area parallel to y-axis. Its area is $2y\,\delta x$ and its CG can be taken as the point $M(x, 0)$.

If $(\bar{x}, \bar{y})$ be the required CG, then by the symmetry $\bar{y}=0$.

Also

$$\bar{x}=\frac{\int_0^{2a} xy\,dx}{\int_0^{2a} y\,dx}=\frac{\int_0^{2a} x\frac{x^{3/2}}{\sqrt{2a-x}}dx}{\int_0^{2a}\frac{x^{3/2}}{\sqrt{2a-x}}dx}.$$

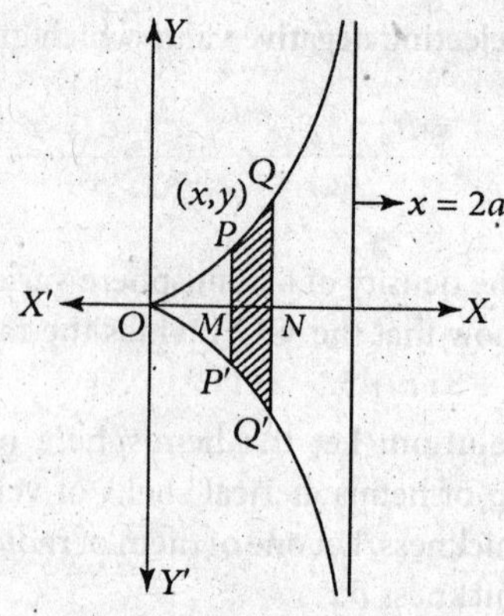

Fig. 7.14

Let $x=2a\sin^2\theta$.

$\therefore\ dx=4a\sin\theta\cos\theta\,d\theta$.

When $x=0, \theta=0$ and when $x=2a$, $\theta=\frac{\pi}{2}$.

$$\therefore\ \bar{x}=\frac{\int_0^{\pi/2}\frac{(2a\sin^2\theta)^{5/2}}{\sqrt{2a-2a\sin^2\theta}}4a\sin\theta\cos\theta\,d\theta}{\int_0^{\pi/2}\frac{(2a\sin^2\theta)^{3/2}}{\sqrt{2a-2a\sin^2\theta}}4a\sin\theta\cos\theta\,d\theta}=\frac{2a\int_0^{\pi/2}\sin^6\theta\,d\theta}{\int_0^{\pi/2}\sin^4\theta\,d\theta}$$

$$=\frac{2a\frac{6-1}{6}\cdot\frac{6-3}{6-2}\cdot\frac{6-5}{6-4}\cdot\frac{\pi}{2}}{\frac{4-1}{4}\cdot\frac{4-3}{4-2}\cdot\frac{\pi}{2}}=\frac{2a\frac{5}{6}\cdot\frac{3}{4}\cdot\frac{1}{2}}{\frac{3}{4}\cdot\frac{1}{2}}=\frac{5a}{3}.$$

Hence, $\bar{x}=\frac{5a}{3}$, $\bar{y}=0$.

Thus, the coordinates of CG is $\left(\frac{5a}{3},0\right)$.

10. The density of a thin circular plate at any point is proportional to the distance of the point from a given tangent to it. Find the position of the CG. [CU(H) 2001, KU(H) 2002]

Solution: Let the tangent and the radius through the point of contact of the tangent be taken as the y- and x-axes. Then the equation of the circular plate is $r=2a\cos\theta$, where a is its radius.

From P, we draw PN perpendicular to OX.

In ΔOPN,

$$\cos\theta=\frac{ON}{OP}=\frac{ON}{r}\Rightarrow ON=r\cos\theta.$$

So, at P density $\rho\ P\propto ON\Rightarrow\rho=k\cdot ON=kr\cos\theta$.

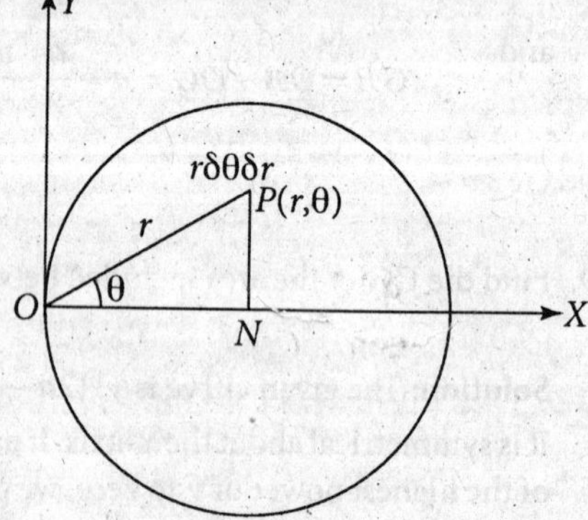

Fig. 7.15

By symmetry, the CG is on the x-axis and

$$\bar{x} = \frac{\int_{\theta=-\pi/2}^{\pi/2}\int_{r=0}^{2a\cos\theta} \rho r\cos\theta\, r\,dr\,d\theta}{\int_{\theta=-\pi/2}^{\pi/2}\int_{r=0}^{2a\cos\theta} \rho r\,dr\,d\theta} \Rightarrow \bar{x} = \frac{2\int_{\theta=0}^{\pi/2}\int_{r=0}^{2a\cos\theta} kr\cos\theta\, r^2\cos\theta\,d\theta\,dr}{2\int_{\theta=0}^{\pi/2}\int_{r=0}^{2a\cos\theta} kr\cos\theta\, r\,dr\,d\theta}.$$

$$= \frac{\int_{\theta=0}^{\pi/2} \frac{(2a\cos\theta)^4}{4}\cos^2\theta\,d\theta}{\int_{\theta=0}^{\pi/2} \frac{(2a\cos\theta)^3}{3}\cos\theta\,d\theta} = \frac{\frac{2a\times 3}{4}\int_{\theta=0}^{\pi/2}\cos^6\theta\,d\theta}{\int_{\theta=0}^{\pi/2}\cos^4\theta\,d\theta}.$$

$$\therefore\ \bar{\theta} = \frac{\frac{3a}{2}\cdot\frac{6-1}{6}\cdot\frac{6-3}{6-2}\cdot\frac{6-5}{6-4}\cdot\frac{\pi}{2}}{\frac{4-1}{4}\cdot\frac{4-3}{4-2}\cdot\frac{\pi}{2}} = \frac{\frac{3a}{2}\cdot\frac{5}{6}\cdot\frac{3}{4}\cdot\frac{1}{2}}{\frac{3}{4}\cdot\frac{1}{2}} = \frac{5a}{4}.$$

Hence, the CG of the circular plate is on its diameter through O and at a distance $\frac{5a}{4}$.

11. An isosceles triangular lamina is such that its mass per unit area at every point is proportional to the product of the distances of the point from the equal sides of the triangle. Prove that the distance of the CG from the vertex is four-fifth of the distance.

[CU(H) 1997]

Solution: Let OAB be the triangle in which $OA = OB$, let $OC = a$ be the attitude and let the angle $AOB = 2\alpha$. We take the axis of x along OC. Let P be the point (x, y) and PM, PN perpendicular to OA, OB.

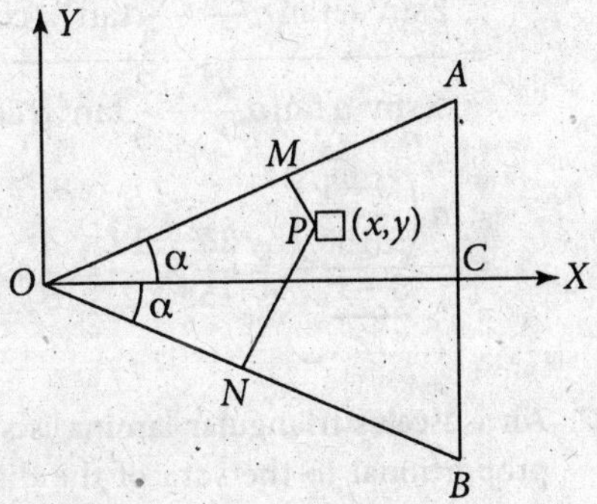

Fig. 7.16

Then equation of OA is $y = x\tan\alpha$.

$\therefore$ PM = length of perpendicular from $P(x, y)$ to OA

$$= \frac{x\tan\alpha - y}{\sqrt{\tan^2\alpha + 1}} = \frac{x\sin\alpha - y\cos\alpha}{\cos\alpha\cdot\sec\alpha} = x\sin\alpha - y\cos\alpha$$

and equation of OB is $y = -x\tan\alpha$.

$\therefore$ PN = length of perpendicular from $P(x, y)$ to OB

$$= \frac{y + x\tan\alpha}{\sqrt{\tan^2\alpha + 1}} = \frac{x\sin\alpha + y\cos\alpha}{\cos\alpha\sec\alpha} = x\sin\alpha + y\cos\theta.$$

The mass per unit area is proportional to $PM\cdot PN$

$$= (x\sin\alpha - y\cos\alpha)(x\sin\theta + y\cos\alpha) = x^2\sin^2\alpha - y^2\cos^2\alpha.$$

So $\rho = k\left(x^2\sin^2\alpha - y^2\cos^2\alpha\right)$.

By symmetry, the CG lies on OX, and its distance from O is given by

$$\bar{x} = \frac{\int_{x=0}^{a}\int_{y=-x\tan\alpha}^{x\tan\alpha} \rho x\,dx\,dy}{\int_{x=0}^{a}\int_{y=-x\tan\alpha}^{x\tan\alpha} \rho\,dx\,dy} = \frac{\int_{x=0}^{a}\int_{y=-x\tan\alpha}^{x\tan\alpha} k\left(x^2\sin^2\alpha - y^2\cos^2\alpha\right) x\,dx\,dy}{\int_{x=0}^{a}\int_{y=-x\tan\alpha}^{x\tan\alpha} k\left(x^2\sin^2\alpha - y^2\cos^2\alpha\right) dx\,dy}$$

$$= \frac{\int_{x=0}^{a} x\left[x^2\sin^2\alpha\cdot y - \frac{y^3}{3}\cos^2\alpha\right]_{y=-x\tan\alpha}^{x\tan\alpha} dx}{\int_{x=0}^{a} \left[x^2\sin^2\alpha\cdot y - \frac{y^3}{3}\cos^2\alpha\right]_{y=-x\tan\alpha}^{x\tan\alpha} dx}$$

$$= \frac{\int_{x=0}^{a} x\left[x^2\sin^2\alpha\cdot 2x\tan\alpha - \frac{2x^3\tan^3\alpha\cdot\cos^2\alpha}{3}\right] dx}{\int_{x=0}^{a} \left[x^2\sin^2\alpha\cdot 2x\tan\alpha - \frac{2x^3\tan^3\alpha\cos^2\alpha}{3}\right] dx}$$

$$\bar{x} = \frac{\int_{0}^{a} \left[2x^4\sin^2\alpha\tan\alpha - \frac{2x^4}{3}\tan^3\alpha\cos^2\alpha\right] dx}{\int_{0}^{a} \left[2x^3\sin^2\alpha\tan\alpha - \frac{2x^3}{3}\tan^2\alpha\cos\alpha\right] dx}$$

$$= \frac{2\sin^2\alpha\tan\alpha\frac{a^5}{5} - \frac{2}{3}\tan^3\alpha\cos^2\alpha\frac{a^5}{5}}{2\sin^2\alpha\tan\alpha\frac{a^4}{4} - \frac{2}{3}\tan^2\alpha\cos\alpha\frac{a^4}{4}} = \frac{2\sin^2\alpha\tan\alpha\left(\frac{1}{5} - \frac{1}{15}\right)}{2\sin^2\alpha\tan\alpha a a^4\left(\frac{1}{4} - \frac{1}{12}\right)}$$

$$= \frac{a\left(\frac{3-1}{15}\right)}{\frac{3-1}{12}} = \frac{2a}{15}\times\frac{12}{2} = \frac{4}{5}a.$$

12. An isosceles triangular lamina is such that its mass per unit area at every point is proportional to the sum of the distances of the point from the equal sides of the triangle. Prove that the distance of the CG from the vertex is three-fourth of the altitude.

[CU(H) 2002, 2006]

Solution: Let OAB be the triangle in which $OA = OB$, let $OC = a$ be the attitude and let the angle $AOB = 2\alpha$. We take the axis of x along OC. Let P be the point (x, y), and PM, PN perpendicular to OA, OB.

Then equation of OA is $y = x\tan\alpha$.

$\therefore$ $PM =$ length of perpendicular from $P(x, y)$ to OA

$$= \frac{x\tan\alpha - y}{\sqrt{\tan^2\alpha + 1}} = \frac{x\sin\alpha - y\cos\alpha}{\cos\alpha\sec\alpha} = x\sin\alpha - y\cos\alpha$$

and equation of OB is $y = -x\tan\alpha$.

$\therefore$ PN = length of perpendicular from $P(x,y)$ to OB

$$=\frac{y+x\tan\alpha}{\sqrt{\tan^2\alpha+1}}=\frac{x\sin\alpha+y\cos\alpha}{\cos\alpha\sec\alpha}$$
$$=x\sin\alpha+y\cos\alpha.$$

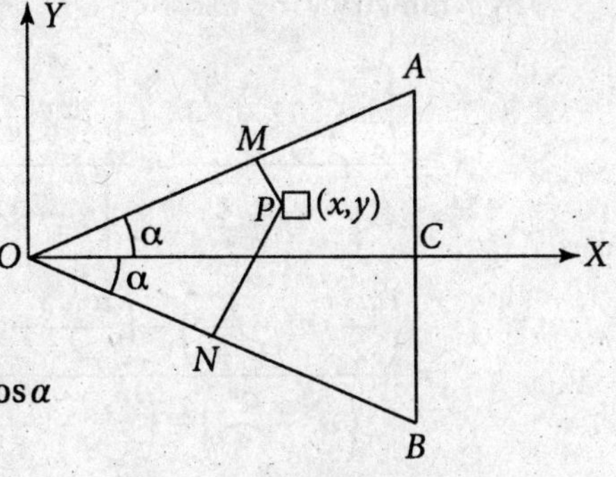

Fig. 7.17

The mass per unit area is proportional to

$$PM+PN=x\sin\alpha-y\cos\alpha+x\sin\alpha+y\cos\alpha$$
$$=2x\sin\alpha.$$

So $\rho=k2x\sin\alpha$.

By symmetry, the CG lies on OX and its distance from O is given by

$$\bar{x}=\frac{\int_{x=0}^{a}\int_{y=-x\tan\alpha}^{x\tan\alpha}\rho x\,dx\,dy}{\int_{x=0}^{a}\int_{y=-x\tan\alpha}^{x\tan\alpha}\rho\,dx\,dy}=\frac{\int_{x=0}^{a}\int_{y=-x\tan\alpha}^{x\tan\alpha}k2x\sin\alpha x\,dx\,dy}{\int_{x=0}^{a}\int_{y=-x\tan\alpha}^{x\tan\alpha}k2x\sin\alpha\,dx\,dy}$$

$$=\frac{\int_{x=0}^{a}x^2 2x\tan\alpha\,dx}{\int_{x=0}^{a}x2x\tan\alpha\,dx}=\frac{\left[\frac{x^4}{4}\right]_0^a}{\left[\frac{x^3}{3}\right]_0^a}=\frac{a^4}{4}\times\frac{3}{a^3}=\frac{3}{4}a.$$

13. Find the CG of the segment of a solid sphere of radius a cut off by a plane situated at a distance $c(<a)$ from the centre. [CU(H) 2003]

Solution: Let the area BCD revolve about DC, the X-axis to generate the spherical segment BCA.

Taking O, the centre of the spherical segment as the origin, the equation of the generating circle is given by

$$x^2+y^2=a^2. \qquad (1)$$

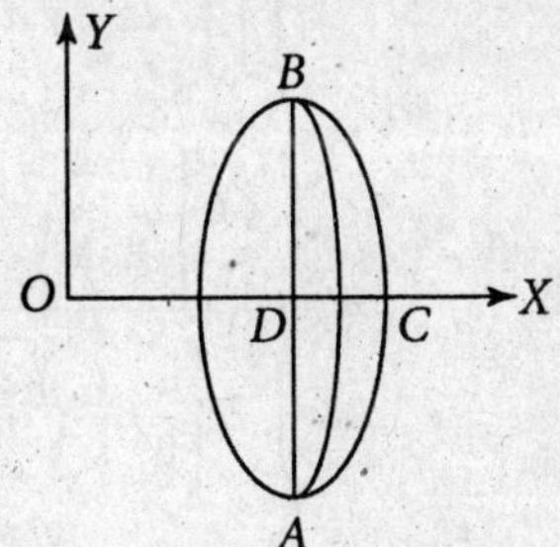

Fig. 7.18

By symmetry $\bar{y}=0$ and

$$\bar{x}=\frac{\int_c^a x\left(\pi y^2\right)dx}{\int_c^a \pi y^2 dx}=\frac{\int_c^a xy^2 dx}{\int_c^a y^2 dx}=\frac{\int_c^a x\left(a^2-x^2\right)dx}{\int_c^a \left(a^2-x^2\right)dx}=\frac{\left[a^2\cdot\frac{x^2}{2}-\frac{x^4}{4}\right]_c^a}{\left[a^2x-\frac{x^3}{3}\right]_c^a}$$

$$=\frac{\left(\frac{a^2}{2}\cdot a^2-\frac{a^4}{4}\right)-\left(\frac{a^2c^2}{2}-\frac{c^4}{4}\right)}{\left(a^3-\frac{a^3}{3}\right)-\left(a^2c-\frac{c^3}{3}\right)}=\frac{\frac{a^4}{4}-\frac{a^2c^2}{2}+\frac{c^4}{4}}{\frac{2}{3}a^3-a^2c+\frac{c^3}{3}}=\frac{\frac{a^4-2a^2c^2+c^4}{4}}{\frac{2a^3-3a^2c+c^3}{3}}$$

$$=\frac{3}{4}\cdot\frac{(a^2-c^2)^2}{[2a^3-2a^2c-a^2c+c^3]}=\frac{2\left(a^2-c^2\right)^2}{4\left[2a^2(a-c)-c(a-c)(a+c)\right]}$$

$$=\frac{3(a-c)^2(a+c)^2}{4(a-c)\left(2a^2-ac-c^2\right)}=\frac{3(a-c)(a+c)^2}{4(2a+c)(a-c)}=\frac{3}{4}\cdot\frac{(a+c)^2}{2a+c}.$$

14. Find the CG of the volume included between the surfaces $\frac{y^2}{b^2}+\frac{z^2}{c^2}-\frac{2x}{a}=0$, $x=2a$, $y=0$ and $z=0$. [CU(H) 1999]

Solution: Limits of z are from 0 to $c\sqrt{\frac{2x}{a}-\frac{y^2}{b^2}}$.

Limits of y are from 0 to $b\sqrt{\frac{2x}{a}}$.

Limits of x are from 0 to $2a$.

$$\therefore\ \bar{x}=\frac{\iiint x\,dx\,dy\,dz}{\iiint dx\,dy\,dz}=\frac{\iint x\,dx\,dy\sqrt{\frac{2x}{a}-\frac{y^2}{b^2}}}{\iint dx\,dy\sqrt{\frac{2x}{a}-\frac{y^2}{b^2}}}$$

$$=\frac{\int xb\left[\frac{1}{2}\cdot\frac{y}{b}\sqrt{\frac{2x}{a}-\frac{y^2}{b^2}}+\frac{1}{2}\cdot\frac{2x}{a}\sin^{-1}\frac{y}{b}\sqrt{\frac{a}{2x}}\right]_0^{b\sqrt{2x/a}}dx}{\int b\left[\frac{1}{2}\cdot\frac{y}{b}\sqrt{\frac{2x}{a}-\frac{y^2}{b^2}}+\frac{1}{2}\cdot\frac{2x}{a}\sin^{-1}\frac{y}{b}\sqrt{\frac{a}{2x}}\right]_0^{b\sqrt{2x/a}}dx}$$

$$=\frac{\int x\left(0+\frac{x}{a}\cdot\frac{\pi}{2}\right)dx}{\int\left(0+\frac{x}{a}\cdot\frac{\pi}{2}\right)dx}=\frac{\int_0^{2a}x^2dx}{\int_0^{2a}x\,dx}=\frac{\frac{8a^3}{3}}{\frac{4a^2}{2}}=\frac{8a^3}{3}\times\frac{2}{4a^2}=\frac{4}{3}a.$$

$$\bar{y}=\frac{\iiint y\,dx\,dy\,dz}{\iiint dx\,dy\,dz}=\frac{\iint y\sqrt{\frac{2x}{a}-\frac{y^2}{b^2}}dx\,dy}{\iint\sqrt{\frac{2x}{a}-\frac{y^2}{b^2}}dx\,dy}.$$

Let

$$t = \frac{2x}{a} - \frac{y^2}{b^2}. \therefore dt = -\frac{2y}{b^2}dy.$$

$$\therefore \int y\left(\frac{2x}{a} - \frac{y^2}{b^2}\right)^{1/2} dy = \int t^{1/2}\left(-\frac{b^2}{2}\right)dt = -\frac{b^2}{2}\cdot\frac{2}{3}t^{3/2}$$

$$= -\frac{b^2}{3}\left[\left(\frac{2x}{a} - \frac{y^2}{b^2}\right)^{3/2}\right]_0^{b\sqrt{\frac{2x}{a}}} = -\frac{b^2}{3}\left[0 - \left(\frac{2x}{a}\right)^{3/2}\right].$$

$$\therefore \bar{y} = \frac{\int \frac{b^2}{3}\left(\frac{2x}{a}\right)^{3/2} dx}{\int b\left(0 + \frac{x}{a}\cdot\frac{\pi}{2}\right)dx}, \text{ as calculated in } \bar{x}$$

$$= \frac{\frac{b^2}{3}\cdot\frac{2\sqrt{2}}{a\sqrt{a}}\cdot\frac{2}{5}\left[x^{5/2}\right]_0^{2a}}{\frac{b\pi}{2a}\left[\frac{x^2}{2}\right]_0^{2a}} = \frac{b\cdot\frac{2\sqrt{2}}{3a\sqrt{a}}\cdot\frac{2}{5}(2a)^{5/2}}{\frac{\pi}{4a}4a^2} = \frac{b\cdot\frac{2\sqrt{2}}{3}\cdot\frac{2}{5}2^{3/2}a}{\pi 4a}$$

$$= \frac{4\cdot 2^3 b}{15\pi} = \frac{32b}{15\pi}.$$

$$\bar{z} = \frac{\iiint z\,dx\,dy\,dz}{\iiint dx\,dy\,dz} = \frac{\frac{c}{2}\iint\left(\frac{2x}{a} - \frac{y^2}{b^2}\right)dx\,dy}{\iint\sqrt{\frac{2x}{a} - \frac{y^2}{b^2}}\,dx\,dy}.$$

$$Nr \text{ of } \bar{z} = \frac{c}{2}\int\left[\frac{2x}{a}y - \frac{y^3}{3b^2}\right]_0^{b\sqrt{\frac{2x}{a}}} dx = \frac{c}{2}\int_{x=0}^{2a}\left[\frac{2}{a}b\sqrt{\frac{2x}{a}}x - \frac{1}{3b^2}\cdot\frac{b^3 2\sqrt{2x}\sqrt{x}}{a\sqrt{a}}\right]$$

$$= \frac{c}{2}\left[\frac{2\sqrt{2}b}{a\sqrt{a}}\cdot\frac{2}{5}x^{5/2} - \frac{b}{3}\cdot\frac{2\sqrt{2}}{a\sqrt{a}}\cdot\frac{2}{5}x^{5/2}\right]_0^{2a}$$

$$= \frac{c}{2}\left[\frac{2\sqrt{2}}{a\sqrt{a}}b\frac{2}{5}(2a)^{5/2} - \frac{2b\sqrt{2}}{3a\sqrt{a}}\cdot\frac{2}{5}(2a)^{5/2}\right]$$

$$= c\left[\frac{2b\sqrt{2}}{5a^{3/2}}(2a)^{5/2} - \frac{2b\sqrt{2}}{3a^{3/2}}\cdot\frac{(2a)^{5/2}}{5}\right]$$

$$= bc\left[\frac{2^4 a}{5} - \frac{2^4 a}{15}\right] = \frac{16abc\times 2}{15} = \frac{32abc}{15}.$$

$$\text{Dr of } \bar{z} \text{ as calculated in } \bar{x} = \int\left(0 + \frac{x}{a}\cdot\frac{\pi}{2}\right)dx = \int_0^{2a}\frac{\pi b}{2a}x\,dy = \frac{\pi b}{2a}\times\frac{(2a)^2}{2} = \pi ba.$$

$$\therefore \bar{z} = \frac{32abc}{15\pi ab} = \frac{32c}{15\pi}.$$

Thus, the coordinates of the CG are $\left(\frac{4}{3}a, \frac{32b}{15\pi}, \frac{32c}{15\pi}\right)$.

15. A uniform board has its inner and outer surfaces concentric hemispheres of radii a and b respectively, prove that the distance of its CG from the centre is $\frac{3}{8} \cdot \frac{(a+b)(a^2+b^2)}{(a^2+ab+b^2)}$.

[KU(H) 2006]

Solution: Let OFD and AFB be the outer and inner surface of the bowl whose centre is O. OEF is the radius of symmetry.

$OA = a$ and $OC = b$.

The bowl can be considered as a solid hemisphere with the hollow one removed.

Let G_1 and G_2 be the centres of gravity of the solid hemisphere and the hollow hemisphere. Weight of the solid hemisphere acting vertically downwards at $G_1 = \frac{2}{3}\pi b^3 W$ and weight of the hollow portion acting vertically upwards at $G_2 = \frac{2}{3}\pi a^3 W$, where W is the weight per unit volume.

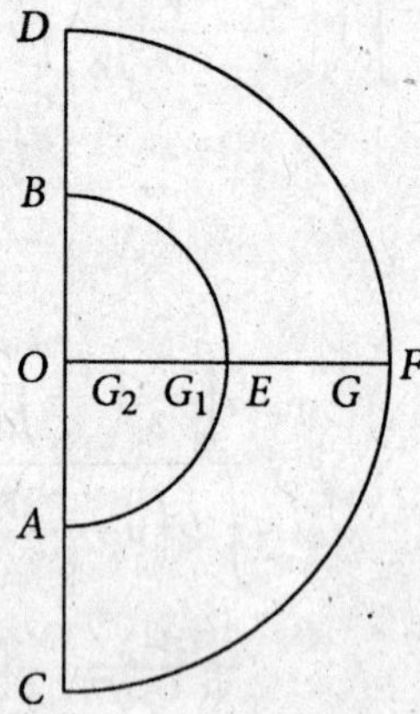

Fig. 7.19

Also $OG_1 = \frac{3}{8}b$ and $OG_2 = \frac{3}{8}a$.

Therefore, if G be the CG of the bowl, then

$$OG = \frac{\frac{2}{3}\pi b^3 W \cdot \frac{3}{8}b - \frac{2}{3}\pi a^3 W \cdot \frac{3}{8}a}{\frac{2}{3}\pi b^3 W - \frac{2}{3}\pi a^3 W} = \frac{3}{8} \cdot \frac{b^3 - a^4}{b^3 - a^3} = \frac{3}{8} \cdot \frac{(b-a)(b+a)(b^2+a^2)}{(b-a)(b^2+ab+a^2)}$$

$$= \frac{3}{8} \cdot \frac{(a+b)(a^2+b^2)}{(a^2+ab+b^2)}.$$

16. Show that the centroid of a circular sector in which the density varies as the nth power of the distance from the centre is on the axis of the sector at a distance $\frac{n+2}{n+3} a \frac{\sin\alpha}{\alpha}$, where 2α is the angle of sector.

[KU(H) 2006]

Solution: Let AOB be the sector that subtends an angle 2α at the centre O and OC is its bisector.

By symmetry the CG lies on the symmetrical radius OC taken as x-axis.

$\therefore\ \bar{y} = 0$.

Also $\bar{x} = \dfrac{\iint \rho r \cos\theta \, r \, d\theta \, dr}{\iint \rho r \, d\theta \, dr}$.

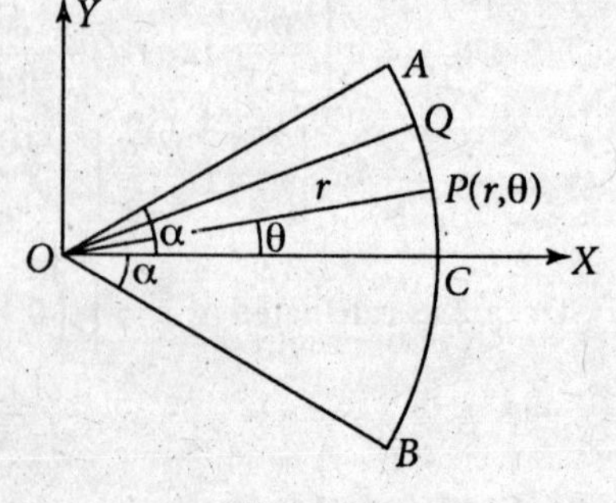

Fig. 7.20

Also $\rho \propto r^n$ (given) $\Rightarrow \rho = kr^n$.

$$\therefore \bar{x} = \frac{\int_{r=0}^{a}\int_{\theta=-\alpha}^{\alpha} kr^n r^2 \cos\theta\, d\theta\, dr}{\int_{r=0}^{a}\int_{\theta=-\alpha}^{\alpha} kr^n r\, d\theta\, dr} = \frac{\int_{r=0}^{a}\int_{\theta=-\alpha}^{\alpha_{n+2}} r^{n+2} \cos\theta\, d\theta\, dr}{\int_{r=0}^{a}\int_{\theta=-\alpha}^{\alpha} r^{n+1} d\theta\, dr}$$

$$= \frac{\left[\frac{r^{n+3}}{n+3}\right]_0^a \left[\sin\theta\right]_{\theta=-\alpha}^{\alpha}}{\left[\frac{r^{n+2}}{n+2}\right]_0^a \left[\theta\right]_{\theta=-\alpha}^{\alpha}} = \frac{n+2}{n+3} \cdot \frac{a^{n+3} 2\sin\alpha}{a^{n+2} \cdot 2\alpha} = \frac{n+2}{n+3} a \frac{\sin\alpha}{\alpha}.$$

17. **From a uniform triangle board a portion consisting of the area of inscribed circle is removed. Show that the distance of the CG of the remainder from any side is $\frac{S}{3as} \cdot \frac{2s^3 - 3\pi aS}{s^2 - \pi S}$, where S is the area and s the semi-spherimeter of the board.**

Solution: Let D be the midpoint of the base BC of a uniform triangular lamin a ABC and G_1 be its CG, where $G_1D = \frac{1}{3}AD$.

Let G_2 be the in centre of ΔABC, then CG of the remainder will lie on the line joining the points G_1 and G_2.

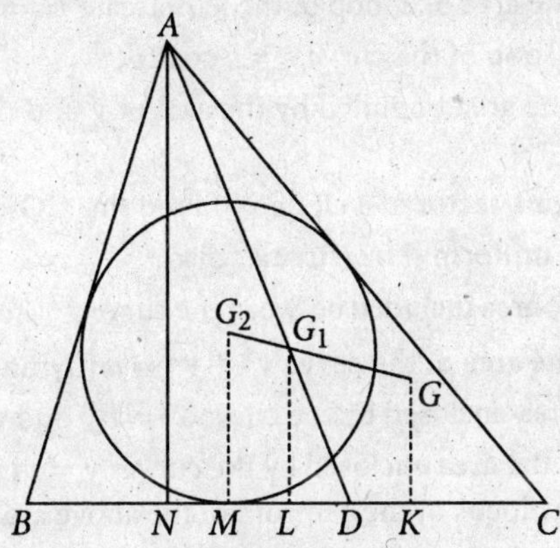

Fig. 7.21

Now area of ΔABC

$$= S = \frac{1}{2} \times BC \times AN = \frac{1}{2} a \cdot AN \Rightarrow AN = \frac{2S}{a}.$$

From similar triangles ΔG_1LD and ΔAND,

$$\frac{G_1L}{AN} = \frac{G_1D}{AD} = \frac{\frac{1}{3}AD}{AD} = \frac{1}{3} \Rightarrow G_1L = \frac{1}{3}AN = \frac{1}{3} \cdot \frac{2S}{a} = \frac{2S}{3a}$$

and $G_2M =$ radius of the incircle $= r = \frac{S}{s}$.

Let w be the weight per unit square of the area, then weight of $\Delta ABC =$ (Area of $\Delta ABC)w = Sw$ and weight of the incircle $=$ (Area of the Incircle) $\cdot w = \pi r^2 w$ $- \pi\left(\frac{S}{s}\right)^2 w$.

So the distance of the CG from the side BC

$$= KG = \frac{(\text{Weight of } \triangle ABC)\cdot G_1L - (\text{Weight of the incircle})\cdot G_2M}{\text{Weight of } \triangle ABC - \text{Weight of the incircle}}$$

$$= \frac{(Sw)\cdot\left(\frac{2S}{3a}\right) - \left[\pi\left(\frac{S}{s}\right)^2 w\right]\cdot\frac{S}{S}}{(Sw) - \left[\pi\left(\frac{S}{s}\right)^2 w\right]} = \frac{S\left[\frac{2}{3a} - \pi\frac{S}{s^3}\right]}{1-\pi\left(\frac{S}{s^2}\right)} = \frac{S}{3as}\left[\frac{2s^3-3\pi aS}{s^2-\pi S}\right].$$

EXERCISE 7

1. Find the position of CG of the arc of the cardioid $r = a\,(1+\cos\theta)$ lying above the initial line.
2. Find the CG of the whole arc of the cardioid $r = a(1+\cos\theta)$.
3. Find the centroid of the arc of the catenary $y = c\cosh\frac{x}{c}$ which is included between the origin and any point (x, y).
4. Find the CG of the arc of the parabola $y^2 = 4ax$ extending from vertex to latus rectum.
5. If O be the pole of lemniscate $r^2 = a^2\cos 2\theta$ and G is the CG of the arc PQ of the curve, prove that OG bisects angle POQ.
6. Find the CG of the area of a loop of the lemniscate $r^2 = a^2\cos 2\theta$.
7. Find the CG of a loop of the curve $r = a\cos 2\theta$.
8. Find the CG of the area bounded by the axis of y and the cycloid $x = a(\theta+\sin\theta)$, $y = a(1-\cos\theta)$.
9. (a) Find the CG of a sector of a circle. (b) Find the CG of a segment of a circle.
10. Find the CG of a uniform semi-circular disc.
11. Find the CG of the area included between the curve $y^2\,(2a-x) = x^3$ and its asymptote.
12. Find the CG of the area of the curve $x^{\frac{2}{3}} + y^{\frac{2}{3}} = a^{\frac{2}{3}}$ lying in the first quadrant.
13. Find the CG of area enclosed by the curve $x^2 = by$ and $y^2 = ax$. [KU(H) 2001]
14. Prove that CG of the area enclosed by the curves $y = mx$ and $y^2 = 4ax$ is $\left(\frac{8a}{5m^2}, \frac{2a}{m}\right)$ and hence, find the locus of the centroid of the above area when m varies.
15. Find the CG of a solid hemisphere.
16. Show that the CG of surface of a spherical segment bisects its height, and hence, deduce the CG of a uniform hemispherical shell.
17. Find the centroid of the volume formed by the revolution of the curve $r = a(1+\cos\theta)$ about the X-axis.
18. Find the centroid of the surface formed by the revolution of the cardioid $r = a(1+\cos\theta)$ about its axis.
19. Find the CG of a solid figure formed by revolving a quadrant of an ellipse about its minor axis.
20. Find the vertical angle of a cone in order that the CG of the whole surface including its plane base may coincide with the CG of its volume.
21. The density at any point of a circular lamina varies as the nth power of the distance from a point on the circumference. Show that the CG of the lamina divides the diameter through O in the ratio $n+2:2$, where n is a positive integer. [KU(H) 2007]

22. A triangular lamina ABC, obtuse angled at C, stands with the side AC in contact with a table. Show that the least weight, which is suspended from B will overturn the triangle, is $\frac{1}{3}w\frac{a^2+3b^2-c^2}{c^2-a^2-b^2}$, where w is the weight of the triangle. Interpret the result, if $c^2 > a^2+3b^2$.

23. Find the CG of one loop of the lemniscate of Bernoulli $r^2 = a^2\cos 2\theta$. [IAS 1988]

24. Find the CG of the solid bounded by the ellipsoid $\frac{x^2}{a^2}+\frac{y^2}{b^2}+\frac{z^2}{c} = 1$ and lying in the positive octant. [IAS 1991]

25. Find the CG of the volume formed by revolving the area bounded by the parabolas $y^2 = 4ax$ and $x^2 = 4by$ about x-axis. [IAS 1992]

26. A circular disc of radius a, whose density is proportional to the distance from the centre, has a hole cut in it bounded by a circle, of diameter b, which passes through the centre. Find the distance from the centre of the disc of the CG of the remaining portion.

27. A portion of an anchor ring is formed by the revolution of a circular area of radius a about a line in its plane at a distance c from its centre, where $c > a$. If 2α be the angle through which it revolves, prove that the CG of the solid is at a distance from the line equal to $\frac{4c^2+a^2}{4c}\frac{\sin\alpha}{\alpha}$.

28. Show that the centre of inertia of a circular disc, the density of which varies inversely as the fourth power of the distance from an external point O in the plane, is the inverse point of O w.r.t. the boundary of the disc.

29. The density at any point of a circular lamina varies as the cube of the distance from a point O on the circumference. Show that the centre of gravity of the lamina divides the diameter through O in the ratio 5 : 2. [CU(H) 2005]

Answers

1. $\left(\frac{4a}{5}, \frac{4a}{5}\right)$. 2. $\left(\frac{4a}{5}, 0\right)$. 3. $\left(x-\frac{c}{s}(y-s), \frac{y}{2}+\frac{cx}{2s}\right)$. 4. $\left\{\frac{a}{4}\left[\frac{3\sqrt{2}-\log(\sqrt{2}+1)}{\sqrt{2}+\log(\sqrt{2}+1)}\right], \frac{4a}{3}\right\}\left[\frac{(2\sqrt{2}-1)}{[\sqrt{2}+\log(\sqrt{2}+1)]}\right]$. 6. $\left(\frac{\pi a}{8}\sqrt{2}, 0\right)$. 7. $\left(\frac{128\sqrt{2}}{105\pi}a, 0\right)$. 8. $\left(\frac{9\pi^2-16}{18\pi}\cdot a, \frac{7}{6}a\right)$. 9. (a) $\left(\frac{2a\sin\alpha}{3\alpha}, 0\right)$; (b) $\left(\frac{2}{3}a\frac{\sin^3\alpha}{\alpha-\sin\alpha\cos\alpha}, 0\right)$. 10. $\left(\frac{4a}{3\pi}, 0\right)$. 11. $\left(\frac{3}{5}a, 0\right)$. 12. $\left(\frac{256a}{315\pi}, \frac{256a}{315\pi}\right)$. 13. $\left[\frac{9a^{1/3}b^{2/3}}{20}, \frac{9a^{2/3}b^{1/3}}{20}\right]$. 14. $y^2 = \frac{5a}{2}x$. 15. $\left(\frac{3a}{8}, 0, 0\right)$. 16. $\left(\frac{a}{2}, 0\right)$, a is the radius. 17. $\left(\frac{4a}{5}, 0\right)$. 18. $\left(\frac{50}{63}a, 0\right)$. 19. $\left(0, \frac{3}{8}b\right)$. 20. $2\sin^{-1}\frac{1}{3}$. 23. $\left[\frac{\pi a\sqrt{2}}{8}, 0\right]$. 24. $\left(\frac{3a}{8}, \frac{3b}{8}, \frac{3c}{8}\right)$. 25. $\left(\frac{20}{9}a^{1/3}b^{2/3}, 0\right)$. 26. $\frac{6b^4}{15\pi a^3-10b^3}$.

Chapter 8

Friction

8.1 Introduction

When one body slides or rolls on another body, it is found from the of experience that a force tending to resist motion is called into *play.* The fact that a body can rest on an inclined plane under no forces except its weight and the reaction of the plane, so that there is a component of the reaction along the plane which balances the component of the weight of the body down the plane. This component of the reaction between the body and the plane, which is in the plane, is called *friction.* Friction is a force which resists sliding of one body over another.

8.2 Definition

If two bodies be in contact with one another the property of the two bodies, by virtue of which a force is exerted between them at their point of contact to prevent one body sliding or rolling on the other is called the *friction* and the force exerted is called the *force of friction.* Its direction is always opposite to the direction of motion of the body. The basic property of the friction is that it is a self-adjusting force and, moreover, it is positive.

8.3 Statical, Dynamical and Limiting Friction [RU(H) 1994]

1. When a body in contact with another body is in the position of equilibrium and the frictional force is just sufficient to maintain the equilibrium, the friction exerted is called the *statical friction.*
2. When a body is just on the point of sliding or rolling on another body, the force of friction exerted reaches its greatest value. The maximum friction exerted is called the *limiting friction.* [RU(H) 2000]
3. When a body actually moves or continues moving on another body, the friction exerted is called the *dynamical friction.*

Clearly the statical and dynamical frictions are less than the limiting friction. We observe that the limiting friction will come into existence when the body is just about to slip or slide.

8.4 Laws of Friction [RU(H) 1998]

The laws of friction can be divided into three groups:

1. The laws of statical friction;
2. The laws of limiting friction;
3. The laws of dynamical friction.

These laws are purely experimental, but they represent a reasonable degree of accuracy in ordinary situations.

Laws of Statical Friction

1. The direction of the friction is opposite to the direction in which the body is urged to move.
2. The magnitude of the friction is, when there is equilibrium, just sufficient to prevent the body from moving.

Laws of Limiting Friction [RU(H) 1991]

1. The magnitude of the limiting friction always bears a constant ratio μ to the normal reaction. The ratio μ depends only on the nature of the substances of which the bodies are made. The constant μ is called the *coefficient of friction.*
2. The magnitude of the limiting friction is independent of the area and the shape of the surfaces of contact.

Laws of Dynamical Friction

1. When motion takes place by one body sliding upon the other, the friction is independent of the velocity and is proportional to the normal reaction. The ratio of the friction to the normal reaction in this case is slightly less than μ.

8.5 Some Definitions

1. **Limiting Equilibrium:** When the two bodies in contact are in equilibrium and one body is just on the point of sliding or rolling over the other, the corresponding equilibrium is called the *limiting equilibrium.*

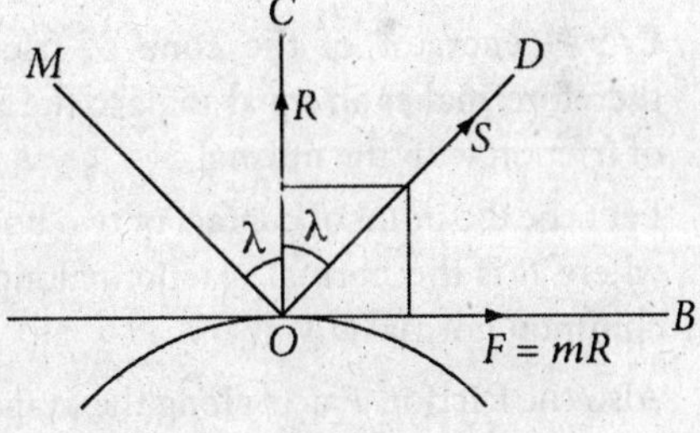

Fig. 8.1

2. **Coefficient of Friction; Angle of Friction:** When the equilibrium is limiting, then the corresponding friction is known as the *limiting friction.* The constant ratio of the limiting friction to the normal reaction is called the *coefficient of friction* and denoted by μ. [RU(H) 1993]

   Let a body, placed on a rough surface, be in limiting equilibrium.

   Let F be the force of friction in the direction of OB, and R be the normal reaction at O along OC.

   F and R can be compounded into a single resultant force which is called the *resultant friction.* It is denoted by S along OD.

   In case of limiting equilibrium, $F = \mu R$.

   The angle COD is called the *angle of friction* and it is denoted by λ.

Resolving forces we have

$$S\cos\lambda = R \tag{1}$$

$$\text{and } S\sin\lambda = F = \mu R. \tag{2}$$

Squaring and adding, we get

$$S^2\left(\cos^2\lambda+\sin^2\lambda\right)=R^2+M^2R^2 \text{ or, } S^2=R^2\left(1+\mu^2\right) \text{ or, } S=R\sqrt{1+\mu^2}.$$

This gives the resultant reaction S in terms of R and μ.
Dividing (2) by (1), we get

$$\frac{S\sin\lambda}{S\cos\lambda}=\frac{\mu R}{R} \text{ or, } \tan\lambda=\mu.$$

Thus, angle of friction is that angle of which the tangent is equal to the coefficient of friction.

In case of ordinary equilibrium we must have $F \le \mu R$.

3. **Cone of Friction:** When a body is in contact with a rough surface, and with the common normal at the point of contact as axes, we describe a right circular cone whose semi-vertical angle is $\lambda = \tan^{-1}\mu$, where μ is the coefficient of friction, this cone is defined as the cone of friction.

[RU(H) 2001]

Every generator of the cone of friction, therefore, makes an equal angle to the angle of friction with the normal.

Let O be the point of contact of two bodies, where R is the normal reaction, along the common normal ON at O.

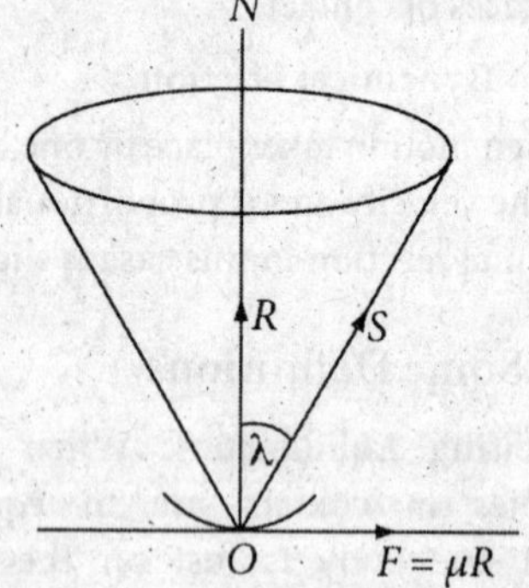

Fig. 8.2

Also the friction F acts along the tangent line OA at O.

If the resultant reaction S makes an angle θ with the reactions R at O, then $R = S\cos\theta$, $F = S\sin\theta$.

Dividing, we get

$$\tan\theta = \frac{F}{R}. \tag{3}$$

But max $F = \mu R$, where μ is the coefficient of friction. Hence, for maximum value of θ, from (1), we have

$$\tan\theta = \frac{\mu R}{R} = \mu = \tan\lambda,$$

where λ is the angle of friction.

Hence, the greatest angle which the direction of the resultant reaction S makes with the normal reaction R is λ. Now we can describe a cone of semi-vertical angle $\lambda = \tan^{-1}\mu$, with ON as its axis and O is the vertex.

The line of action of S, the resultant reaction will be within, or at the most, on the cone. [RU(H) 1993]

This cone is called the *cone of friction.*

Definition 1. *The right cone described with its vertex at the point of contact of two rough bodies and having the common normal at the point of contact as axis and angle of friction as the semi-vertical angle is called the cone of friction.*

8.6 Equilibrium of a Body on a Rough Inclined Plane under its Own Weight. [IAS 1996; RU(H) 1992]

A body is placed on a rough plane inclined to the horizon at an angle greater than the angle of friction and is supported by a force, acting in a vertical plane through the line of greatest slope. To find the limits between which the force must lie.

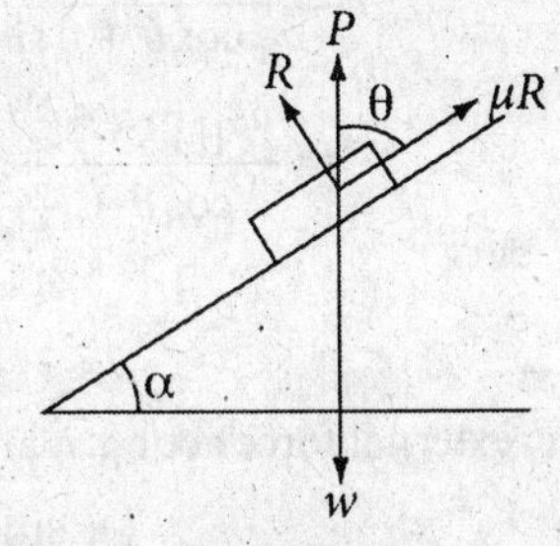

Fig. 8.3

Let a heavy body of weight w be placed on a rough inclined plane of inclination α to the horizon, and be acted on by a force P at an angle θ to the plane. Let μ be the coefficient of friction.

Case I. Let the body be just supported, i.e., just on the point of slipping down.

If R be the normal reaction of the plane, the friction in this case, which is limiting, is μR up the plane.

Hence, for equilibrium, resolving along and perpendicular to the plane

$$P\cos\theta + \mu R = w\sin\alpha$$

$$P\sin\theta + R = w\cos\alpha$$

$$R = w\cos\alpha - P\sin\theta. \tag{1}$$

From (1),

$$P\cos\theta + \mu(w\cos\alpha - P\sin\theta) = w\sin\alpha$$

$$\text{or, } P(\cos\theta - \mu\sin\theta) = w(\sin\alpha - \mu\cos\alpha).$$

$$\therefore P = \frac{w(\sin\alpha - \mu\cos\alpha)}{\cos\theta - \mu\sin\theta}. \tag{2}$$

If λ be the angle of friction, so that $\mu = \tan\lambda$, then from (2) we have

$$P = \frac{w(\sin\alpha - \tan\lambda\cos\alpha)}{\cos\theta - \tan\lambda\sin\theta} = \frac{w(\sin\alpha - \frac{\sin\lambda}{\cos\lambda}\cos\alpha)}{\cos\theta - \frac{\sin\alpha}{\cos\lambda}\sin\theta}$$

$$= \frac{w(\sin\alpha\cos\lambda - \sin\lambda\cos\alpha)}{\cos\theta\cos\lambda - \sin\theta\sin\lambda} = \frac{w\sin(\alpha-\lambda)}{\cos(\theta+\lambda)}, \tag{A}$$

giving necessary value of P in the given direction just to support the weight on the plane.

Case II. Let the body be on the point of sliding up.

In this case, the limiting friction μR is down the plane, and hence, the equations for equilibrium give

$$P\cos\theta - \mu R = w\sin\alpha \tag{3}$$

$$\text{and } P\sin\theta + R = w\cos\alpha,$$

$$R = w\cos\alpha - P\sin\theta.$$

So from (3), $P\cos\theta - \mu(w\cos\alpha - P\sin\theta) = w\sin\alpha$

$$P(\cos\theta + \mu\sin\theta) = w(\sin\alpha + \mu\cos\alpha)$$

$$P = \frac{w(\sin\alpha + \mu\cos\alpha)}{\cos\theta + \mu\sin\theta} = \frac{w[\sin\alpha + \tan\lambda\cos\alpha]}{\cos\theta + \tan\lambda\sin\theta}$$

$$= \frac{w\left[\sin\alpha + \frac{\sin\lambda}{\cos\lambda}\cos\alpha\right]}{\cos\theta + \frac{\sin\lambda}{\cos\lambda}\sin\theta} = \frac{w(\sin\alpha\cos\lambda + \sin\lambda\cos\alpha)}{\cos\theta\cos\lambda + \sin\lambda\sin\theta}.$$

$$P = \frac{w\sin(\lambda + \alpha)}{\cos(\theta - \lambda)}. \tag{B}$$

Hence, external force acting at an angle θ with line of the greatest slope must lie between

$$\frac{w\sin(\alpha - \lambda)}{\cos(\theta + \lambda)} \text{ and } \frac{w\sin(\alpha + \lambda)}{\cos(\theta - \lambda)}$$

in order to keep the body at rest on the inclined plane.

Special Case

1. The least force to pull the body up is $w\sin(\alpha + \lambda)$.
   From equation (B), we have P as least when $\cos(\theta - \lambda)$ as the greatest, i.e., $\cos(\theta - \lambda) = 1$, i.e., $\theta - \lambda = 0$ or $\theta = \lambda$.
   Then the least force is $w\sin(\lambda + \alpha)$.
2. The least force to pull the body down is $w\sin(\alpha - \lambda)$.
   From equation (A), we have P as least when $\cos(\theta + \lambda)$ as greatest, i.e., $\cos(\theta + \lambda) = 1$.
   Then $P = w\sin(\alpha - \lambda)$.
3. The limits of the force is $w\sin(\alpha \pm \lambda)$ when the force acts along.
   Taking $\theta = 0$ in (A) and (B), we get
   $$P = \frac{w\sin(\alpha \pm \lambda)}{\cos\lambda}.$$
4. The least force necessary to move a load w on a rough horizontal plane.
   From equation (B) we get by taking $\alpha = 0$.
   $$P = \frac{w\sin\alpha}{\cos(\theta - \lambda)}.$$
   P is least when $\cos(\theta - \lambda)$ is the greatest, i.e., $\cos(\theta - \lambda) = 1$.
   $$P = w\sin\alpha.$$

» **Example 1.** *To find the conditions of Equilibrium of a Particle constrained to rest on a Rough Curve under any given forces.* **[RU(H) 2001]**

Let X, Y be the resolved parts of external forces parallel to coordinate axes in any position P of the particle. Let R be the normal reaction measured inwards of the curve on the particle and F the friction, which, if the particle is in limiting equilibrium, is equal to μR, where μ is the coefficient of friction for the curve. If the tangent at P makes an angle ψ with the x-axis, then we have

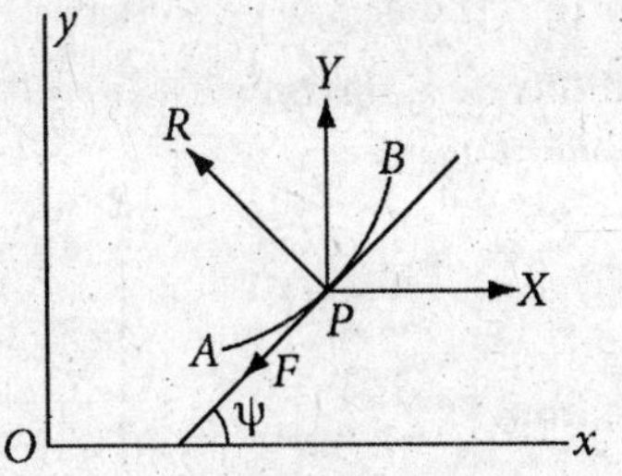

Fig. 8.4

$$\tan\psi = \frac{dy}{dx}, \ \sin\psi = \frac{dy}{ds}, \ \cos\psi = \frac{dx}{ds}.$$

Resolving all the forces along the normal, we get for equilibrium

$$X\sin\psi - Y\cos\psi = R \text{ or, } X\frac{dy}{ds} - Y\frac{dx}{ds} = R. \qquad (1)$$

Similarly, resolving all the forces along the tangent, we obtain

$$X\cos\psi + Y\sin\psi = F \text{ or, } X\frac{dx}{ds} + Y\frac{dy}{ds} = F.$$

Assuming that as in the figure friction acts downwards, i.e., on the assumption that the tendency of the particle is to move upwards.

Now in any case the numerical value of F should not be greater than μR. So, for equilibrium, $F \le \mu R$;

$$\text{or, } \left|X\frac{dx}{ds} + Y\frac{dy}{ds}\right| \le \mu\left|X\frac{dy}{ds} - Y\frac{dx}{ds}\right| \text{ or, } \left(X\frac{dx}{ds} + Y\frac{dy}{ds}\right)^2 \le \mu^2\left(X\frac{dy}{dx} - Y\right),$$

where $\frac{dy}{dx}$ is given by the known equation of the curve.

WORKED-OUT EXAMPLES

1. A uniform beam of length $2h$, can rest with one end on a rough horizontal plane and against the top of a wall of height h, in a vertical plane perpendicular to the wall and at any inclination to the wall is geometrically possible. Show that the angle of friction between the beam and both wall and ground supposed to be equally rough, must not be less than $\frac{1}{2}\sin^{-1}\left(\frac{4}{3\sqrt{3}}\right)$. [IAS 1981]

   Solution: Let θ be the inclination of the beam to the wall where $AB = 2h$, $AG = h$, $CD = h$.

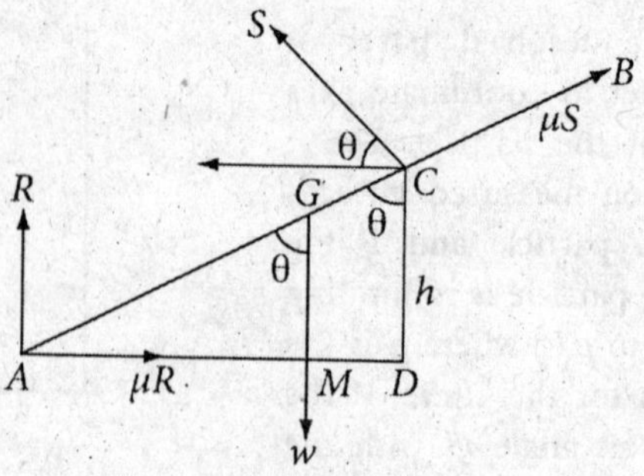

Fig. 8.5

In $\triangle ACD$,

$$\cos\theta = \frac{CD}{AC} = \frac{h}{AC};\ AC = h\sec\theta.$$

In $\triangle AGM$,

$$\sin\theta = \frac{AM}{AG} = \frac{AM}{h};\ AM = h\sin\theta.$$

As the beam rests at any inclination we consider its limiting equilibrium. Resolving horizontally and vertically, we get

$$\mu R + \mu S\sin\theta = S\cos\theta \qquad (1)$$

$$S\sin\theta + \mu S\cos\theta + R = w$$

$$\text{or, } \mu S\sin\theta + \mu^2 S\cos\theta + \mu R = \mu w. \qquad (2)$$

(1)–(2) gives

$$-\mu^2 S\cos\theta = S\cos\theta - \mu w \text{ or, } \mu w = S\cos\theta\left(\mu^2+1\right). \qquad (3)$$

Taking moment about A, we get

$$SAC = wAM \text{ or, } Sh\sec\theta = wh\sin\theta \text{ or, } S\sec\theta = w\sin\theta. \qquad (4)$$

Dividing (3) by (4), we get

$$\frac{S\cos\theta(\mu^2+1)}{S\sec\theta} = \frac{\mu w}{w\sin\theta} \text{ or, } \cos^2\theta(\mu^2+1) = \frac{\mu}{\sin\theta}$$

$$\text{or, } \frac{2\mu}{\mu^2+1} = 2\cos^2\theta\sin\theta \text{ or, } \sin 2\lambda = 2\cos^2\theta\sin\theta.$$

Now λ will be max. or min. according as $\sin 2\lambda$ is max. or min.
Let $z = \sin 2\lambda$, $z = 2\cos^2\theta\sin\theta$.

$$\frac{dz}{d\theta} = 2\left(\cos^3\theta - 2\cos\theta\sin^2\theta\right).$$

For max or min z,

$$\frac{dz}{d\theta} = 0 \text{ or, } 2\cos\theta\left(\cos^2\theta - \sin^2\theta\right) = 0.$$

Now $\cos\theta$ cannot be zero, for in that case $\theta = \frac{\pi}{2}$, which is not possible geometrically.

$$\therefore \cos^2\theta - 2\sin^2\theta = 0 \text{ or, } 1 - 3\sin^2\theta = 0 \text{ or, } \sin^2\theta = \tfrac{1}{3}.$$

$$\therefore \cos^2\theta = \tfrac{2}{3}.$$

Hence,

$$\sin 2\lambda = 2\cos^2\theta \sin\theta = 2\cdot\frac{2}{3}\frac{1}{\sqrt{3}} = \frac{4}{3\sqrt{3}}.$$

$$\therefore 2\lambda = \sin^{-1}\left(\frac{4}{3\sqrt{3}}\right);\ \lambda = \frac{1}{2}\sin^{-1}\left(\frac{4}{3\sqrt{3}}\right).$$

2. A uniform rod rests in limiting equilibrium within a rough hollow sphere; if the rod subtends an angle 2α at the centre of the sphere and if λ be the angle of friction, show that the angle of inclination of the rod to the horizon is $\tan^{-1}\left[\dfrac{\sin 2\lambda}{\cos 2\alpha + \cos 2\lambda}\right]$.

[IAS 1988, RU(H) 1996]

Solution: Let AB be the rod in limiting equilibrium. Given $AB = 2a$. The two forces R and S (resultant reactions of A and B) meet in C. Therefore, the third force W must also pass through C. The resultant reactions makes angles λ with the directions of the normal reactions.

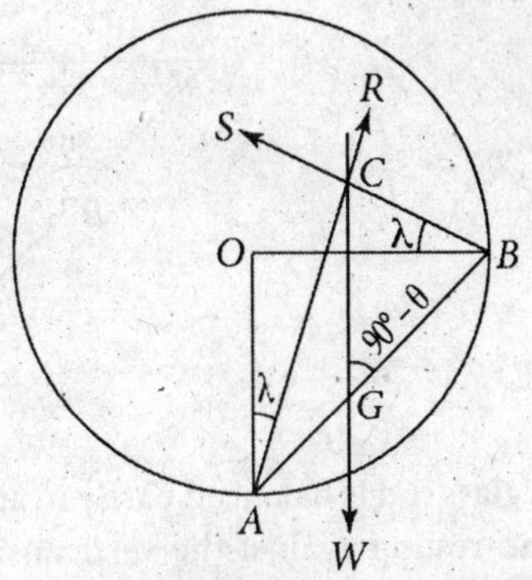

Fig. 8.6

$$\therefore \angle BAC = 90 - (\alpha + \lambda) = 90 - \alpha - \lambda$$

$$\angle ABC = 90 - \alpha + \lambda \text{ and}$$

$$\angle BGC = 90 - \theta.$$

Using $m\cdot n$ formula of Trigonometry on ΔABC, we get

$$(AG + BG)\cot(90 - \theta) = BG \cot BAC - AG \cot ABC,$$

$$\text{i.e., } 2a\tan\theta = a\cot(90^\circ - \alpha - \lambda) - a\cot(90^\circ - \alpha + \lambda),$$

$$\text{i.e., } 2\tan\theta = \tan(\alpha+\lambda) - \tan(\alpha-\lambda) = \frac{\sin(\alpha+\lambda)}{\cos(\alpha+\lambda)} - \frac{\sin(\alpha-\lambda)}{\cos(\alpha-\lambda)}$$

$$= \frac{\sin[\alpha+\lambda-(\alpha-\lambda)]}{\cos(\alpha+\lambda)\cos(\alpha-\lambda)}$$

$$\text{or, } 2\tan\theta = \frac{2\sin 2\lambda}{2\cos(\alpha+\lambda)\cos(\alpha-\lambda)} \text{ or, } 2\tan\theta = \frac{2\sin 2\lambda}{\cos 2\alpha + \cos 2\lambda}$$

$$\text{or, } \tan\theta = \frac{\sin 2\lambda}{\cos 2\alpha + \cos 2\lambda},\ \therefore \theta = \tan^{-1}\left[\frac{\sin 2\lambda}{\cos 2\alpha + \cos 2\lambda}\right] \textbf{ (Proved.)}$$

3. One end of a heavy uniform rod AB can slide along a rough horizontal rod AC to which it is attached by a ring, B and C are joined by a string. If ABC be a right angle

when the rod is on the point of sliding, μ the coefficient of friction and α the angle between AB and the vertical, show that $\mu = \dfrac{\tan\alpha}{\tan^2\alpha + 2}$.

Solution: Let S be the resultant of reaction R and μR inclined at an angle λ to the direction of normal reaction. Let the vertical through G and CB produced (along which will act tension T) meet at P, then for equilibrium, line of action of S will also pass through P. Now we apply mn theorem of Trigonometry on ΔABP in which $\angle BGP = \alpha$ and $BG = AG$, we get

$$(AG + BG)\cot\alpha = AG\cot\lambda - BG\cot(90° - \alpha) \text{ or, } 2\cot\alpha = \cot\lambda - \tan\alpha$$

$$\text{or, } \cot\lambda = \frac{2}{\tan\alpha} + \tan\alpha = \frac{2 + \tan^2\alpha}{\tan\alpha}. \therefore \mu = \tan\alpha = \frac{\tan\alpha}{\tan^2\alpha + 2} \text{ (Proved).}$$

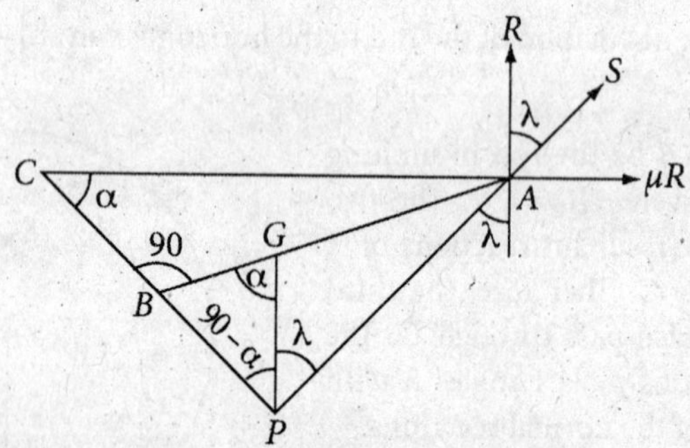

Fig. 8.7

4. A glass rod is balanced partly in and partly out of a cylindrical tumbler with the lower end resting against the vertical side of the tumbler. If α and β be the greatest and least angles which the rod can make with the vertical, show that angle of friction is $\frac{1}{2}\tan^{-1}\frac{\sin^3\alpha - \sin^3\beta}{\sin^2\alpha\cos\alpha + \cos\beta\sin^2\beta}$.

Solution: If the rod AB is making the greatest angle α with the vertical, then its end A is on the point of moving upwards along the vertical side of the tumbler and if CG is outside the tumbler. The forces have been marked in the figure.

Resolving these forces horizontally and vertically, we have

$$R = S\cos\alpha + \mu S\sin\alpha \quad (1)$$

$$\text{and } S\sin\alpha = \mu R + \mu S\cos\alpha + w. \quad (2)$$

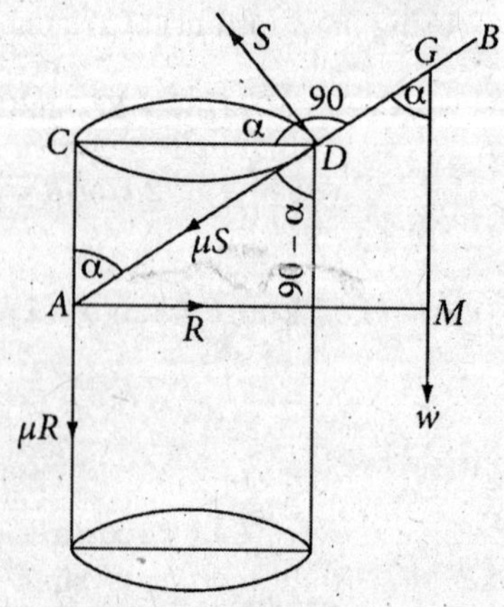

Fig. 8.8

$$S\sin\alpha = \mu(S\cos\alpha + \mu S\sin\alpha) + \mu S\cos\alpha + w$$

$$\text{or, } w = S[\sin\alpha - \mu(\cos\alpha + \mu\sin\alpha) - \mu\cos\alpha]$$

$$w = S[(1 - \mu^2)\sin\alpha - 2\mu\cos\alpha]. \quad (3)$$

Also taking moment about A, we have $S \cdot AD = w \cdot AM$.

In $\triangle ACD$ and $\triangle AGM$,

$$\sin\alpha = \frac{CD}{AD} = \frac{2r}{AD} \text{ [let } CD = 2r]$$

$$AD = 2r \operatorname{cosec}\alpha.$$

In $\triangle AGM$,

$$\sin\alpha = \frac{AM}{AG} = \frac{AM}{a}$$

$$AM = a\sin\alpha \text{ [Let } AB = 2a,\ AG = GB = a]$$

Now $S \cdot AD = w \cdot AM$

$$\therefore S.2r \operatorname{cosec}\alpha = wa\sin\alpha$$

$$S.2r \operatorname{cosec}\alpha = a\sin\alpha S\left[\left(1-\mu^2\right)\sin\alpha - 2\mu\cos\alpha\right]$$

$$\text{or, } 2r = a\sin^2\alpha[(1-\mu^2)\sin\alpha - 2\mu\cos\alpha]. \quad (4)$$

Again, when the rod AB is making the least angle β with the vertical, then its end A is on the point of slipping downwards along the vertical side of tumbler and G inside the tumbler.

For this case substituting β for α and μ for μ in (4), we get

$$2r = a\sin^2\beta\left[\left(1-\mu^2\right)\sin\beta + 2\mu\cos\beta\right]. \quad (5)$$

From (4) and (5), we have

$$a\sin^2\alpha\left[\left(1-\mu^2\right)\sin\alpha - 2\mu\cos\alpha\right] = a\sin^2\beta\left[\left(1-\mu^2\right)\sin\beta + 2\mu\cos\beta\right]$$

$$\text{or, } \left(1-\mu^2\right)\left(\sin^3\alpha - \sin^3\beta\right) = 2\mu\left(\sin^2\beta\cos\beta + \sin^2\alpha\cos\alpha\right)$$

$$\text{or, } \frac{2\mu}{1-\mu^2} = \frac{\sin^3\alpha - \sin^3\beta}{\sin^2\alpha\cos\alpha + \sin^2\beta\cos\beta} \text{ or, } \tan 2\lambda = \frac{\sin^3\alpha - \sin^3\beta}{\sin^2\alpha\cos\alpha + \sin^2\beta\cos\beta}$$

$$\text{or, } \lambda = \frac{1}{2}\tan^{-1}\left[\frac{\sin^3\alpha - \sin^3\beta}{\sin^2\alpha\cos\alpha + \sin^2\beta\cos\beta}\right] \text{ (Proved).}$$

5. A sphere whose radius is a and whose CG is at a distance c from the centre rests in limiting equilibrium on a rough plane inclined at any angle α to the horizon. Show that it may be turned through an angle $2\cos^{-1}\left(\frac{a\sin\alpha}{c}\right)$ and still be in equilibrium.

Solution: The normal reaction and the force of friction pass through A; hence, the weight must also act along the vertical through A and let it meet the horizontal through O in L. If G_1 be the position of CG, then $OG_1 = c$.

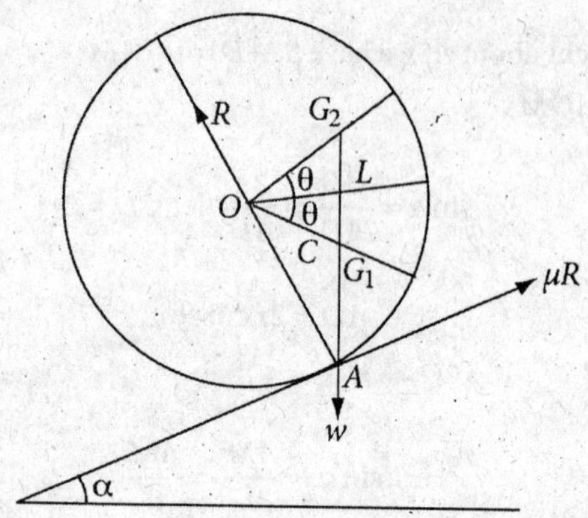

Fig. 8.9

In ΔOG_1L,

$$\cos\theta = \frac{OL}{OG_1} = \frac{OL}{c}.$$

$$\left.\begin{aligned} OL &= c\cos\theta \\ OG_1 &= c \end{aligned}\right\}. \qquad (1)$$

Resolving along and perpendicular to the plane, we get

$$\mu R = w\sin\alpha \text{ and } R = w\cos\alpha. \qquad (2)$$

Taking moment about O, we get $\mu Ra = wOL$.

$$\mu Ra = wc\cos\theta \text{ [from (1)]}$$

$$\text{or, } w\sin\alpha = wc\cos\theta \text{ [from (2)]}$$

$$\text{or, } \cos\theta = \frac{a}{c}\sin\alpha \text{ or, } \theta = \cos^{-1}\left(\frac{a}{c}\sin\alpha\right). \qquad (3)$$

The above equation of equilibrium remains unchanged when θ is replaced by $-\theta$ or G_1 takes the position G_2 and hence the sphere remains in equilibrium when it is rotated through an angle 2θ or

$$2\cos^{-1}\left(\frac{a}{c}\sin\alpha\right) \text{ [from (3)].}$$

6. A heavy uniform elliptical wire of semi-axes a, b is hung over a small rough peg. Show that if the wire can be in equilibrium with any point in contact with the peg, the coefficient of friction cannot be less than $\frac{a^2-b^2}{2ab}$. [CU(H) 2007]

Solution: Since the equilibrium is possible for all positions of P, the force of friction will not be maximum. The weight w acts vertically downwards and hence, the resultant S of normal reaction R and force of friction (not max) F should act vertically up and let it be inclined at an angle θ to R, then clearly λ = angle of friction $\geq \theta$ or, $\tan\lambda \geq \tan\theta$ or, $\mu \geq \tan\theta$.

$\therefore \mu \geq$ max value of $\tan\theta$ for all positions of P.

Now

$$\tan\theta = \tan(\alpha-\beta) = \frac{\tan\alpha - \tan\beta}{1+\tan\alpha\tan\beta};$$

$$\tan\theta = \frac{m_1 - m_2}{1+m_1 m_2}. \quad (1)$$

Equation of normal at $P(a\cos\phi, b\sin\phi)$ is $ax\sec\phi - by\operatorname{cosec}\phi = a^2 - b^2$ and its slope is

Fig. 8.10

$$m_1 = \tan\alpha = \frac{a\sec\phi}{b\operatorname{cosec}\phi} = \frac{a}{b}\tan\phi;$$

$$m_2 = \tan\beta = \text{slope of } CP = \frac{b\sin\phi}{a\cos\phi} = \frac{b}{a}\tan\phi.$$

So from (1),

$$\tan\theta = \frac{\left(\frac{a}{b} - \frac{b}{a}\right)\tan\phi}{1+\tan^2\phi} = \frac{a^2-b^2}{ab}\cdot\frac{\sin^2\phi}{2}.$$

Now max value of $\sin 2\phi = 1$.

$\therefore$ max value of $\tan\theta = \frac{a^2-b^2}{ab}$.

$$\therefore \mu \geq \frac{a^2-b^2}{2ab}.$$

7. A hemisphere shell rests on a rough plane, whose angle of frictions is λ, show that the inclination of the plane base of the rim to the horizontal cannot be greater than $\sin^{-1}(2\sin\lambda)$. **[RU(H) 1989]**

Solution: Let θ be the angle which the plane base makes with the horizontal, G the CG of the shell, O the centre of the plane base through which normal reaction R passes and A the point of contact.

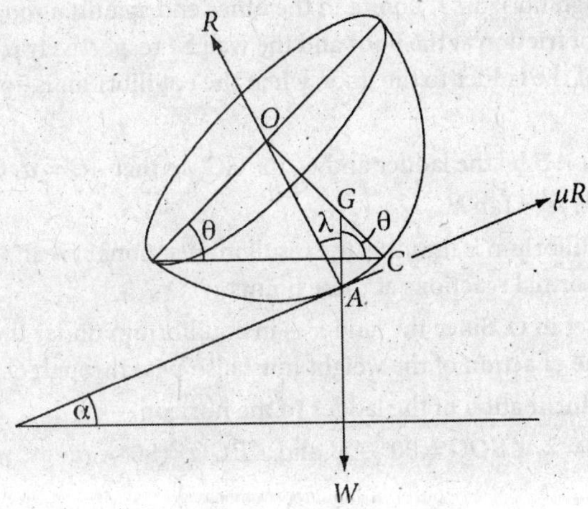

Fig. 8.11

Then $\angle AGC = \theta$ and $\angle OAG = \lambda = \alpha =$ inclination of the plane to the horizon.
From ΔAOG,

$$\frac{AO}{\sin(\pi-\theta)} = \frac{OG}{\sin\lambda} \Rightarrow \frac{a}{\sin\theta} = \frac{a}{2\sin\lambda} \Rightarrow \sin\theta = 2\sin\lambda \Rightarrow \theta = \sin^{-1}(2\sin\lambda)\ \textbf{(Proved)}.$$

8. A solid hemisphere of weight W rests in limiting equilibrium with its curved surface on a rough inclined plane and its plane face is kept horizontal by a weight P attached to a point on its rim. Prove that the coefficient of friction is $\frac{P}{\sqrt{W(2P+W)}}$. **[RU(H) 2000]**

Solution: Let the plane be inclined at an angle α to the horizon, the reaction and the force of friction at the point of contact being R and μR. Resolving horizontally,

$$R\sin\alpha = \mu R\cos\alpha \Rightarrow \mu = \tan\alpha.$$

If O be the centre, A the point of contact and OB vertical, then $\angle AOB = \alpha$, hence if a be the radius of the hemisphere, we have on taking moments about A,

$$W \cdot a\sin\alpha = P(a - a\sin\alpha) \Rightarrow \sin\alpha = \frac{P}{W+P}.$$

$$\therefore\ \mu = \tan\alpha = \frac{\sin\alpha}{\sqrt{1-\sin^2\alpha}}$$

$$= \frac{P}{\sqrt{(W+P)^2 - P^2}}$$

$$= \frac{P}{\sqrt{W(W+2P)}}\ \textbf{(Proved)}.$$

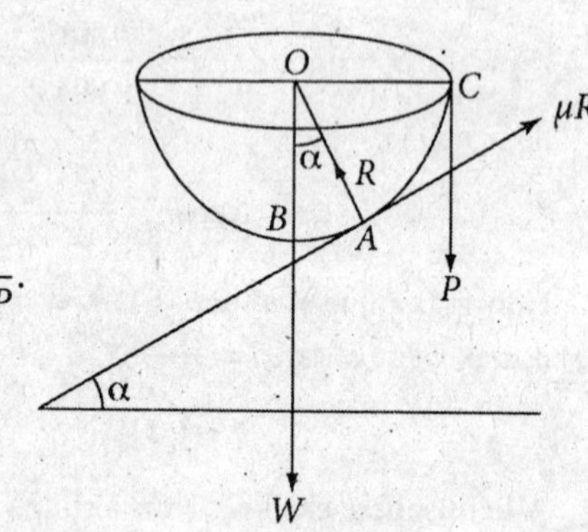

Fig. 8.12

9. A ladder whose CG divides it into two portions of lengths a and b rests with one end on a rough horizontal floor and the other end against a rough vertical wall. If the coefficients of friction at the floor and the wall be respectively μ and μ', show that the inclination of the ladder to the floor, when the equilibrium is limiting is $\tan^{-1}\frac{a-b\mu\mu'}{(a+b)\mu}$.

[RU(H) 1998]

Solution: Let AB be the ladder and G, its CG, so that $AG = a$, $GB = b$.
Let $\mu = \tan\lambda$, $\mu' = \tan\lambda'$.
Since the equilibrium is limiting, the resultant reactions at A and B make angles λ and λ' with the normal reactions at these points.
Let them meet in O. Since the ladder is in equilibrium under the action of only three forces the line of action of the weight must also pass through O.
Let α be the inclination of the ladder to the horizon.
Then $\angle AOG = \lambda$, $\angle BOG = 90° - \lambda'$ and $\angle BGO = 90° - \alpha$.

By mn theorem in ΔAOB, we have

$$(a+b)\cot(90^\circ-\alpha)=a\cot\lambda-b\cot(90^\circ-\lambda')$$

$$\Rightarrow (a+b)\tan\alpha=a\cot\lambda-b\tan\lambda'$$

$$\Rightarrow (a+b)\tan\alpha=a.\frac{1}{\mu}-b\mu'$$

$$\Rightarrow (a+b)\tan\alpha=\frac{a-b\mu\mu'}{\mu}$$

$$\Rightarrow \tan\alpha=\frac{a-b\mu\mu'}{\mu(a+b)}$$

$$\Rightarrow \alpha=\tan^{-1}\frac{a-b\mu\mu'}{(a+b)\mu}$$ **(Proved).**

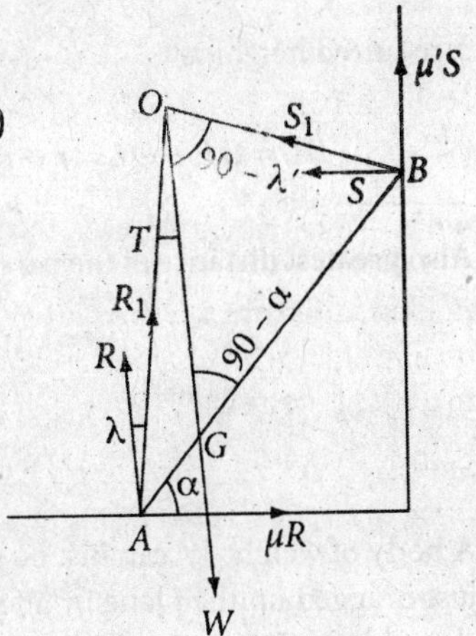

Fig. 8.13

10. How high can a particle rest inside a hollow sphere of radius r, μ being the coefficient of friction α. Also show that the greatest distance of the particle from the vertical diameter is $\frac{\mu r}{\sqrt{1+\mu^2}}$. **[RU(H) 1999]**

Solution: Let O be the centre and W be the weight of the sphere. Let P be the limiting position of equilibrium of the particle and R be the normal reaction between the particle and the sphere acting in the direction PO, as all the normals of a sphere passes through its centre.

Also as the particle is on the point of slipping downwards, so the force of friction μR is acting upwards at right angles to OP, i.e., tangentially to the sphere.

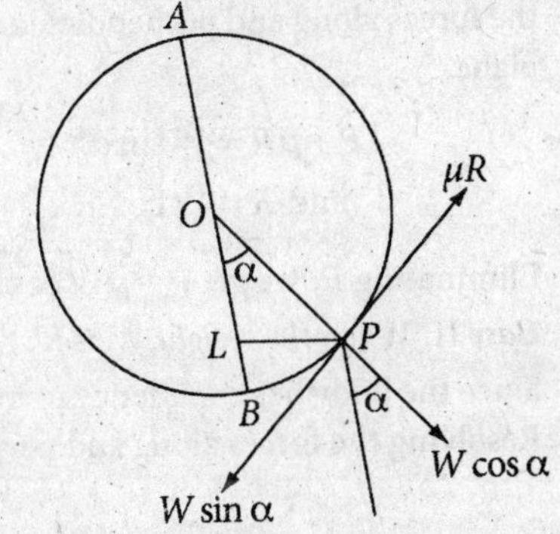

Fig. 8.14

From P, we draw PL perpendicular to the vertical diameter AB of the sphere. $W\cos\alpha$ and $W\sin\alpha$ are the components of W in the normal and tangential directions at P.

Let $\angle POB=\alpha$. Resolving forces along and perpendicular to PO, we get

$$R=W\cos\alpha \tag{1}$$

$$\text{and } \mu R=W\sin\alpha. \tag{2}$$

Dividing (2) by (1), we get $\mu=\tan\alpha$.

Now in ΔOPL,

$$\cos\alpha=\frac{OL}{OP}=\frac{OL}{r}\Rightarrow OL=r\cos\alpha.$$

$\therefore$ required height is

$$BL = OB - OL = r - r\cos\alpha = r\left[1 - \frac{1}{\sec\alpha}\right] = r\left[1 - \frac{1}{\sqrt{1+\mu^2}}\right].$$

Also greatest distance of the particle from vertical diameter $= PL = OP\sin\alpha = r\sin\alpha$.

$$\because \tan\alpha = \mu \Rightarrow \sin\alpha = \frac{\mu}{\sqrt{1+\mu^2}},$$

$$\therefore PL = \frac{\mu r}{\sqrt{1+\mu^2}}.$$

11. A body of weight W can just be sustained on a rough inclined plane by a force P and just dragged up the plane by a force Q; P, Q both acting along the line of the greatest slope, show that the coefficient of the friction is $\frac{Q-P}{\sqrt{4W^2-(P+Q)^2}}$. [KU(H) 2001]

Solution: *Case* I. *When the weight W is just supported by the force P.*

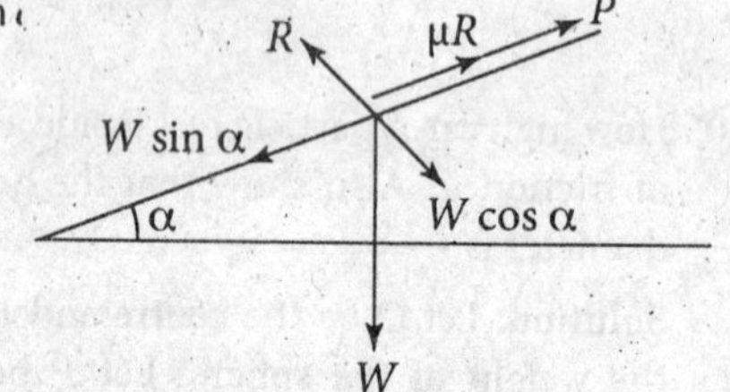

Fig. 8.15

Since the body has a tendency to move down, the friction μR acts upwards, where R is the normal reaction. Resolving the forces along and perpendicular to the plane.

$$P + \mu R = W\sin\alpha \qquad (1)$$

$$\text{and } R = W\cos\alpha. \qquad (2)$$

Eliminating R, we get $P + \mu W\cos\alpha = W\sin\alpha \Rightarrow P = W(\sin\alpha - \mu\cos\alpha)$.

Case II. *When the weight W is just dragged by by the force Q.*

Since the body has a tendency to move up the plane, the friction μR acts downwards. Resolving the forces along and perpendicular to the plane,

$$Q = \mu R + W\sin\alpha \qquad (4)$$

$$\text{and } R = W\cos\alpha. \qquad (5)$$

Eliminating R, we get

$$Q = \mu W\cos\alpha + W\sin\alpha$$

$$\Rightarrow Q = W(\mu\cos\alpha + \sin\alpha). \qquad (6)$$

Adding and subtracting (3) and (6), we get

$$P + Q = 2W\sin\alpha, \quad \frac{Q-P}{\mu} = 2W\cos\alpha.$$

Squaring and adding, we get

$$(P+Q)^2 + \frac{(Q-P)^2}{\mu^2} = 4W^2\left(\sin^2\alpha + \cos^2\alpha\right) \Rightarrow \frac{(Q-P)^2}{\mu^2} = 4W^2 - (P+Q)^2$$

$$\Rightarrow \mu^2 = \frac{(Q-P)^2}{4W^2-(P+Q)^2} \Rightarrow \mu = \frac{Q-P}{4W^2-(P+Q)^2}.$$

12. Two equal uniform ladders are jointed at one end and stand with the other ends on a rough horizontal plane. A man whose weight is equal to that of the ladders ascends one of them. Prove that the other will slip first. Supposing that it slips when he has ascended a distance x, prove that the coefficient of friction is $\left(\frac{a+x}{2a+x}\right)\tan\alpha$, a being the length of each ladder, and α the angle which each makes with the vertical.[CU(H) 2001]

Solution: Let M be the position of the man on the ladder AB at any instant when $BM=x, R, R'$ the normal reactions on the ground and F, F' the frictions at the instant at B and C respectively.

Now considering the two ladders AB and AC as forming one system action and reaction at A on the two neutralise each other. Hence, resolving horizontally and vertically and taking moment about B for the equilibrium of the combined system, we get

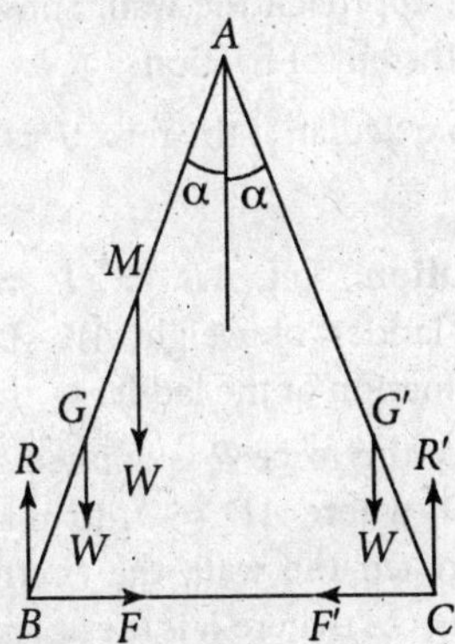

Fig. 8.16

$$F=F' \qquad (1)$$

$$R+R'=3W \qquad (2)$$

and

$$R'\cdot 2a\sin\alpha = W\cdot\tfrac{1}{2}a\sin\alpha + W\cdot x\sin\alpha + W\cdot\tfrac{3}{2}a\sin\alpha. \qquad (3)$$

Again, considering the equilibrium of the ladder AC separately, taking moment about A whereby the unknown action at A of AB on AC will not enter the equation, we get

$$F'\cdot a\cos\alpha = R'\cdot a\sin\alpha - W\cdot\tfrac{a}{2}\sin\alpha. \qquad (4)$$

From (3),

$$R'=\frac{W}{2\cdot 2a}[a+2x+3a] \Rightarrow R'=\frac{2W(2a+x)}{2\cdot 2a} \Rightarrow R'=\frac{W(2a+x)}{2a}.$$

Then from (2),

$$R=3W-R'-3W-\frac{W(2a+x)}{2a}=\frac{W(6a-2a-x)}{2a}=\frac{W(4a-x)}{2a}.$$

Hence, from (1) and (4),

$$Fa\cos\alpha = a\sin\alpha\left[R'-\frac{W}{2}\right] \Rightarrow F=\tan\alpha\left[\frac{W(2a)+x}{2a}-\frac{W}{2}\right]$$

$$\Rightarrow F=\frac{W\tan\alpha}{2a}[2a+x-a] \Rightarrow F=\frac{W\tan\alpha(a+x)}{2a}.$$

Thus,

$$\frac{F}{R}=\left(\frac{a+x}{4a-x}\right)\tan\alpha \text{ and } \frac{F}{R'}=\tan\alpha\left(\frac{a+x}{2a+x}\right).$$

Now $(4a-x)-(2a+x)=2(a-x)$ is positive, for $x<a$.

Hence, $\frac{F}{R'} > \frac{F}{R}$ for all values of x.

One of the ladders will slip when either $\frac{F}{R}$ or $\frac{F'}{R'}$ is equal to the coefficient of friction μ and as $\frac{F'}{R'}$ is greater, this will attain the value μ first. Thus, the other ladder AC will slip first and the coefficient of friction is connected to x by the relation $\mu = \frac{F'}{R'} = \frac{a+x}{2a+x}\tan\alpha$.

13. A uniform ladder of length l and weight W rests with its foot on a rough ground and its upper end against a smooth vertical wall, its inclination to the vertical being θ. A force P is applied to it horizontally at a distance c from the foot so as to make the foot approach the wall. Show that P must exceed $\frac{Wl}{l-c}\left(\mu + \frac{1}{2}\tan\theta\right)$, where μ is the coefficient of friction.

In particular, if the force P is applied at the foot, show that P is $W\left(\mu + \frac{1}{2}\tan\theta\right)$.

[CU(H) 1985]

Solution: Let $AB = l =$ length of the ladder of weight W. $\angle ABO = \theta$, inclination of the ladder to the vertical.

Since the force P is applied horizontally at D, where $AD = c$, to make the foot approach the wall, the friction μR acts along OA, where R is the normal reaction. Resolving the forces along vertical,

$$W = R. \tag{1}$$

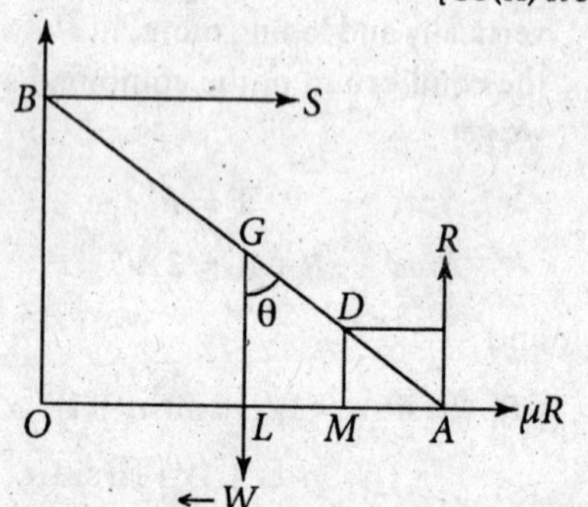

Fig. 8.17

Taking moment about B, we have

$$R\cdot OA + \mu R\cdot OB = W\cdot OL + P\cdot OM$$

$$\Rightarrow R\cdot l\sin\theta + \mu R\cdot l\cos\theta = W\cdot\frac{l}{2}\sin\theta + P(l-c)\cos\theta \tag{2}$$

$$\Rightarrow Wl\sin\theta + \mu Wl\cos\theta = W\cdot\frac{l}{2}\sin\theta + P(l-c)\cos\theta$$

$$\Rightarrow \frac{Wl}{2}\sin\theta + \mu Wl\cos\theta = P(l-c)\cos\theta$$

$$\Rightarrow P(l-c) = \frac{Wl}{2}\tan\theta + \mu Wl$$

$$\Rightarrow P(l-c) = Wl\left(\mu + \frac{1}{2}\tan\theta\right)$$

$$\Rightarrow P = \frac{Wl}{l-c}\left(\mu + \frac{1}{2}\tan\theta\right). \tag{3}$$

For the second part: If P is applied horizontally at the foot, we have $c = 0$, so that from (3)

$$P = W\left(\mu + \tfrac{1}{2}\tan\theta\right).$$

14. A heavy rod of length $2a$ lines over a rough peg, with one extremity leaning against a rough wall, θ being its inclination to the wall, if c be the distance of the peg from the wall and λ the angle of friction both at the peg and at the wall, show that

(i) If the point of contact of the rod and the wall be below the peg, then the rod is on the point of slipping downwards when $\sin^2 \alpha \sin(\theta + 2\lambda) = \frac{c}{a}\cos^2\lambda$, and when it is on the point of slipping upwards, then $\sin^2\theta \cdot \sin(\theta - 2\lambda) = \frac{c}{a}\cos^2\lambda$.

(ii) If the point of contact of the rod with the wall is above the peg, then the rod is on the point of sliding downwards when $\sin^3\theta = \frac{c}{a}\cos^2\lambda$.

(iii) If a be the distance of the peg from the wall, show that $\sin^3\theta = \frac{1}{1+\mu^2}$, where μ is the coefficient of friction.

Solution: (i) Let AB be a heavy rod of length $2a$ which lies on a rough peg at P. If the rod is on the point of slipping down, then the forces of friction are marked in the direction along which they will act.

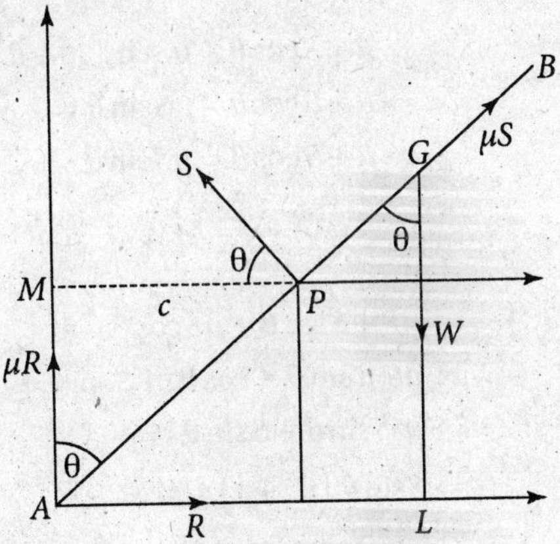

$MP = c$.

In $\triangle APM$,

$$\sin\theta = \frac{PM}{AP} = \frac{C}{AP} \Rightarrow AP = c\,\text{cosec}\,\theta$$

and $AG = a$.

In $\triangle AGL$,

$$\sin\theta = \frac{AL}{AG} = \frac{AL}{a} \Rightarrow AL = a\sin\theta.$$

Fig. 8.18

Resolving horizontally and vertically, we get

$$R + \mu S\cos(90° - \theta) = S\cos\theta \Rightarrow R + \mu S\sin\theta = S\cos\theta. \quad (1)$$

and

$$\mu R + S\sin\theta + \mu S\cos\theta = W. \quad (2)$$

Eliminating R between (1) and (2), we get

$$S\mu(\cos\theta - \mu\sin\theta) + S\sin\theta + \mu S\cos\theta = W$$

$$\Rightarrow S\sin\theta\left(1+\mu^2\right) + 2\mu S\cos\theta = W$$

$$\Rightarrow S\frac{\sin\theta\,(1+\mu^2)}{1+\mu^2} + \frac{2\mu S\cos\theta}{1+\mu^2} = \frac{W}{1+\mu^2}$$

$$\Rightarrow S\sin\theta\frac{1+\tan^2\lambda}{1+\tan^2\lambda} + \frac{2\tan\lambda}{1+\tan^2\lambda}S\cos\theta = \frac{W}{1+\tan^2\lambda}$$

$$\Rightarrow S\sin\theta\cos 2\lambda + S\cos\theta\sin 2\lambda = W\cos^2\lambda$$

$$\Rightarrow S\sin(\theta + 2\lambda) = W\cos^2\lambda. \quad (3)$$

Taking moment of all the forces about A, we get

$$S \cdot AP = W \cdot AL \Rightarrow Sc\,\text{cosec}\,\theta = W \cdot a\sin\theta \Rightarrow S = \frac{Wa}{c}\sin^2\theta. \quad (4)$$

Putting the value of S from (4) in (3), we get

$$\frac{Wa}{c}\sin^2\theta\sin(\theta+2\lambda)=W\cos^2\lambda\Rightarrow\sin^2\theta\sin(\theta+2\lambda)=\frac{c}{a}\cos^2\lambda. \qquad (5)$$

When the rod is on the point of slipping upwards, then forces of frictions will act in opposite directions. Then either proceeding as above or changing λ into $-\lambda$, we get from (5),

$$\sin^2\theta\sin(\theta-2\lambda)=\frac{c}{a}\cos^2\lambda.$$

(ii) Resolving the forces horizontally and vertically, we get

$$R+S\cos\theta+\mu S\cos(90+\theta)=0$$
$$\Rightarrow R+S\cos\theta-\mu S\sin\theta=0$$
$$\Rightarrow R+S\cos\theta=\mu S\sin\theta \qquad (6)$$

and

$$\mu R+S\sin\theta+\mu S\cos\theta=W$$
$$\Rightarrow \mu S(\mu\sin\theta-\cos\theta)+S\sin\theta+\mu S\cos\theta=W$$
$$\Rightarrow \mu^2 S\sin\theta+S\sin\theta=W$$
$$\Rightarrow S\sin\theta\left(\mu^2+1\right)=W. \qquad (7)$$

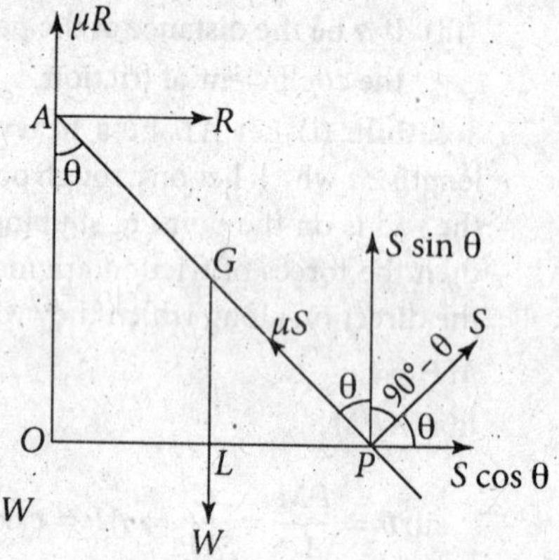

Fig. 8.19

Taking moment of all the forces about A, we get

$$S\cdot AP-W\cdot ML=0\Rightarrow Sc\operatorname{cosec}\theta=W\cdot a\sin\theta\Rightarrow S=\frac{Wa}{c}\sin^2\theta. \qquad (8)$$

From (7) and (8), we get

$$\frac{Wa}{c}\sin^3\theta\left(1+\tan^2\lambda\right)=W\Rightarrow\frac{Wa}{c}\sin^3\theta\sec^2\lambda=W\Rightarrow\sin^3\theta=\frac{c}{a}\cos^2\lambda.$$

(iii) If $c=a$, then

$$\sin^2\theta=\cos^2\lambda=\frac{1}{1+\tan^2\lambda}=\frac{1}{1+\mu^2}.$$

15. An equilateral triangle of uniform thickness and density rests with one end of its base on a rough horizontal plane and the other against a smooth vertical wall. Show that the least angle that its base can make with the horizontal plane is given by the equation $\cot\theta=2\mu+\frac{1}{\sqrt{3}}$, μ being the coefficient of friction. [CU(H) 1998]

Solution: AB is the base of the triangle ABC which is resting on a rough horizontal plane and smooth vertical wall. AB will make the least angle with the horizontal when the triangle ABC is on the point of motion and the force of friction μR will be as marked in the figure, where R is the normal reaction at A. Let R' be the resultant reaction at A making an angle λ with R. The resultant reaction R' at A, the reaction S at B and the weight W of the triangle meet at O. Let the line of action of W acting at G, the CG of $\triangle ABC$ meet AB in K.

Then $\angle GKB = 90^\circ - \theta$.

Also $CF = AC\sin 60^\circ = a\sqrt{3}$, where $AB = 2a$ (say) and F is the midpoint of AB.

$$\therefore\ GF = \frac{1}{3}CF = \frac{1}{3}a\sqrt{3} = \frac{a}{\sqrt{3}}.$$

In ΔGKF,

$$KF = GF\tan\theta = \frac{a}{\sqrt{3}}\tan\theta.$$

$$\therefore\ AK = AF - KF = a - \frac{a}{\sqrt{3}}\tan\theta.$$

Fig. 8.20

Also $\angle AOK = \lambda$ and $\angle BOK = 90^\circ$.

Applying $m-n$ formula for ΔAOB, we have

$$(AK+KB)\cot(90^\circ-\theta) = AK\cot\lambda - Bk\cot 90$$

$$\Rightarrow AB\tan\theta = \left[a - \frac{a}{\sqrt{3}}\tan\theta\right]\cot\lambda \Rightarrow 2a\tan\theta = a\left(1-\frac{1}{\sqrt{3}}\tan\theta\right)\frac{1}{\tan\lambda}$$

$$\Rightarrow 2\tan\theta = \left(1-\frac{1}{\sqrt{3}}\tan\theta\right)\frac{1}{\mu} \Rightarrow \left(2+\frac{1}{\mu\sqrt{3}}\right)\tan\theta = \frac{1}{\mu}$$

$$\Rightarrow \left(2\mu+\frac{1}{\sqrt{3}}\right) = \frac{1}{\tan\theta} = \cot\theta \Rightarrow \cot\theta = 2\mu + \frac{1}{\sqrt{3}}.$$

16. A straight uniform beam of length $2h$ rests in limiting equilibrium, in contact with a rough vertical wall of height h with one end on a rough horizontal plane and with the other end projecting beyond the wall. If both the wall and the plane be equally rough, prove that λ the angle of friction is given by $\sin 2\lambda = \sin 2\alpha \sin\alpha$, α being the inclination of the beam to the horizon. **[IAS 2008]**

Solution: Let W be the weight of the beam acting at the middle point O of the beam, which rests at A and B on the horizontal plane and the vertical wall respectively. If R and S be the reactions of the plane and the wall respectively, the forces of friction μR and μS act along the horizontal plane and the beam respectively.

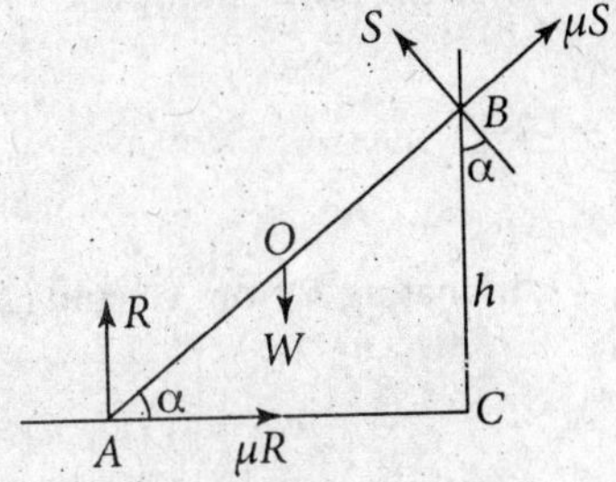

Fig. 8.21

Resolving horizontally and vertically, we get

$$\mu R + \mu S\cos\alpha \quad S\sin\alpha - 0 \qquad (1)$$

$$\text{and } R + \mu S\sin\alpha + S\cos\alpha = W. \qquad (2)$$

Taking moments about A, we get

$$W\cdot h\cos\alpha=S\cdot AB\Rightarrow Wh\cos\alpha=S\cdot h\operatorname{cosec}\alpha\Rightarrow S=\frac{W}{2}\sin 2\alpha. \quad (3)$$

Eliminating R from (1) and (2), we have

$$\mu[W-\mu S\sin\alpha-S\cos\alpha]+\mu S\cos\alpha-S\sin\alpha=0\Rightarrow\mu(W-\mu S\sin\alpha)-S\sin\alpha=0$$

$$\Rightarrow\mu W=S\left(\mu^2+1\right)\sin\alpha\Rightarrow\mu W=\frac{W}{2}\sin 2\alpha\left(\mu^2+1\right)\sin\alpha \text{ [from (3)]}$$

$$\Rightarrow\sin 2\alpha\sin\alpha=\frac{2\mu}{\mu^2+1}=\frac{2\tan\lambda}{1+\tan^2\lambda}=\sin 2\lambda.$$

17. A ladder rests on a horizontal floor leans against a vertical wall. The coefficients of friction of the floor and the wall with the ladder are μ and μ_1 respectively. If a man, whose weight is n times that of the ladder, want to climb up the ladder, find the minimum safe angle of the ladder with the floor. [IAS 2003]

Solution: Let W be the weight of the ladder AB which acts at its midpoint G.

$\therefore AB=\frac{1}{2}AB.$

Let normal reactions at A and B be R and S respectively and a man of weight nW when reaches at B, the top of the ladder, then it is about to slip.

In the limiting equilibrium the force of friction μR at A will act in the direction of AC and the force of friction μ_1, S at B will act in the direction of CB. Let the inclination of the ladder with the horizontal in this case be θ.

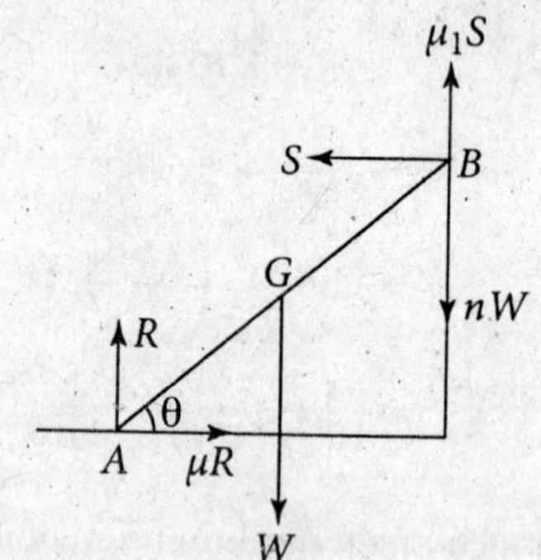

Fig. 8.22

Resolving horizontally and vertically all the forces, we get

$$\mu R=S \quad (1)$$

$$\text{and } R+\mu_1 S=W+nW. \quad (2)$$

Eliminating R from (1) and (2), we get

$$\frac{S}{\mu}+\mu_1 S=W(1+n)\Rightarrow S\left(1+\mu\mu_1\right)=\mu(1+n)W\Rightarrow S=\frac{\mu(1+n)W}{1+\mu\mu_1}. \quad (3)$$

Now taking moment of all the forces about A, we get

$$W\cdot AG\cos\theta+nW\cdot AB\cos\theta=S\cdot AB\sin\theta+\mu_1 S\cdot AB\cos\theta$$

$$\Rightarrow W\cdot\tfrac{1}{2}AB\theta+nW\cdot AB=S\cdot AB\tan\theta+\mu_1 S\cdot AB$$

$$\Rightarrow W(1+2n)=2S\left(\tan\theta+\mu_1\right)$$

$$\Rightarrow W(1+2n)=\frac{2\mu(1+n)W}{1+\mu\mu_1}(\tan\theta+\mu_1)\ \text{[from (3)]}$$

$$\Rightarrow (1+2n)(1+\mu\mu_1)=2\mu(1+n)(\tan\theta+\mu_1)$$

$$\Rightarrow (1+2n)(1+\mu\mu_1)=2\mu(1+n)\tan\theta+2\mu\mu_1(1+n)$$

$$\Rightarrow 1+\mu\mu_1+2n+2n\mu\mu_1-2\mu\mu_1-2n\mu\mu_1=2\mu(1+n)\tan\theta$$

$$\Rightarrow \frac{(1+2n-\mu\mu_1)}{2\mu(1+n)}=\tan\theta \Rightarrow \theta=\tan^{-1}\left[\frac{1+2n-\mu\mu_1}{2\mu(1+n)}\right].$$

EXERCISE 8

1. A uniform rod AB is lying in a horizontal position upon two parallel horizontal equally rough rods and is at right angles to them. The points of contact are C and D, C being nearest to A. A gradually increasing force is applied to the rod AB at A parallel to the other two rods. Show that if $\frac{4}{AB} < \frac{1}{AC} + \frac{1}{AC}$ equilibrium will be ultimately broken by the rod turning round D and slipping at C.
2. A horizontal lamina has two points A, B projecting from it which can slide in two straight slots at right angles to one another. If a couple be applied to the lamina, prove that it will not move unless the sum of the angles of friction at A and B falls short of 90°.
3. A uniform ladder of length 70 ft rests against a vertical wall with which it makes an angle of 45°, the coefficient of friction between the ladder and the wall is $\frac{1}{3}$ and that between the ladder and the floor is $\frac{1}{2}$. If a man whose weight is one half that of the ladder ascends it, how high will be he when the ladder slips?
4. A uniform ladder rests at an angle of 45° with the horizon, with its extremity against a rough vertical wall and its lower extremity on the ground. μ and μ_1 are the coefficients of limiting friction between the ladder and ground and the ladder and the wall, respectively. Show that the least horizontal force which will move the lower extremity towards the wall is $\frac{1}{2}w\frac{1+2\mu-\mu\mu'}{1-\mu'}$.
5. A ladder on a horizontal floor leans against a vertical wall. Show that if its CG is at the middle point of the ladder, the greatest inclination to the wall which is consistent with equilibrium is $\tan^{-1}\frac{2\mu}{1-\mu^2}$, where μ is the coefficient of friction for contact both with the floor and the wall. Or show that the inclination of the ladder to the wall is twice the angle of friction. Prove further that the lines of action of resultant forces acting at the foot and the top of the ladder are at right angles to one another.
6. A ladder of length $2l$ is in contact with a vertical wall and a horizontal floor, the angle of friction being λ at each contact. If the weight of the ladder acts at a point, distant kl below the mid-point, prove that its limiting inclination θ to the vertical is given by $\cot\theta=\cot 2\lambda-k\operatorname{cosec}2\lambda$.
7. A uniform rod AB is supported at an inclination α to the horizontal with its lower end B on a rough horizontal plane by a light string attached to A. Prove that the greatest inclination θ of the string to the vertical is given by $\cot\theta=\frac{1}{\mu}\pm 2\tan\alpha$.
8. A uniform rod rests at two rough pegs A and B. AB is inclined at angles α to the horizontal and the coefficients of friction are μ_1 and μ_2, respectively. If the rod is in limiting equilibrium and the centre of the rod divides AB into two parts a and b, then prove that $\tan\alpha=\frac{\mu_1 b+\mu_2 a}{a+b}$ and $\frac{a}{b}=\frac{\mu_1-\tan\alpha}{\tan\alpha-\mu_2}$.

9. A uniform rod MN has its ends in two fixed straight rough grooves OA and OB, in the same vertical plane, which makes angles α and β with the horizontal. Show that when the end M is on the point of slipping in the direction AO, the tangent of the angle of inclination to MN of the horizontal is $\frac{\sin(\alpha-\beta-2\lambda)}{2\sin(\beta+\lambda)\sin(\alpha-\lambda)}$, where λ is the angle of friction. Also find the normal reactions of the planes.
10. A uniform rod rests in a vertical plane within a fixed hemispherical bowl whose radius is equal to the length of the rod. μ is the coefficient of friction between the rod and the bowl. Show that in limiting equilibrium the inclination of the rod to the horizontal is $\tan^{-1}\frac{4\mu}{3-\mu^2}$.
11. A uniform rectangular board whose sides are $2a$ and $2b$ rests in limiting equilibrium in contact with two rough pegs in the same horizontal line at a distance d apart. Show that the inclination θ of the side $2a$ to the horizontal is given by the equation $d\cos\lambda\cos(\lambda+2\theta)=a\cos\theta-b\sin\theta$, where λ is the angle of friction.
12. A uniform rod of length $2a$ and weight w rests with its middle point upon a rough horizontal cylinder whose axis is perpendicular to the rod. Show that the greatest weight that can be attached to one end of the rod without sliding it off the cylinder is $\frac{b\lambda}{a-b\lambda}w$, where b is the radius of the cylinder and λ the angle of friction.
13. If a hemisphere rests in equilibrium with its curved surface in contact with a rough plane inclined to the horizon at an angle α, show that the inclination of the plane base of the hemisphere to the horizontal is $\sin^{-1}\left(\frac{8}{3}\sin\alpha\right)$, provided that α is less than $\sin^{-1}\frac{3}{5}$ and also less than the angle of friction. [CU(H) 2005]
14. A perfectly rough plane is inclined at an angle α to the horizon. Show that the least eccentricity of the ellipse which can rest on the plane is $\sqrt{\frac{2\sin\alpha}{1+\sin\alpha}}$. [IAS 1999, CU(H) 2008]
15. A rough cycloid (coefficient of friction μ) is placed with its axis vertical and vertex downwards. Find the lowest point on it from the base where a heavy particle will rest in equilibrium. The equations of the cycloid are $x=a(\theta+\sin\theta)$, $y=a(1-\cos\theta)$ referred to vertex as origin, the axis of y drawn vertically upwards. [CU(H) 1993, VU(H) 2004]
16. A cone rests on a rough table and a cord fastened to the vertex of the cone passes over a smooth pulley at the same height as the top of the cone and supports a weight. Show that, if the weight be continually increased, the cone will turn over or slide according as the coefficient of friction is $>$ or $<$ $\tan\alpha$, where α is the semi-vertical angle of the cone.
17. A flat heavy circular disc lies on a rough plane and can turn freely about a pin in its circumference. Show that it will rest in any position if the coefficient of friction is $> \frac{(9\pi\tan\alpha)}{32}$, where α is the inclination of the plane to the horizontal, it being assumed that the weight of the disc is uniformly distributed over its area.
18. A rod resting on a rough inclined plane, whose inclination α to the horizon is greater than the angle of friction λ, is free to turn about one of its ends, which is attached to the plane. Show that, for equilibrium, the greatest possible inclination of the rod to the line of greatest slope is $\sin^{-1}(\tan\lambda\cot\alpha)$. [RU(H) 1990]

Answers

3. $25\sqrt{2}$ ft. 9. $\frac{w\sin(\beta+\lambda)\cos\lambda}{\sin(\alpha+\beta)}$, $\frac{w\sin(\alpha-\lambda)\cos\lambda}{\sin(\alpha+\beta)}$. 15. $\frac{2a}{1+\mu^2}$.

Chapter 9

Bending Moment

9.1 Introduction

A structural element which is subjected to loads transverse to its axis is known as a *beam.* In general, a beam is either free from any axial force or its effect is negligible. Analysis of beams involves the determination of shear force, bending moment and depletion S at various sections.

Usually, a beam is considered horizontal and the loads vertical. Other cases are considered as exceptions. A concentrated or point load is assumed to act at a point, though in practice it may be disturbed over a small area.

A distributed load is one which is spread over some length of the beam. The rate of loading may be uniform or may vary from point of point.

9.2 Types of Supports and Beams

A beam may have the following kinds of supports:

I. Roller Support

When a beam rests on a sliding surface such as a roller or any flat surface like a masonry wall, the support is known as a *roller support.* A roller support can sustain as the possible movement on the supporting surface does not allow any resistance along that direction. Thus, if it has a reaction, normal to the surface only and the reaction along the rolling surface is zero. A roller support allows the rotation of the body. The conventional methods of showing a roller support are shown in Fig. 9.1(a).

II. Hinged Support and Hinged Joint

In a hinged support, no translational displacement of the beam is possible, however, it is free to rotate. A hinged support can sustain reactions in any direction, i.e., it can have components in vertical as well as horizontal directions. As a hinged support can rotate, in beams with hinged joints, the bending moment at the hinge is taken to be zero. Sometimes, beams are also provided with hinged joints. In such cases there is also no reaction which is there in case of a hinged support.

The conventional methods of showing a hinged support and a hinged joint are shown in Fig. 9.1(b).

III. Fixed or Encastre or Built-in-Support

A beam built into a rigid support, which does not allow any type of movement or rotation is known as *fixed* or *encastrè* or *built-in-support* [Fig. 9.1(c)]. A fixed support exerts a fixing moment and a reaction on the beam.

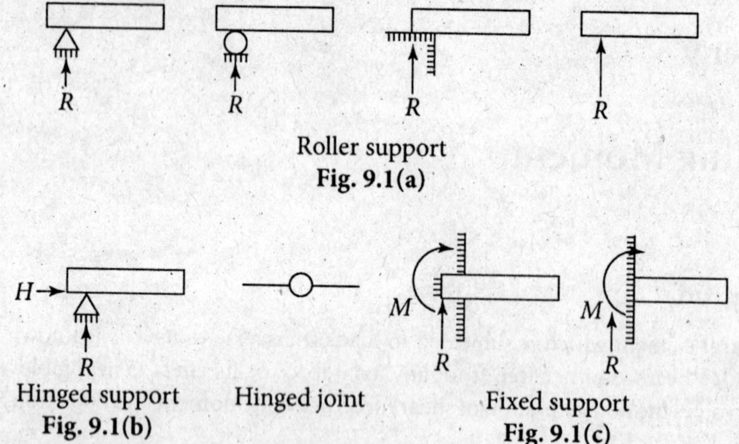

Roller support
Fig. 9.1(a)

Hinged support
Fig. 9.1(b)

Hinged joint

Fixed support
Fig. 9.1(c)

Depending upon the type of supports, the beams are classified as follows:

(i) **Simply supported beam.** When both supports of beams are roller supports or one support is roller and the other hinged, the beam is known as a *simply supported beam* [Fig. 9.2(a)].

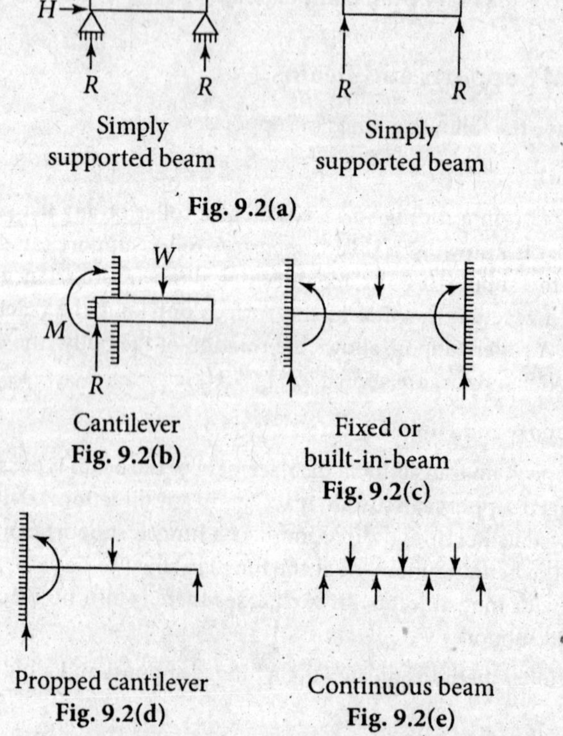

Simply supported beam

Simply supported beam

Fig. 9.2(a)

Cantilever
Fig. 9.2(b)

Fixed or built-in-beam
Fig. 9.2(c)

Propped cantilever
Fig. 9.2(d)

Continuous beam
Fig. 9.2(e)

(ii) **Cantilever beam.** A beam with one end fixed and the other end free is called a *cantilever* [Fig. 9.2(b)]. There is a vertical reaction and moment at the fixed end (known as *fixing moment*).

(iii) **Fixed beam.** A beam with both ends fixed is known as fixed beam [Fig. 9.2(c)].

(iv) **Propped cantilever beam.** Beams with one end fixed and the other simply supported are known as *propped cantilever beam* [Fig. 9.2(c)].

(v) **Continuous beam.** Beams being supported at more than two sections are known as *continuous beams* [Fig. 9.2(e)].

Generally, beams with more than two reaction components cannot be analysed using the equations of static equilibrium alone and are known as *statically indeterminate beams.*

9.3 Shear Force

Shear force is the unbalanced vertical force on one side (to the left or right) of a section of a beam and is the sum of all the normal forces on one side of the section. It also represents the tendency of either portion of the beam to slide or shear laterally relative to the other. Remember that a *force* at a section means the sum of all the forces on one side of the section.

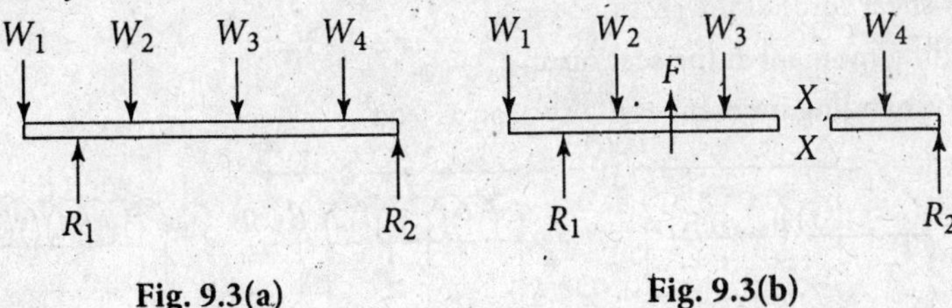

Fig. 9.3(a) **Fig. 9.3(b)**

Consider the beam as shown in the Fig. 9.3(a). It is simply supported at two points and carries four loads. The reactions at the supports are R_1 and R_2. Now if the beam is imagined to cut a section $x-x$ into two portions [Fig. 9.3(b)], the resultant of all the forces (loads as well as reaction of the support) to the left of the section is F (assuming upwards). Also as the beam is in equilibrium, the resultant of the forces to the right of $x-x$ must also be F downwards. Therefore, F is known as *shear* or *shearing force* (SF).

Shear force is considered positive when the resultant of the forces to the left of a section is upwards or to the right downwards.

A shear force diagram (SFD) *shows the variation of shear force along the length of a beam.*

9.4 Bending Moment

Bending moment at a section of a beam is defined as the algebraic sum of moments all the forces on one side of the section.

If the moment M about the section $x-x$ of all the forces to the left is clockwise (Fig. 9.4), then for the equilibrium, the moment of the forces to the right of $x-x$ must be M counter-clockwise.

Bending moment is considered positive if the moment on the left portion is clockwise or on the right portion counter-clockwise. This is usually referred to as sagging bending moment since it tends to cause concavity upwards. A bending moment causing convexity upwards is taken as negative bending moment and is called *hogging bending moment.*

A bending moment diagram (BMD) shows the variation of bending moment along the length of a beam.

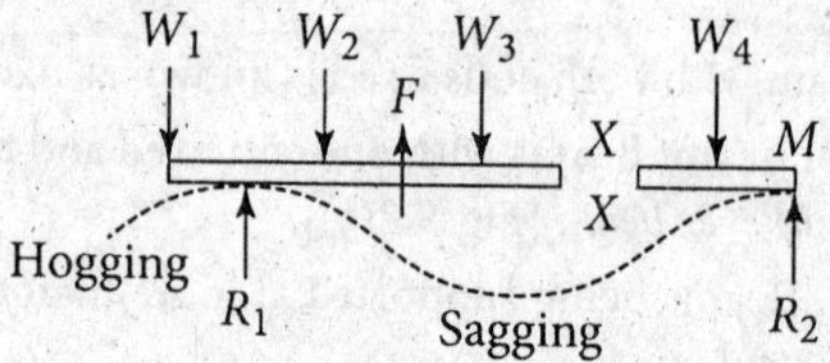

Fig. 9.4

9.5 Relation between Load, Shear Force and Bending Moment

Consider a small length δu cut out from a loaded beam at a distance x from a fixed origin O (Fig. 9.5). Let

W = mean rate of loading on the length δx

F = shear force at the section x

$F+\delta F$ = shear force at the section $x+\delta x$

M = bending moment at the section x

$M+\delta M$ = bending moment at the section $x+\delta x$

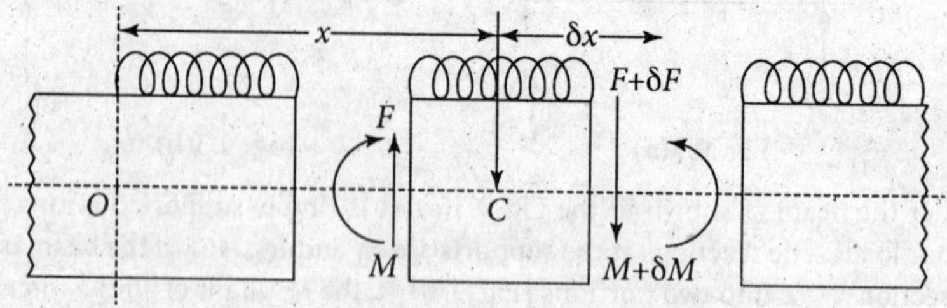

Fig. 9.5

Total load on the small length $= w\cdot\delta x$ acting approximately through the centre C (if the load is uniformly distributed it will be exactly acting through C).

For equilibrium of the element of length δx, equating vertical forces,

$$F=W\delta x+(F+\delta F)\Rightarrow W=-\frac{dF}{dx}\ \text{(taking limit when } \delta x\to 0), \qquad (1)$$

i.e., the *rate of change of shear force (or slope of the shear force curve) is equal to intensity of loading.*

Taking moment about C, we have

$$M+F\cdot\frac{\delta x}{2}+(F+\delta F)\cdot\frac{\delta x}{2}-(M+\delta M)=0.$$

Neglecting the product and squares of small quantities and taking limit when $\delta x\to 0$, we get

$$F=\frac{dM}{dx}, \qquad (2)$$

i.e., *rate of change of bending moment curve is equal to the shear force.*

Also from (1) and (2), we get

$$W = -\frac{d}{dx}\left(\frac{dM}{dx}\right) = -\frac{d^2M}{dx^2}.$$

Note. (1) At a point C, where $\frac{dM}{dx} = 0$, i.e., shear force is zero, the bending moment will have the maximum value.

(2) The point of zero bending moment, i.e., where the type of bending changes from sagging to hogging is called a *point of inflection* or *contra flexure*.

Integrating (1), between two values of x at two sections A and B, we have

$$F_a - F_b = \int_a^b w\,dx,$$

which is the area under the load distribution diagram.

Similarly, integrating (2) between two values of x at two sections A and B

$$M_b - M_a = \int_a^b F\,dx.$$

This shows that the variation of bending moment between two sections is equal to the area under the shear force diagram.

9.6 Bending of Rods

Consider S to be any section of a thin rod RSQ as a curved line. Now, let the components of the stress force along the tangent and normal at S be T and U, and L represents the mutual action of the two parts RS, QS of the rod on each other. Here T is the tension at S, positive towards the right; U, the shear positive towards the inward normal and L, the stress couple, or the bending moment, positive in the anticlockwise sense.

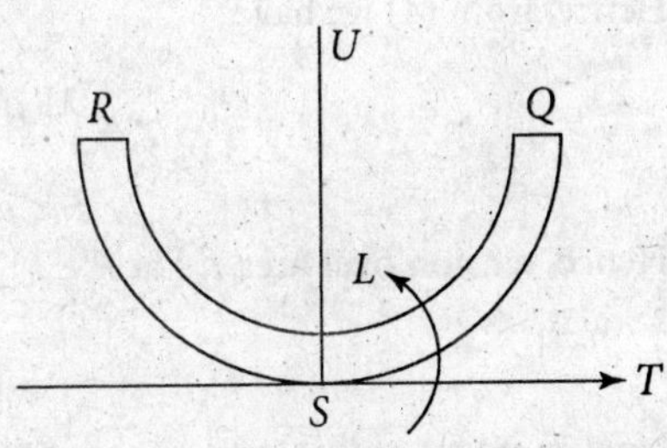

Fig. 9.6

But by experimental law, we have

$$L = k\left(\frac{1}{\rho} - \frac{1}{\rho_0}\right), \tag{1}$$

where k = constant, called the flexural rigidity, ρ_0 = radius of curvature of the rod when not deformed, ρ = that when deformed.

When the rod is initially straight, we have

$$\rho_0 = \infty \Rightarrow L = \frac{k}{\rho} \text{ [from (1)]} \tag{2}$$

The stress developed by the deformation must be some function of the deformation. Let any point be displaced through a distance s, then we have

F = stress = some function of $s = f(s) = f(0) + sf'(0) + \ldots$ (by Maclaurin's theorem).

But, when $s = 0$,

$$F = 0 \Rightarrow f(0) = 0 \Rightarrow F = sf'(0) \text{ (approx.)}$$

This shows that the displacement is a linear function.

In other words, all points lying in a place originally must lie on a plane after deformation.

When we confine our attention to a small portion only, it will have a definite radius of curvature, and the upper boundary will be expanded, while the lower boundary will get contracted. So there must exist some intermediate curve, which is neither expanded, nor contracted of separation on the surface.

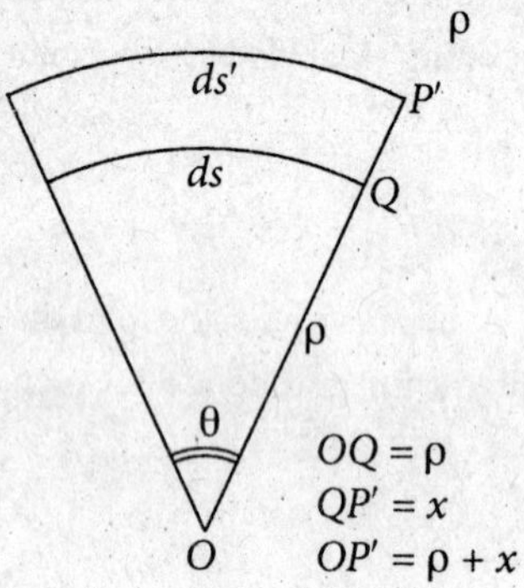

In an intermediate section P', we have

$$T = \frac{\lambda(ds' - ds)}{ds}. \quad (4)$$

ρ ds′ P′ ds Q ρ θ O OQ = ρ QP′ = x OP′ = ρ + x

Fig. 9.7

Now let O be the centre of curvature of the section Q such that $OQ = \rho$ and let $P'Q = x$ so that $ds' = (\rho + x)\theta, ds = \rho\theta$.

Hence, from (4) we have

$$T = \frac{\lambda[(\rho + x)\theta - \rho\theta]}{\rho\theta} = \frac{\lambda x}{\rho}.$$

Hence, tension over area ds at P'

$$= \frac{\lambda x}{\rho} ds. \quad (5)$$

Now, moment of this tension about Q

$$= \lambda \frac{x ds}{\rho} x = \frac{\lambda x^2}{\rho} ds \Rightarrow L = \Sigma\lambda \frac{x^2}{\rho} ds = \frac{\lambda}{\rho} \Sigma x^2 ds = \frac{\lambda I}{\rho}, \quad (6)$$

where $I = \Sigma x^2 ds$ is the moment of inertia of the cross section.

If $k = \lambda I$, then from (6), $L = \frac{k}{\rho}$.

9.7 To find the work done in twisting a rod

Let the form of the rod be a plane curve so that the plane of the couple is also the plane of the curve itself.

Now consider a length ds of the rod. Let ψ_0, ψ be the angles of contingence at the original position and at any other instant; then we have

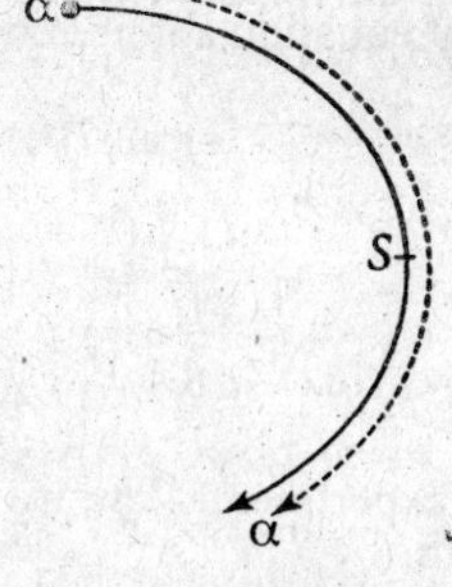

Fig. 9.8

$$\rho_0 = \frac{ds}{\psi_0},\ \rho = \frac{ds}{\psi} \Rightarrow L = k\left(\frac{1}{\rho} - \frac{1}{\rho_0}\right)$$

$$\Rightarrow L = \frac{k}{ds}(\psi - \psi_0).$$

Hence, the work done in twisting the rod from ψ_0 to ψ

$$= \int_{\psi_0}^{\psi} L d\psi = \frac{k}{ds}\int_{\psi_0}^{\psi}(\psi - \psi_0)\, d\psi = \frac{k}{2ds}(\psi - \psi_0)^2$$

$$= \frac{k}{2ds}\left(\frac{1}{\rho} - \frac{1}{\rho_0}\right)^2 (ds)^2 = \frac{k}{2}\left(\frac{1}{\rho} - \frac{1}{\rho_0}\right)^2 ds.$$

Hence, the total work done in bending the whole rod

$$= \int_0^s \frac{k}{2}\left(\frac{1}{\rho} - \frac{1}{\rho_0}\right)^2 ds.$$

9.8 To find the General Equation of Equilibrium of a Bend Rod

Figure is self-explanatory. The element $PQ = ds$ of the bent rod is in equilibrium under the stresses of the two portions on the two sides and external forces. The stresses constitute the extreme tension T, $T + dT$, shears N, $N + dN$ and the couples L, $L + dL$. Let F and G be the components of the external force per unit length of the rod along the tangent and normal at P as shown in the figure.

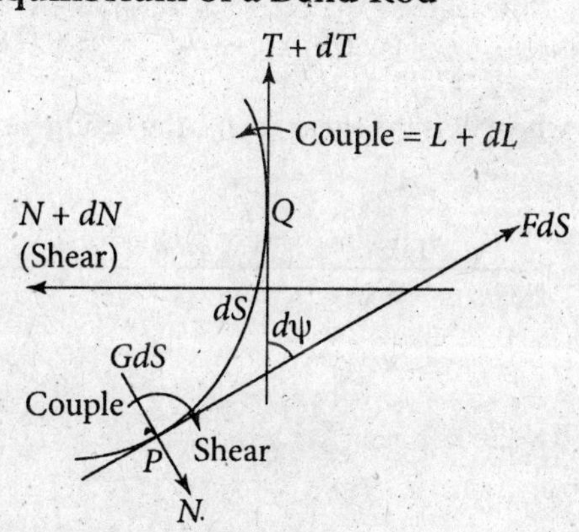

Fig. 9.9

Resolving tangentially and normally at P,

$$-T + (T + dT)\cos d\psi - (N + dN)\sin d\psi + F ds = 0 \text{ (tangentially)} \quad (1)$$

$$\Rightarrow -T + T + dT - (N + dN)d\psi + FdS = 0 \Rightarrow \frac{dT}{dS} - N\frac{d\psi}{dS} + F = 0$$

$$\Rightarrow \frac{dT}{dS} - \frac{N}{\rho} + F = 0 \quad (2)$$

and $-N + (N + dN)\cos d\psi + (T + dT)\sin d\psi + GdS = 0$ (normally) (3)

$$\Rightarrow -N + (N + dN) + (T + dT)d\psi + G ds = 0$$

$$\Rightarrow \frac{dN}{dS} + T\frac{d\psi}{dS} + G = 0 \Rightarrow \frac{dN}{dS} + \frac{T}{\rho} + G = 0. \quad (4)$$

Moment equation about P gives

$$-L+L+dL+(T+dT)ds\sin d\psi+(N+dN)ds=0 \tag{5}$$

$$\Rightarrow dL+(T+dT)ds\,d\psi+(N+dN)ds=0$$

$$\Rightarrow \frac{dL}{dS}+N=0, \tag{6}$$

where by experimental law, we have $L=k\left(\frac{1}{\rho}-\frac{1}{\rho_0}\right)$.

9.9 Clapeyron's Equation of the Three Moments

Statement. If L_1, L_2, L_3 be the bending moments at three points of supports A, B, C of a heavy rod, the

$$L_1a+2L_2(a+b)+L_3b+\frac{w}{4}\left(a^2+b^3\right)+\frac{W}{b}(b-c)(2b-c)c=6k\left(\frac{y_1}{a}+\frac{y_2}{b}\right),$$

where horizontal distance between A, B and B, C are a, b; y_1, y_2 are altitudes of A and C above B; w the weight per unit length of the rod; W the weight suspended from a point between B and C at a horizontal distance c from B.

Proof. For the portion BP, taking moment about P, we get

$$L+\frac{wx^2}{2}+Nx+W(x-c)-L_2=0, \tag{1}$$

where N is the shear at B, L the bending moment at P, $L_2=$ bending moment at B.

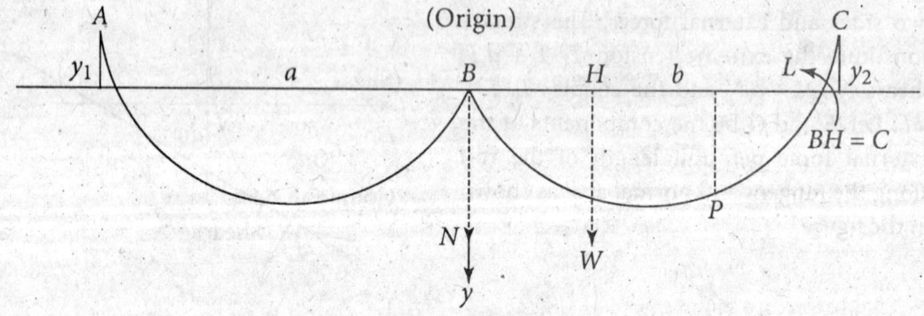

Fig. 9.10

So (1) becomes

$$L=L_2-Nx-\frac{wx^2}{2}-W(x-c)=-k\frac{d^2y}{dx^2},$$

whence integrating, we obtain

$$L_2x-\frac{Nx^2}{2}-\frac{wx^3}{6}-\frac{W(x-c)^2}{2}=-k\left(\frac{dy}{dx}-\beta\right).$$

At B, when $x=0$, let $\frac{dy}{dx}=\text{constant}=\beta$, the limits of x being from 0 to x and of $(x-c)$ from c to x.

Now integrating all the terms from 0 to x and the last term from c to x, we obtain

$$-k(y-\beta x)=L_2\frac{x^2}{2}-\frac{Nx^3}{6}-\frac{wx^4}{24}-\frac{W}{6}(x-c)^3$$

$$\Rightarrow -6k(y-\beta x)=3L_2x^2-Nx^3-\frac{wx^4}{4}-W(x-c)^3$$

$$\Rightarrow -6k(y-\beta x)=3L_2x^2-x^2\left[L_2-L-\frac{wx^2}{2}-W(x-c)\right]-\frac{wx^4}{4}-W(x-c)^3$$

[using (1)]

$$\Rightarrow -6k(y-\beta x)=(2L_2+L)\,x^2+\frac{wx^4}{2}-\frac{wx^4}{4}+W(x-c)x^2-W(x-c)^3$$

$$\Rightarrow -6k(y-\beta x)=(2L_2+L)\,x^2+\frac{wx^4}{4}+W(x-c)\left[x^2-\left(x^2+c^2-2xc\right)\right]$$

$$\Rightarrow -6k(y-\beta x)=(2L_2+L)\,x^2+\frac{wx^4}{4}+W(x-c)c(2x-c). \tag{2}$$

Again, at C, $L=L_3$, $y=-y_2$, $x=b$.
So, from (2), we have

$$-6k\left(-y_2-\beta b\right)=(2L_2+L_3)\,b^2+\frac{wb^4}{4}+W(b-c)c(2b-c). \tag{3}$$

In order to find L_1, we put $W=0$, $y=-y_1, x=-a, L=L_1$ in (2)

$$-6k\left(-y_1+\beta a\right)=(2L_2+L_1)a^2+\frac{wa^4}{4}. \tag{4}$$

Now multiplying both the sides of (3) by a, we get

$$-6k\left(-ay_2-ba\beta\right)=(2L_2+L_3)\,b^2a+\frac{wb^4a}{4}+W(b-c)(2b-c)ca. \tag{5}$$

Also multiplying both the sides of (4) by b, we get

$$-6k\left(-by_1+ab\beta\right)=(2L_2+L_1)\,a^2b+\frac{wa^4b}{4}. \tag{6}$$

Adding (5) and (6), we get

$$-6k\left(-ay_2-by_1\right)=2L_2ab(b+a)+ab(aL_1+bL_3)+\frac{wab}{4}\left(a^3+b^3\right)$$
$$+W(b-c)(2b-c)ca$$

$$\Rightarrow 6k\left(ay_2+by_1\right)=2L_2ab(a+b)+ab(aL_1+bL_3)+\frac{wab}{4}\left(a^3+b^3\right)$$
$$+W(b-c)(2b-c)ca.$$

Dividing both the sides by ab, we get

$$6k\left(\frac{y_1}{a}+\frac{y_2}{b}\right)=aL_1+2L_2(a+b)+bL_3+\frac{w}{4}\left(a^3+b^3\right)+\frac{W}{b}(b-c)(2b-c)c.$$

Particular Cases

1. If A, B, C lie in the same horizontal line, we then have

$$0 = L_1 a + 2L_2(a+b) + L_3 b + \frac{w}{4}\left(a^2+b^2\right) + \frac{W}{b}(b-c)(2b-c)c.$$

2. If $W = 0$, we have

$$0 = L_1 a + 2L_2(a+b) + L_3 b + \frac{w}{4}\left(a^2+b^2\right).$$

3. If A, C are free extremities, we have $L_1 = 0, L_3 = 0$ and then

$$L_2 = -\frac{w}{8}\cdot\frac{a^3+b^3}{a+b} = -\frac{w}{8}\left(a^2-ab+b^2\right).$$

4. If $a = b$, in this case, we have $L_2 = -\frac{w}{8}a^2$.

WORKED-OUT EXAMPLES

1. A uniform plank of length $2l$ and weight w per unit length is supported in a horizontal position at its ends. A boy whose weight is nw attempts to walk across the plank. If M is the maximum bending moment the plank will stand, show that the boy can only cross the plank, if $n < \left(\frac{2M}{lw} - 1\right)$. [NBU(H) 2006]

Solution: Let AB be the given uniform plank of length $2l$ and the boy walks from A to B. Let R and R' be the reactions of the supports at the ends.

Let the boy be at P after time t, where $AP = x$.

The forces acting on the system are:

1. Weight $2lw$ of the plank acting at the midpoint C of AB.

2. Weight nW of the boy acting at P.

3. The reaction R and R' of the supports acting at A and B respectively.

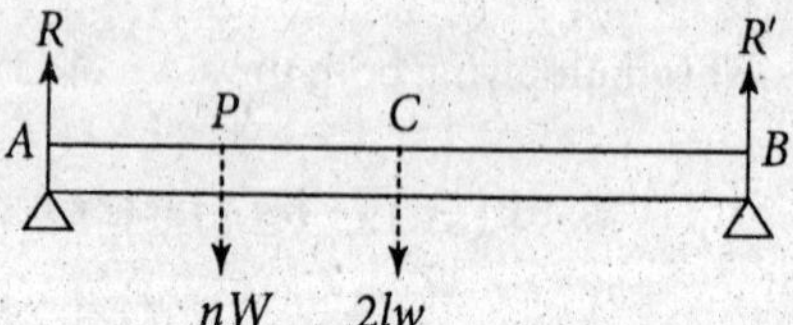

Fig. 9.11

Now taking moments of all the forces about the point B, we get

$$-R\cdot 2l + nwPB + 2lwCB = 0 \Rightarrow R\cdot 2l = nw(AB-AP) + 2lw\cdot l$$

$$\Rightarrow R2l = nw(2l-x) + 2lw\cdot l \Rightarrow R = nw\left(1-\frac{x}{2l}\right) + wl. \qquad (1)$$

The bending moment at C is

$$M_c = Rx - \frac{wx^2}{2} = nw\left(1-\frac{x}{2l}\right)x + wlx - \frac{wx^2}{2} \Rightarrow M_c = nw\left(x-\frac{x^2}{2l}\right) + wlx - \frac{wx^2}{2}.$$

$$\therefore \frac{dM_c}{dx} = nw\left(1-\frac{x}{l}\right) + wl - wx \qquad (2)$$

and

$$\frac{d^2M_c}{dx^2} = -\frac{nw}{l} - w.$$

For maximum or minimum M_c,

$$\frac{dM_c}{dx}=0\Rightarrow nw\left(1-\frac{x}{l}\right)+wl-wx=0\Rightarrow w(l+n)=wx\left(1+\frac{n}{l}\right)$$

$$\Rightarrow w(l+n)=\frac{wx}{l}(l+n)\Rightarrow n=l.$$

At $x=l$,

$$\frac{d^2M_c}{dx^2}=-w\left(1+\frac{n}{l}\right)<0.$$

Hence, the bending moment is maximum at $x=l$.

So

$$\max M_c=nw\left(l-\frac{l^2}{2l}\right)+wl^2-\frac{wl^2}{2}=nw\frac{l}{2}+\frac{wl^2}{2}=\frac{wl}{2}(n+l).$$

Since M is the maximum bending moment, so, we have

$$\frac{wl}{2}(n+l)<M\Rightarrow n+l<\frac{2M}{wl}\Rightarrow n<\frac{2M}{wl}-l.$$

2. A stiff horizontal beam AB is supported at one end A and at some other point C. If the greatest possible uniformly distributed weight is to be placed upon it without breaking it show that C must divide the beam in the ratio $\sqrt{2}+1:1$. [NBU(H) 2005]

Solution: Let w be the weight per unit length of beam AB. Then total weight of the beam is $2lw$ which acts at the midpoint G of AB. Let R_1 and R_2 be the reactions of the supports at A and C respectively.

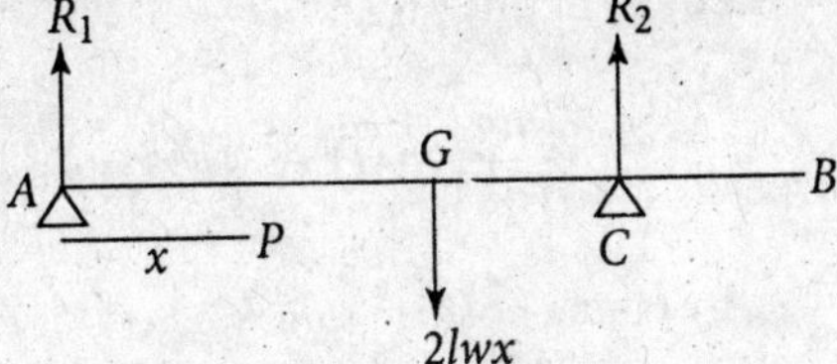

Fig. 9.12

Let $AB=2l, AC=y, AP=x$.

Taking moment of all the forces about A, we get

$$-2lw\cdot AG+R_2AC=0\Rightarrow 2lwl=R_2y\Rightarrow R_2=\frac{2wl^2}{y}. \tag{1}$$

Again, taking moment of all the forces about C, we get

$$-R_1AC+2lwGC=0\Rightarrow R_1AC=2lwGC\Rightarrow R_1y=2lw(AC-AG)$$

$$\Rightarrow R_1y=2lw(y-l)\Rightarrow R_1=2lw\left(\frac{y-l}{y}\right). \tag{2}$$

We take a point P on AC so that $AP=x$. Then the bending moment at P is

$$M=R_1x-\frac{wx^2}{2}=2lw\frac{(y-l)}{y}x-\frac{wx^2}{2}\ \text{[using (2)]}$$

For maximum or minimum value of M,

$$\frac{dM}{dx}=0.$$

Hence, from (3)

$$2wl\frac{(y-l)}{y}-wx=0\Rightarrow x=\frac{2l}{y}(y-l)=2l\left(1-\frac{l}{y}\right).$$

Also,

$$\frac{d^2M}{dx^2}=-w<0.$$

So M is maximum at $x=2l\left(1-\frac{l}{y}\right)$ and therefore,

$$\max M=2lw\frac{(y-l)}{y}\cdot\frac{2l(y-l)}{y}-\frac{w}{2}4l^2\left(1-\frac{l}{y}\right)^2=\frac{4l^2w}{y^2}(y-l)^2-\frac{2l^2w(y-l)^2}{y^2}$$
$$=\frac{2l^2w}{y^2}(y-l)^2. \quad (4)$$

Similarly, the bending moment for the part CB is maximum at, the maximum value being

$$\frac{w}{2}CB^2=\frac{w}{2}(2l-y)^2. \quad (5)$$

The maximising bending moment would be as small as possible when (4) and (5) are equal. In this case,

$$\frac{2l^2w}{y^2}(y-l)^2=\frac{w}{2}(2l-y)^2\Rightarrow\frac{4l^2}{y^2}=\left(\frac{2l-y}{y-l}\right)^2\Rightarrow\frac{2l}{y}=\pm\left(\frac{2l-y}{y-l}\right). \quad (6)$$

By taking positive sign, we get

$$\Rightarrow 2ly-y^2=2ly-2l^2\Rightarrow y^2=2l^2\Rightarrow y=\sqrt{2}l.$$

In this case,

$$CB=2l-y=2l-\sqrt{2}l=l\left(2-\sqrt{2}\right).$$

Hence,

$$\frac{AC}{CB}=\frac{y}{2l-y}=\frac{\sqrt{2}l}{l\left(2-\sqrt{2}\right)}=\frac{\sqrt{2}}{2-\sqrt{2}}=\frac{1}{\sqrt{2}-1}=\sqrt{2}+1.$$

If we take negative sign in (6), then

$$2l-y=-2l\left(1-\frac{l}{y}\right)\Rightarrow 2l-y=-2l+\frac{2l^2}{y}$$
$$\Rightarrow 2yl-y^2=-2yl+2l^2\Rightarrow y^2-4yl+2l^2=0$$
$$\Rightarrow y=\frac{4l\pm\sqrt{16l^2-8l}}{2}\Rightarrow y=\frac{4l\pm2\sqrt{2}l}{2}$$
$$\Rightarrow y=\left(2\pm\sqrt{2}\right)l.$$

But $l\le y<2l$. So this value of y is not possible.

3. A beam AB of of uniform material projects horizontally from a wall at A. It is of length l and its cross section topers uniformly from a^2 at A and b^2 at B. Prove that the greatest bending moment is $\frac{wl^2}{12}(a^2+2ab+3b^2)$, where w is the density of the material. [NBU(H) 2007]

Solution: Let AB be the centre line of the beam fixed in a wall at A. The cross section of the beam is taken to be a square, so that the face $CD=a$ and $EF=b$.

Let the lines CE and DF when produced meet at O and let $\angle COA=\angle DOA=\alpha$.

Then $AO=\frac{a}{2}\cot\alpha$ and $BO=\frac{b}{2}\cot\alpha$.

Now

$$AB=l\Rightarrow OA-OB=l$$

$$\Rightarrow \frac{a}{2}\cot\alpha-\frac{b}{2}\cot\alpha=l$$

$$\Rightarrow \frac{1}{2}\cot\alpha(a-b)=l$$

$$\Rightarrow \cot\alpha=\frac{2l}{a-b}. \qquad (1)$$

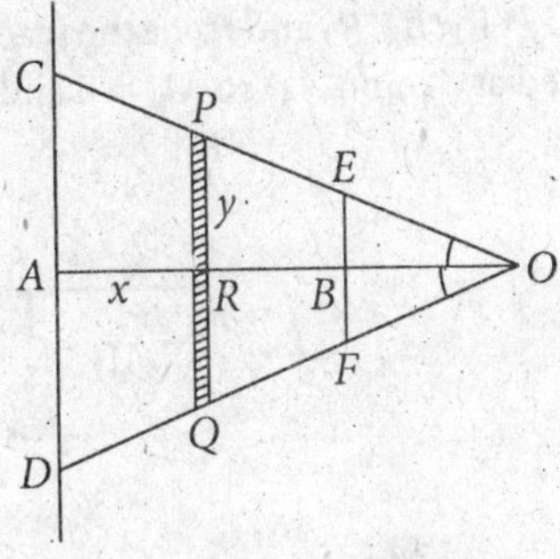

Fig. 9.13

Take a thin slice of the beam of thickness dx at a distance x from A.

In ΔOPR,

$$\tan\alpha=\frac{PR}{OR}=\frac{PR}{OA-AR}=\frac{PR}{\frac{a}{2}\cot\alpha-x}$$

$$\Rightarrow PR=\left(\frac{a}{2}\cot\alpha-x\right)\tan\alpha=\frac{a}{2}-x\tan\alpha=\frac{1}{2}(a-2x\tan\alpha).$$

So width of the thin slice $=PQ=2PR=a-2x\tan\alpha$.

Hence, area of the square section at $P=PQ^2=(a-2x\tan\alpha)^2$.

Therefore, the weight of this section $=(a-2x\tan\alpha)^2 w\,dx$ and its bending moment about A due to this weight $=(a-2x\tan\alpha)^2 wx\,dx$.

Obviously, this bending moment will be greatest. Then the greatest bending moment due to the weight of the whole beam

$$=\int_0^l (a-2x\tan\alpha)^2 wx\,dx=w\int_0^l\left(a^2-4ax\tan\alpha+4x^2\tan^2\alpha\right)x\,dx$$

$$=w\left[\frac{a^2l^2}{2}-4a\tan\alpha\frac{l^3}{3}+4\tan^2\frac{l^4}{4}\right]=\frac{wl^2}{6}\left[3a^2-8a\tan\alpha l+6l^2\tan^2\alpha\right]$$

$$=\frac{wl^2}{6}\left[3a^2-8al\frac{(a-b)}{2l}+6l^2\frac{(a-b)^2}{4l^2}\right]=\frac{wl^2}{12}\left[6a^2-8a(a-b)+3(a-b)^2\right]$$

$$=\frac{2l^2}{12}\left[6a^2-8a^2+8ab+3a^2-6ab+3b^2\right]=\frac{wl^2}{12}\left(a^2+2ab+3b^2\right).$$

4. A uniform rod A_1A_4 is supported at the ends A_1 and A_4, and at points A_2, A_3 the supports being at the same horizontal level, find the thrusts on the rod and the bending moments at the points $A_1A_2A_3$ and A_4 in terms of weight W. Also find the equation of the curve in the middle portion. [CU(H) 1991]

Solution: Let l and W be the length and weight of the rod A_1A_4. We take two points A_2 and A_3 on the rod such that $A_1A_2 = A_2A_3 = A_3A_4 = a$. Let w be the weight per unit length of the rod.

So

$$w = \frac{W}{l} = \frac{W}{3a} \Rightarrow W = 3aw. \tag{1}$$

Let R_1, R_2, R_3 and R_4 be the reactions at A_1, A_2, A_3, A_4 respectively. As the rod is supported at A_1 and A_4, so $M_1 = 0$ and $M_4 = 0$.

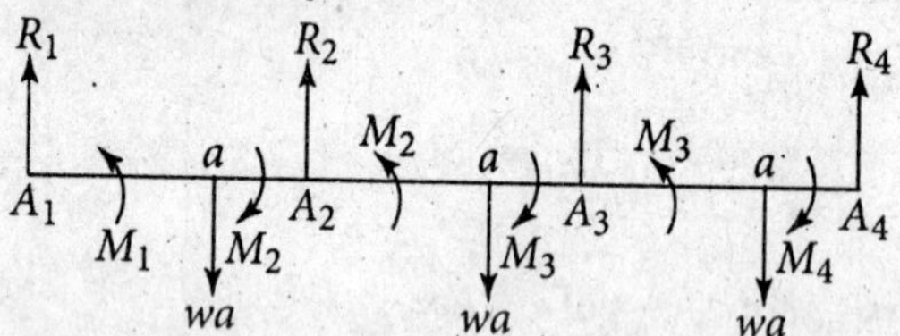

Fig. 9.14

Now, using Clapeyron's equation of the three moments to the portion $A_1A_2A_3$ and $A_2A_3A_4$, we get respectively.

$$M_1 \cdot a + 2M_2(a+a) + M_3 \cdot a = \frac{w}{4}\left(a^3+a^3\right) \Rightarrow 4M_2 + M_3 = \frac{wa^2}{2} \quad (\because M_1 = 0) \tag{2}$$

and

$$M_2a + 2M_3(a+a) + M_4a = \frac{w}{4}\left(a^3+a^3\right) \Rightarrow M_2 + 4M_3 = \frac{wa^2}{2} \quad (\because M_4 = 0). \tag{3}$$

Solving (2) and (3), we get

$$M_2 = M_3 = \frac{wa^2}{10} = \frac{wa}{30}. \tag{4}$$

Now, consider the portion A_1A_2, taking moment about A_2, we get

$$M_2 + R_1 \cdot a = wa \cdot \frac{a}{2} \Rightarrow \frac{wa^2}{10} + R_1a = \frac{wa^2}{2} \text{ [using (4)]}$$

$$\Rightarrow R_1 = \frac{wa}{2} - \frac{wa}{10} = \frac{4wa}{10} = \frac{2wa}{5} = \frac{2W}{15} \text{ [using (1)]}$$

So, from symmetry, $R_2 = \frac{2W}{15}$.

Considering the portion A_2A_3, taking moment about A_3, we get

$$M_3 + M_2 - R_2a + wa \cdot \frac{a}{2} = 0 \Rightarrow R_2a = \frac{wa^2}{2} + 2\frac{wa^2}{10}$$

$$\Rightarrow R_2 = \frac{wa}{2} + \frac{wa}{5} = \frac{7wa}{10} = \frac{7W}{30} \text{ [using (1)]}$$

Thus, from symmetry $R_3 = \frac{7W}{30}$.

Taking origin at A_1, x-axis in the horizontal and y-axis in the vertically downward direction, the equation of the middle portion is given by

$$-EI\frac{d^2y}{dx^2}=R_1x+R_2(x-a)-\frac{wx^2}{2}$$
$$\Rightarrow EI\frac{d^2y}{dx^2}=-\frac{2}{15}Wx-\frac{11}{30}W(x-a)+\frac{x^2}{2}\cdot\frac{W}{3a}$$
$$\Rightarrow EI\frac{d^2y}{dx^2}=\frac{W}{30a}\left[-4xa-11(x-a)a+5x^2\right]=\frac{W}{30a}\left[5x^2-15ax+11a^2\right].$$

Integrating, we have

$$EI\frac{dy}{dx}=\frac{W}{30a}\left[\frac{5x^3}{3}-15\frac{ax^2}{2}+11a^2x\right]+C_1.$$

Integrating, again we get

$$EIy=\frac{W}{30a}\left[\frac{5x^4}{12}-\frac{5ax^3}{2}+11\frac{a^2x^2}{2}\right]+c_1x+c_2. \tag{5}$$

Also $y=0$, when $x=a$ or, $x=2a$.

Hence,

$$0=\frac{W}{30a}\left[\frac{5a^4}{12}-\frac{5aa^3}{2}+\frac{11a^2a^2}{2}\right]+c_1a+c_2. \tag{6}$$

and

$$0=\frac{W}{30a}\left[\frac{80a^4}{12}-\frac{40a^4}{2}+\frac{44a^4}{2}\right]+c_12a+c_2. \tag{7}$$

Subtracting (5) from (7), we get

$$0=c_1a+\frac{W}{30a}\left[\frac{75a}{12}-\frac{35a^4}{2}+\frac{33a^4}{2}\right]\Rightarrow -c_1a=\frac{W}{30\times 12}[75-210+198]$$
$$\Rightarrow c_1=-\frac{Wa^2}{30\times 12}[63]=-\frac{7Wa^2}{40}.$$

From (5),

$$-c_2=\frac{Wa^3}{30}\left[\frac{5}{12}-\frac{5}{2}+\frac{11}{2}\right]-\frac{7Wa^3}{40}\Rightarrow c_2=a^3W\left[\frac{7}{40}-\frac{1}{30}\left(\frac{5+6\times 6}{12}\right)\right]$$
$$\Rightarrow c_2=a^3W\left[\frac{7}{40}-\frac{41}{360}\right]\Rightarrow c_2=\frac{a^3W}{360}[63-41]=\frac{a^3W\times 22}{360}\Rightarrow c_2=\frac{a^3W11}{180}.$$

Putting the values of c_1 and c_2 in (5), we get

$$EIy=\frac{W}{30a}\left[\frac{5x^4}{12}-\frac{5ax^3}{2}+\frac{11a^2x^2}{2}\right]-\frac{7Wa^2x}{40}+\frac{a^3W11}{180}$$
$$=\frac{W}{360a}\left[5x^4-30ax^3+66a^2x^2\right]-\frac{7Wa^2x}{40}+\frac{a^3W11}{180}.$$
$$=\frac{W}{360a}\left[5x^4-30ax^3+66a^2x^2-63a^3x+a^422\right]$$
$$=\frac{W}{360a}\left[(x-a)(x-2a)\left(5x^2-15ax+11a^2\right)\right]. \tag{8}$$

Thus, both y and d^2y are zero, when

$$x = a\left(\frac{3}{2} \pm \frac{\sqrt{5}}{10}\right).$$

Thus, the middle portion has two points of inflexion both of which lies on the line joining the supports.

5. An elastic rod, clamped at one end so that it is horizontal there, is bent by its own weights. Show that the deflection of its extremity is $\frac{3}{8}$th of what it would be if the deflection were caused by a weight equal to its own weight hung on at its extremity. [CU(H) 1992]

Solution: Let O be fixed point and y be the deflection by its own weight W.

Let $OP' = x, OA' = l = OA$. Taking moments about P', we get

$$B\frac{d^2y}{dx^2} = W(l-x),$$

where $B = El$.

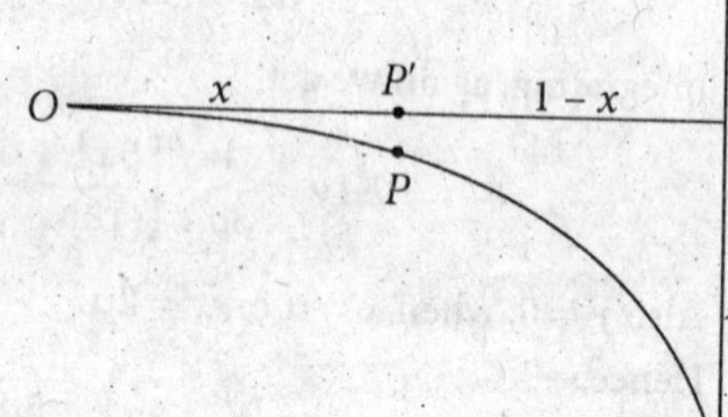

Fig. 9.15

Integrating, we get

$$B\frac{dy}{dx} = W\left(lx - \frac{x^2}{2}\right) + c_1. \quad (1)$$

At O, $x = 0$, $\frac{dy}{dx} = 0$. So from (1), $c_1 = 0$.

Hence, from (1),

$$B\frac{dy}{dx} = W\left(lx - \frac{x^2}{2}\right).$$

Integrating again, we get

$$By = W\left(\frac{lx^2}{2} - \frac{x^3}{6}\right) + c_2. \quad (2)$$

At $0, x = 0, y = 0$. So from (2), $c_2 = 0$.

Hence, from (2),

$$By = W\left(\frac{lx^2}{2} - \frac{x^3}{6}\right).$$

Therefore, deflection at A is

$$y_A = \frac{W}{B}\left(\frac{l^3}{2} - \frac{l^3}{6}\right) = \frac{Wl^3}{3B}.$$

Also, taking moments about P, we get

$$B\frac{d^2y}{dx^2} = (l-x)\cdot\frac{(l-x)}{2}w,$$

w being weight per unit length.

Integrating, we get

$$B\frac{dy}{dx}=-\frac{w}{6}(l-x)^3+c_3. \tag{3}$$

Integrating again, we get

$$By=\frac{w}{24}(l-x)^4+c_3x+c_4. \tag{4}$$

When $x=0, y=0, \frac{dy}{dx}=0.$

Hence, from (3) and (4), we get

$$c_3=w\frac{l^3}{6} \text{ and } c_4=-\frac{w}{24}l^4.$$

Putting the values of c_3 and c_4 in (4), we get

$$By=\frac{w}{24}(l-x)^4+\frac{wl^3}{6}x-\frac{wl^4}{24}\Rightarrow y=\frac{w}{B}\left[\frac{(l-x)^4}{24}+\frac{l^3}{6}x-\frac{l^4}{24}\right].$$

Deflection is

$$y'_A=\frac{w}{B}\left[0+\frac{l^4}{6}-\frac{l^4}{24}\right]=\frac{w}{B}\cdot\frac{3l^4}{24}=\frac{wl^4}{8B}=\frac{W}{l}\cdot\frac{l^4}{8B}=\frac{Wl^3}{8B}.$$

Hence,

$$\frac{Y'_A}{Y_A}=\frac{Wl^3/8B}{Wl^3/3B}=\frac{3}{8}\Rightarrow y'_A=\frac{3}{8}y_A.$$

EXERCISE 9

1. A uniform heavy rod AB is supported at each end. If w be the weight per unit length prove that the bending moment at any point P is $\frac{1}{2}wAP\cdot BP$.
2. A light beam of length l is supported at its ends and carries a load W concentrated at a point at a distance a from one end. Find the shearing stress and bending moment.
3. AB is a stiff uniform beam of weight W and length $2l$, and is supported at its end so as to be horizontal, a man of weight W' stands on it at P, where $AP=\xi(<l)$. Show that the curve of bending moment consists of two arcs of parabolas of equal latus rectum and that the bending moment is greatest at a point distant $\left(1-\frac{W'}{W}\xi\right)$ from A.
4. A uniform girder is supposed at its ends and the load is concentrated at its middle. Show that the maximum bending moment is twice as great as it would be if the load were uniformly distributed.
5. A uniform beam AB, of length l, is supported at its ends and loaded with a weight w at a point Q. Where $AQ=a$. If the weight of the beam be neglected, show that the equation of AQ is

$$Ely=\frac{W(l-a)}{16}\left[(2l-a)x-x^2+a^2\right]x$$

and that of BQ is

$$Ely=\frac{Wa}{16}\left[\left(l^2-a^2\right)(l-x)-(l-a)^3\right].$$

6. A uniform slightly elastic rod, of length $a+b$, rests on three supports in a horizontal line, situated at its ends A, B and at a point C distant a from A. Show that the points of inflexion I and J on the rod are situated in the segments AC and CB respectively, and are such that $a \cdot IC = b \cdot JC = \frac{1}{4}(a^2 - ab + b^2)$.

Answers

2. When $0 < x < a, M = W\left(1 - \frac{a}{l}\right)x, S = W\left(1 - \frac{a}{l}\right); a < x < l, M = W\left(1 - \frac{x}{l}\right)a, S = \frac{Wa}{l}$.

OBJECTIVE QUESTIONS

Fill in the Blanks

1. If the displacement of the point of application of the force is perpendicular to the direction of the force, then $W=$ —.
2. Work done by a couple during a small displacement is independent of the — but depends upon — only.
3. The equation $\vec{u}\cdot\vec{R}+\vec{e}\cdot\vec{G}=0$ is called the equation of —.
4. The work done by the tension of an inextensible string during a small displacement is —.
5. The work done by the thrust of an inextensible rod during a small displacement is —.
6. The work done by the tension T of an inextensible string of length l during a small displacement is —.
7. The work done by the thrust T of an extensible rod of length l during a small displacement is —.
8. A uniform beam of thickness $2b$ rests symmetrically on a rough horizontal cylinder of radius a, the position of the beam will be stable when —.
9. The intrinsic equation of the catenary is —. [MU(H) 2008]
10. The Cartesian equation of a common catenary is —.
11. The relation between yards is —.
12. $y=c$ —.
13. If the centre of gravity of a body is below the point of support, the body is in — equilibrium.
14. If the centre of gravity of a body is above the point of support, then the body is in — equilibrium.
15. If the centre of gravity of a body is at the point of support, then the body is in — equilibrium.
16. If the work function of the body is maximum, then the body is in — equilibrium.
17. If the work function of the body is minimum, then the body is in — equilibrium.
18. In the position of equilibrium, if the potential energy of the body is minimum, then the body is in — equilibrium.
19. If the height of centre of gravity of the body is minimum, then the body is in — equilibrium.
20. If $\frac{1}{h}>\frac{1}{r}+\frac{1}{R}$, then the body is in — equilibrium.
21. If $\frac{1}{h}\leq\frac{1}{r}+\frac{1}{R}$, then the body is in — equilibrium.
22. The radius of curvature of the plane is —.

True or False

1. Mutual reaction between two bodies is omitted in forming the equation of virtual work.
2. If T be the thrust of an inextensible rod of length l, then the work done by T during a small displacement is $-T\delta l$.
3. If $\vec{r}$ is the position vector of a point P and if it is given a small displacement $d\vec{r}$ consisting of a small translation $\vec{u}$ and small rotation $\vec{e}$, then $d\vec{r}=\vec{u}+\vec{e}\times\vec{r}$.
4. For a common catenary, $s=c\log(\sec\psi+\tan\psi)$.
5. For a common catenary, $T=wy$.
6. The catenary, when it is very tightly stretched, becomes a parabola to a first approximatio
7. If $\frac{1}{h}<\frac{1}{r}+\frac{1}{R}$, then the body is in stable equilibrium.
8. If $\frac{1}{h}+\frac{1}{r}+\frac{1}{R}$, then the body is in unstable equilibrium.
9. If the potential energy of the body is minimum, then the position of equilibrium is stable.
10. If the depth of centre of gravity of the body is maximum, then the position of equilibrium of the body is stable.
11. The intrinsic equation of the catenary is $s=c\tan\psi$. [MU(H) 2011]
12. The equation of the line of action of the resultant force of a system of coplanar forces is $xY+yX+G=0$. [MU(H) 2011]
13. For a system in equilibrium under conservative forces, the potential energy is maximum for stable and minimum for unstable equilibrium.
14. A system consisting of a force and couple such that the line of action of the force is orthogonal to the axis of the couple is called a wrench.
15. The equilibrium is stable if $\frac{1}{h}>\frac{1}{r}-\frac{1}{R}$. [MU(H) 2007]
16. The algebraic sum of the virtual works done by any system of forces is equal to the virtual work done by their resultant. [MU(H) 2013]
17. A system of coplanar forces acting upon a rigid body cannot be reduced to a single force acting at an arbitrarily chosen point together with a couple.
18. Virtual work is the real work done by a force.
19. The reaction of a smooth surface upon a body which rolls without sliding on the surface can be omitted in forming the equation of virtual work.

Multiple Choice Questions

1. Work done by the thrust T of the rod of length

   (a) zero (b) $-T\delta l$ (c) $T\delta l$ (d) none of these.
2. Cartesian equation of catenary is

   (a) $s=c\tan\psi$ (b) $y+s=ce^{x/c}$ (c) $T/wl=\sec\psi$ (d) $y=c\cosh\frac{x}{c}$.

3. Tension at any point (x, y) of a catenary is

   (a) $c \sec \psi$ (b) $w\sqrt{s^2+c^2}$ (c) $\cosh\left(\frac{x}{c}\right)$ (d) $c \tan \psi$.

4. Tension at the lowest point of a catenary is

   (a) wc (b) $\sqrt{y^2-c^2}$ (c) $(y-c)$ (d) $x^2/2c$.

5. The positions of minimum potential energy are the positions of

   (a) stable equilibrium (b) unstable equilibrium (c) neutral equilibrium (d) none of these. [MU(H) 2010]

6. The equation of the line of action of the resultant force of a coplanar force-system is

   (a) $xY+yX+G=0$ (b) $xY-yX+G=0$ (c) $yX-xY+G=0$. [MU(H) 2009]

7. For a rigid body, the principle of virtual work helps to determine the position of

   (a) resultant (b) instability (c) equilibrium. [MU(H) 2009]

Answers

Fill in the Blanks

1. 0. **2.** Translation, Rotation. **3.** Virtual work. **4.** 0, 5, 0. **6.** $-T\delta l$. **7.** $T\delta l$. **8.** $b < a$. **9.** $s = c\tan\psi$. **10.** $y = c\cosh\frac{x}{c}$. **11.** $y^2 = c^2+s^2$. **12.** $c\sec\psi$. **13.** Stable. **14.** Unstable. **15.** Neutral. **16.** Stable. **17.** Unstable. **18.** Stable. **19.** Stable. **20.** Stable. **21.** Unstable. **22.** ∞.

True or False

1. T. **2.** F. **3.** T. **4.** F. **5.** T. **6.** T. **7.** F. **8.** T. **9.** T. **10.** T. **11.** T. **12.** F. **13.** F. **14.** F. **15.** T. **16.** T. **17.** F. **18.** F. **19.** T.

Multiple Choice Questions

1. (c). **2.** (d). **3.** (b). **4.** (a). **5.** (a). **6.** (c). **7.** (c).